Agricultural Communications

in Action

A Hands-On Approach

RICKY TELG • TRACY ANNE IRANI

Delmar Cengage Learning
is proud to support
FFA activities

Join us on the web at
www.cengage.com/community/agriculture

Agricultural Communications

in Action

A Hands-On Approach

RICKY TELG • TRACY ANNE IRANI

DELMAR
CENGAGE Learning™

Australia • Brazil • Japan • Korea • Mexico • Singapore • Spain • United Kingdom • United States

**Agricultural Communications in Action:
A Hands-On Approach**
Ricky Telg, PhD, Tracy Anne Irani, PhD

Vice President, Editorial: Dave Garza

Director of Learning Solutions: Matthew Kane

Senior Acquisitions Editor: Sherry Dickinson

Managing Editor: Marah Bellegarde

Senior Product Manager: Juliet Steiner

Editorial Assistant: Scott Royael

Vice President, Marketing: Jennifer Baker

Marketing Director: Deborah Yarnell

Marketing Coordinator: Erin D'Angelo

Senior Production Director: Wendy Troeger

Production Manager: Mark Bernard

Senior Content Project Manager: Katie Wachtl

Senior Art Director: David Arsenault

Production Technology Analyst: Chris Catalina

Design Credits:
© iStockphoto/bonnie Jacobs
© iStockphoto/savas keskiner
© iStockphoto/Ricky Corey
© iStockphoto/Petrovich9

For product information and technology assistance, contact us at
Cengage Learning Customer & Sales Support, 1-800-354-9706

For permission to use material from this text or product,
submit all requests online at **www.cengage.com/permissions.**

Further permissions questions can be e-mailed to
permissionrequest@cengage.com

Example: Microsoft ® is a registered trademark of the Microsoft Corporation.

Library of Congress Control Number: 2011936567

ISBN-13: 978-1-1113-1714-0

ISBN-10: 1-1113-1714-3

Delmar
5 Maxwell Drive
Clifton Park, NY 12065-2919
USA

Cengage Learning is a leading provider of customized learning solutions with office locations around the globe, including Singapore, the United Kingdom, Australia, Mexico, Brazil, and Japan. Locate your local office at:
international.cengage.com/region

Cengage Learning products are represented in Canada by Nelson Education, Ltd.

To learn more about Delmar, visit **www.cengage.com/delmar**

Purchase any of our products at your local college store or at our preferred online store **www.cengagebrain.com**

Notice to the Reader

Publisher does not warrant or guarantee any of the products described herein or perform any independent analysis in connection with any of the product information contained herein. Publisher does not assume, and expressly disclaims, any obligation to obtain and include information other than that provided to it by the manufacturer. The reader is expressly warned to consider and adopt all safety precautions that might be indicated by the activities described herein and to avoid all potential hazards. By following the instructions contained herein, the reader willingly assumes all risks in connection with such instructions. The publisher makes no representations or warranties of any kind, including but not limited to, the warranties of fitness for particular purpose or merchantability, nor are any such representations implied with respect to the material set forth herein, and the publisher takes no responsibility with respect to such material. The publisher shall not be liable for any special, consequential, or exemplary damages resulting, in whole or part, from the readers' use of, or reliance upon, this material.

Printed in the United States of America
1 2 3 4 5 6 7 15 14 13 12 11

DEDICATIONS

Ricky W. Telg:

This book is dedicated to my wife Jodie, our adult children, and our parents.

Tracy A. Irani:

This book is dedicated to my husband Jim and our three children, Sarah, Jessica, and Andrew.

We wish to thank everyone who contributed to this textbook and the supplemental educational materials, such as sidebars, videos, and professional communications material examples. We greatly appreciate your assistance.

Contents

Preface

Most, if not all, managers who regularly interview job candidates will say that they are looking for two primary skill sets in a new hire:

- Knowledge and skills in a given subject or technical area, and
- Effective communication abilities.

The first skill set is fairly easy to come by in school. In mathematics, science, career and technical education, English, and elective courses, students are tested on their technical knowledge with standardized exams to indicate their proficiency in these subject areas. But how can students develop communication skills?

We, the authors of *Agricultural Communications in Action: A Hands-On Approach,* believe that students can learn these skills, provided they have the tools to do so. This textbook features content to help students become better equipped as effective communicators.

CONCEPTUAL APPROACH

Agricultural communications is a fast-growing segment of agricultural education. At the university level, "traditional" departments of agricultural education have added and embraced agricultural communications courses in their curriculum. At the high school level, more agricultural education programs are adding full courses or a series of lessons in agricultural communications. In addition, in 2000 the National FFA Organization created the Agricultural Communications Career Development Event to highlight students' agricultural communications abilities.

Agricultural communications is a wide and varied field; it includes news writers, graphic designers, video and radio reporters and producers, special events planners, photographers, Web designers, advertising and public relations practitioners, and many more types of professional communicators. This book showcases the variety of communication skills, knowledge, and styles needed to be a successful agricultural communicator, making it a "one-stop shop" for instructors who have had little or no communications training. It provides a breadth and depth of information so that students can learn communications development concepts quickly and then apply them immediately. Because so many types of communications software programs exist, the book is not designed to teach students how to use a specific type of software program—for example, Adobe's Photoshop or Dreamweaver—but students can apply the concepts learned in the textbook to these and other similar software programs.

ORGANIZATION

The book is divided into seven sections, for a total of 18 chapters. The major book sections are as follows:

Introduction and Message Development. In this section, students will receive an overview of agricultural communications and its history, agricultural communications organizations, and agricultural communications careers. This section also provides the basis for the rest of the book—message development. Knowing how to create an understandable and effective message is essential to the communications process and to the skills covered in the remainder of the book. This section also features research methods used in communications, including how to create surveys, polls, and focus groups.

Writing and Document Design. The chapters in this section focus on business writing, news media writing, and document design. Students will learn how to craft business letters, memos, and electronic mail, as well as how to write journalistically and to think like a news reporter. Students also will learn the basics of document design and how to develop newsletters and brochures.

Visual Communication. Because our society is increasingly emphasizing visual communication, effective communicators also need to know how to develop a visually appealing message. In this section, students will learn how to design posters, fliers, displays, and projected presentations, and how to shoot and edit digital photographs. How to speak in public—another form of visual communication—also is included in this section.

Video and Online Communication. Two of the fastest-growing segments of communication are video and online communication. In this section, techniques used in video and audio production will be emphasized. Students will also learn basic principles of Web page design and how to integrate "new media," including social media, into effective communications messages.

Working with the Media. This section discusses how to work with news media to get information out about events and activities. The chapter on risk and crisis communication provides important information on how to prepare for potential crises and how to communicate effectively during a crisis situation.

Putting It Together. In this section, students will be shown how to put together the communication skills learned in the book to achieve a specific goal or objective. This section covers persuasive messages, special events planning, and communication campaign development.

The Future. Five years ago, who knew what Facebook was? Ten years ago, who knew that almost every youth and adult would have a cell phone? Twenty years ago, the word "web" usually referred to a spider's web. Now, "web" means the World Wide Web. So what will the future of agricultural communications look like? What technologies will be used and what issues will agriculture face in the near future? This chapter reviews current trends and peers into what the future could hold for agricultural communications.

FEATURES

Current communication trends are integrated into this instructional and functional textbook. The text is a practical, "how-to" approach for students. Key features include:

- Real-life examples and sidebars from professionals in various agricultural industries illustrate how professionals tackle communication issues and problems.

- Chapter Objectives and Key Terms set the stage for student learning and are excellent study and review tools.
- Specific written communication topics covered include message development, research methods, writing, and document design.
- Coverage of visual communication discusses how to design posters, fliers, displays, and projected presentations; how to shoot and edit digital photographs; and how to speak in public.
- Cutting-edge information on video and online communication emphasizes basic principles of Web page design and how to integrate "new media"—including social media—into effective communications messages.
- A working with the media section discusses how to work with news media to get information out about events and activities and how to communicate during a crisis situation.
- The text is rich with photos and figures illustrating communication concepts and demonstrating design techniques.
- Chapter Exercises provide application opportunities to apply concepts covered in the chapter, as well as questions such as multiple choice, matching, and fill in the blank to test mastery of information.

TEACHING AND LEARNING PACKAGE

The complete supplements package for *Agricultural Communications in Action* was developed to achieve two goals:

- To assist students in learning the information presented in the text, and
- To assist instructors in planning and implementing their course for the most efficient use of time and other resources.

ClassMaster DVD-ROM

ISBN 10:1-1113-1715-1

ISBN 13:978-1-11131-715-7

This robust supplement provides the instructor with valuable resources to simplify the planning and implementation of the instructional program. Materials provided on this DVD-ROM include the following.

Instructor's Guide

An electronic Instructor's Guide provides excellent tools to help the instructor create a dynamic and engaging learning experience for the student. The Instructor's Guide contains the tools listed here but can be downloaded and modified to meet individual instructional goals:

- Teaching Tips: This section provides engaging ideas and tips for the instructor to use in conjunction with the chapter topics.
- Discussion Topics: These excellent strategies for stimulating class discussion can be used to challenge student critical thinking and to create an interactive classroom experience.
- Additional Assignments: These additional ready-to-use activities and assignments are tailored to provide thoughtful activities directly supporting the individual chapters.
- Additional Resources: This section provides additional websites, organizations, references, and more to help the instructor gain additional information supporting each chapter.
- Answers to Chapter Exercises: Answers and intended outcomes for the end-of-chapter questions are provided to assist the instructor in grading and evaluation.

Video Interviews

- Thirteen videos are provided featuring professional agricultural communicators' discussions about topics that relate to book chapters.
- Discussion activities related to the videos are also provided.

Printed Tutorials

- Step-by-step tutorials take students through some of the agricultural communications industry's most-used software programs: Adobe's Dreamweaver, InDesign, and Photoshop, and Apple's Final Cut Express video editing program.
- Work files are included so that students can work with the files that are presented in the tutorials.

Communications Campaign Plan Book Example

- An actual student-developed example of a communications campaign plan book is included to provide students with a model to utilize to develop a communications campaign plan.
- Step-by-step worksheets covering each section of a communications campaign plan are provided that instructors can print off and hand out to students as homework or a small-group activity.

AP Quizzes

- Quizzes focusing on Associated Press Style, grammar, punctuation, and social media listings from *The Associated Press Stylebook* are provided.
- The quizzes will help students hone their Associated Press Style skills, which is the standard writing format for journalists and public relations professionals.
- For more information about Associated Press Style and *The Associated Press Stylebook*, visit http://www.apstylebook.com.

Computerized Testbank in ExamView™

- Includes 540 multiple-choice, matching, completion, and true/false questions that test students on retention and application of material in the text.
- All questions provide the correct answers.
- Instructors can create custom tests by mixing questions from the 18 chapters of questions, modifying existing questions, and even adding additional questions to meet individual instructional needs.

Instructor Slides Created in PowerPoint

- A comprehensive offering of over 350 instructor slides created in PowerPoint outlines the concepts from the text to assist the instructor with lectures.
- Ideas are presented to stimulate discussion and critical thinking.
- Wherever possible, figures and photos from the text have been included to help engage visual learners.

About the Authors

The authors of *Agricultural Communications in Action: A Hands-On Approach* have more than 30 years of combined experience in communications.

University of Florida (UF) professor **Ricky W. Telg** has taught communication courses on digital media development and news writing for more than 15 years. He has a bachelor's degree and doctorate from Texas A&M University and a master's degree from the University of North Texas. Before joining UF's Department of Agricultural Education and Communication in 1995, Telg worked as a television reporter, radio personality, and newspaper reporter and editor. At UF, Telg advises the collegiate organization Agricultural Communicators and Leaders of Tomorrow and serves on several college and university committees that focus on enhancing undergraduate education and communication skills.

Telg has served in leadership roles in several national organizations, including the National Agricultural Communicators of Tomorrow (ACT), where he served as National ACT adviser for 5 years, and the Association for Communication Excellence in Agriculture, Natural Resources, and Life and Human Sciences (ACE). Telg has received UF's College of Agricultural and Life Sciences' Undergraduate Advisor of the Year award and the Undergraduate Teacher of the Year award twice, as well as being selected as the North American Colleges and Teachers of Agriculture's (NACTA) Southern Region Outstanding Teacher in 2008. Telg was named one of two national United States Department of Agriculture Food and Agricultural Sciences Excellence in Teaching Award recipients in 2010. Telg has given numerous paper presentations at regional, national, and international meetings and has published articles on distance education-and agricultural communications-related topics.

Telg's extensive experience in writing news stories, developing digital media, and creating targeted messages to various audiences is integrated into *Agricultural Communications in Action: A Hands-On Approach.*

Tracy A. Irani is a professor at the University of Florida, teaching courses in communication campaigns, public relations, Web design, and critical and creative thinking. Prior to teaching at UF, Irani worked in marketing, public relations, and advertising. She brings a wealth of knowledge in communication methods.

Irani received a bachelor of arts degree in Journalism and Communications from Point Park College, a master's degree in Corporate Communications from Duquesne University, and a doctorate in Mass Communications from the University of Florida. Irani joined UF's Department of Agricultural Education and Communication's faculty in 1999; she holds a teaching and research appointment in the area of agricultural communications. Irani's research interests focus on critical thinking, problem solving, and decision making with respect to controversial science, communications, and technology issues. Irani is a certified administrator of the Kirton Adoption Innovation (KAI) inventory, one of the most widely used assessments of cognitive style and problem solving. She is a past chair of the ACE Research and Academic Programs Special Interest Groups, and is a past president of UF's chapter of Gamma Sigma Delta, the international agricultural honorary society. In addition, she is past head of the Association for Education in Journalism and Mass Communication's Science Communication Interest Group.

She has authored or co-authored numerous research articles and made presentations at local, national, and international meetings on marketing communications, new and social networking media, branding, problem solving and decision making, and distance education.

Irani is currently development director for the Florida Center for Public Issues Education in Agriculture and Natural Resources in UF's Department of Agricultural Education and Communication.

Introduction and Message Development

Chapter 1
- Introduction to Agricultural Communications

Chapter 2
- Effective Communication and Message Development

Chapter 3
- Research Methods

1

Introduction to Agricultural Communications

OBJECTIVES

After completing this chapter, the student will be able to:

- Define *agricultural communications*.
- Identify major agricultural communications organizations.
- List possible careers in the agricultural communications field.

INTRODUCTION

In the past century, U.S. agriculture has changed dramatically. The Farm Credit HORIZONS project and the United States Department of Agriculture (USDA) reported that the percentage of the U.S. population living on a farm decreased from around 40 percent in 1900 to less than 5 percent in 2000 (United States Farm Credit System, 2006). The percentage of the U.S. workforce employed in agriculture also has shifted radically from the beginning of the twentieth century to the start of the twenty-first century. In 1900, around 40 percent of the U.S. workforce was employed in agriculture. By 1945, agriculture's share of the U.S. workforce had dropped to around 16 percent. In 2000, 1.9 percent of the U.S. labor force worked in agriculture (Dimitri, Effland, & Conklin, 2005).

One of the results of Americans moving from rural areas to urban settings and being removed from their agrarian heritage is that their knowledge about and understanding of agriculture has declined over generations (Figure 1-1). As a result, many people do not understand agricultural industries and how these industries impact people's everyday lives. In addition, some information does not portray agriculture in a positive light, and unless someone communicates a positive message about agriculture, no one will know about the contributions agriculture provides. Therefore, the need for good agricultural communicators is as great now as it ever has been. In addition to communicating information to farmers and ranchers about improved agricultural practices, good agricultural communications professionals also communicate targeted messages about agriculture to an urban audience.

FIGURE 1-1: Because of the move from rural to urban settings, twenty-first-century Americans are less informed about how agricultural industries impact their everyday lives. (*Courtesy of USDA Agricultural Research Service/Photo by Brian Prechtel*)

This textbook will help prepare you to be a competent and effective agricultural communicator. You will learn how to develop messages using effective communication methods, including document design, Web design, video production, news writing, and other methods, so you can better communicate with your target audience. In this chapter, you will define agricultural communications, understand its history, and explore careers in the field.

DEFINING AGRICULTURAL COMMUNICATIONS

Communications is the process of transferring information from a sender to a receiver with the use of a medium—such as newspapers, your voice, radio, television, the Web, or other media—in which the communicated information is understood by both the sender and the receiver. **Agriculture** is the process of producing food, feed, and fiber through the raising of plants and animals. Agricultural communications blends the definitions of communications and agriculture. **Agricultural communications** is the exchange of information about the agricultural and natural resources industries through effective and efficient media, such as newspapers, magazines, television, radio, and the Web, to reach appropriate audiences.

Jim Bret Campbell

Senior Director of Marketing and Publications, American Quarter Horse Association

Why is agricultural communications important?

Most Americans do not know where their food comes from, how it is processed, and why they spend so little of their income on food. Agricultural communications is vital to informing consumers, regulators, and legislators about the many advantages of our agricultural system and its importance to this country.

© Cengage Learning 2012

Jennifer Latzke

Associate Editor, High Plains Journal

Why is agricultural communications important?

Take any top story on tonight's news and you can trace it back to an agriculture connection. From the price of a bag of groceries to the property taxes needed to keep public schools open, and from the cost of foreign petroleum supplies to the latest in conservation efforts, farming and ranching have a key role in our everyday lives. However, fewer and fewer people are choosing careers in agriculture, or they're generations removed from farming and ranching and don't understand our modern production methods.

The closest some people come to agriculture is the meat counter or a trip to the county fair. They have messages from all sorts of radical environmentalists and animal rights activists telling them that modern farming and ranching are bad for people and the environment. But, they need the other point of view. They need to be educated about what it takes to feed and clothe and fuel an ever-growing world population.

When you don't understand agriculture and the scientific concepts behind advancements in biotechnology, animal reproduction, or economies of scale, it's easy to have a skewed view of modern agriculture.

That's where agricultural communications is key.

Charlie Arnot, chief executive officer of the Center for Food Integrity, once explained it like this: "Consumers envision a pastoral 'farm,' with a barn and chickens and cows, but it's like comparing a new car to a 1957 Chevy. New cars have seat belts and airbags and computers and all the refinements. But, the public is stuck in 1957. They're in the grocery store thinking like a 1957 Chevy, and we zoom by in our 2008 Taurus" (keynote address to the Nebraska Ag Classic, Dec. 15, 2008).

So, as our farming and ranching methods become more modern and as we advance in science to ensure that we can feed, clothe, and fuel an ever-growing world population, it's vital that we have communicators to spread our message to members of the general public, who are our ultimate consumers.

BRIEF HISTORY OF AGRICULTURAL COMMUNICATIONS

Agricultural communications is not a new field. Agricultural communications in the United States can be traced back to colonial times, and even though the communications media have changed over time, agricultural communicators still develop important messages for targeted audiences. This section provides a brief overview of some of the important events in the history of American agricultural communications.

Colonial Days through the 1800s

In the late 1700s during the United States' colonial era, society was predominantly agrarian. Much of the communication was word of mouth, from farmer to farmer, and most of the information about how to grow crops came from the colonists' backgrounds in Europe (Boone, Meisenbach, & Tucker, 2000).

Several magazines and newspapers that included information on American farming practices began publication soon after the American Revolution. New Jersey's first newspaper, the *New Jersey Gazette*, which began in December 1777, was one of the first of the colonial-era newspapers to encourage articles on farming. In 1785, the Philadelphia Society for Promoting Agriculture was organized and was the first agricultural society to publish results of agricultural experimental work. Two of its most notable members were George Washington and Benjamin Franklin (Fletcher, 1976). *The Farmer's Almanac*, which still is published but now is titled *The Old Farmer's Almanac*, began publication in 1792 (Figure 1-2). It published tips to improve agricultural practices, weather forecasts, planting charts, tide tables, astronomical information, recipes, and trends in fashion. *The Old Farmer's Almanac* is the oldest continuously published periodical in North America (*The Old Farmer's Almanac*, 2010).

The first American agricultural magazine, the *Agricultural Museum*, began publication in 1810, but lasted only two years (Mott, 1930). In 1819, the *American Farmer* journal started, published by Baltimore postmaster

John Stuart Skinner, who is considered the father of American agricultural journalism. The *American Farmer* was the first agricultural journal in the United States to attain prominence, running from 1819 to 1897. Skinner was also a prominent person during the American Revolution. Skinner was with Francis Scott Key on the mission at Baltimore's Fort McHenry that inspired Key to write *The Star-Spangled Banner* (Molotsky, 2000). Skinner's *American Farmer* showcased information on every branch of agriculture, allowing his readers to be better farmers and livestock producers. Skinner also started several other agriculture-related publications, including the *American Turf Register and Sporting Magazine*, the *Farmer's Library and Monthly Journal of Agriculture*, and the *Plough, the Loom, and the Anvil*. Skinner's periodicals helped spark interest in agriculture during the mid-1800s (Virtualology.com, 2001) However, many other agricultural publications that started in the early 1800s went out of business quickly because making a profit from these early publications proved difficult (Boone et al., 2000). Figure 1-3 is an 1878 illustration from the

Kenneth V. Pilon/www.Shutterstock.com

FIGURE 1-2: Several farmer's almanacs from the early to mid-1800s are shown here. Farmer's almanacs and other agriculture-related publications provided information to help farmers grow crops and raise livestock.

FIGURE 1-3: An 1878 engraving entitled "The Condition of the Turkey" by P.R. Morris and engraver A. Bellenger as published in Cassell, Petter, Galpin & Co.'s *Magazine of Art Illustrated*. (*Courtesy of www.fromoldbooks.org*)

© Cengage Learning 2012

Scott Wallin

Director, Producer Communications, Dairy Management Inc.

Why is agricultural communications important?

I think any aspect of agriculture has a great story to tell. We're all united in agriculture, and that's a great thing to see when farmers from all industries get together and they all have the same goal of providing a safe, abundant supply of food for consumers. I enjoy seeing other agricultural organizations that share those same goals, and I can't stress it enough that we all need to bond together. We all need to tell our story and make people understand that there is a farmer at the start of that food source. It's something that I think we take for granted an awful lot in this country—people really don't understand what it takes to make food in the United States. There's an awful lot of hard work and preparation that goes behind it, and we have such a safe, abundant supply of food, and that's a story that you can never tell too many times. I think it's critical that we are the voice of those farmers because there's no way that they can get off the farm every day and tell the story about themselves. They need communication organizations to learn about their industries and go forward and tell that story to people.

Magazine of Art Illustrated, providing an interesting glimpse into agricultural representations in print from this time.

By the 1840s and 1850s, some scientists in colleges of agriculture began writing for agricultural publications (Boone et al., 2000). In some cities, newspapers, such as the *Chicago Tribune* and *The New York Times,* employed farm writers. Some magazines that were started in the mid- to late 1800s are still being published, such as *Drovers Journal* and *Farm Journal.*

1900s to Today

The early 1900s saw growth in agricultural magazines and rural weekly newspapers that featured farming news. Between 1880 and 1920, the number of farm magazines and newspapers grew from 157 to 400, and circulation of agricultural publications also grew, from about 1 million in 1880 to around 17 million by 1920 (Boone et al., 2000). The growth in farm publications also coincided with the expansion of agricultural lands in the West and Midwest, as Americans moved further west.

During this time, agricultural journalism was recognized as an academic field. The first course in agricultural journalism was taught at Iowa State University in 1905 (Boone et al., 2000), and the first department of agricultural journalism was formed at the University of Wisconsin in 1908 (University of Wisconsin-Madison, 2010).

Beginning in the early 1920s, radio became a major force to communicate agricultural news. Listeners received up-to-the-minute weather and market reports. State agricultural colleges also used radio to broadcast lectures from their Extension agents to rural audiences. The *National Farm and Home Hour,* which was co-produced by the USDA and NBC, was one of the most successful early radio programs. The *National Farm and Home Hour* ran from 1929 to 1944 and was heard six days a week, with a mix of music and celebrity guests. The USDA used the program as a way to present agriculture to the public. After World War II in the late 1940s and early 1950s, television increased in popularity as a way to communicate agricultural topics

Bob Stobaugh

Public Affairs Director, United States Department of Agriculture, Natural Resources Conservation Service

Why is agricultural communications important?

Agricultural communications takes many forms. From a federal government perspective, it is critical to the financial bottom line of many farmers to be informed of government cost-share programs, subsidies, and farm loans available to them. With the number of farmers dwindling, it is important that the ones left have all of the information available to them in order to make informed decisions on the financial well-being of their farms.

to rural and urban audiences. Many of the agriculture-related television programs were broadcast early in the day or at noon when farmers were in from their fields (Baker, 1981).

Agriculture-related television programs are produced at regional and national levels. *AgDay* and *U.S. Farm Report* are current national agricultural television news programs and cover agriculture-related topics of national scope and importance. *AgDay* runs daily, and *U.S. Farm Report* airs once a week. *Market to Market* in Iowa, *Farmweek* in Mississippi, and *Sunup* in Oklahoma are examples of weekly regional agricultural television programs that provide news to their local agriculture audiences. In 2000, RFD-TV, a satellite television channel that focuses on the interests of rural America, was launched. RFD-TV airs programs on agricultural news, music, entertainment, and rural living (RFD-TV, 2010).

Agricultural magazines have become more specialized in content. These publications target specific types of farmers by income, geography, and the agricultural commodity they produce, such as beef, dairy, or vegetable crops. By the mid-1980s, most large newspapers had dropped farm reporters and farm news, due to Americans' migration from rural areas and towns to the cities. Farm reporters were no longer seen as a viable use of a large newspaper's resources. Large newspapers' coverage of agriculture tended to focus on crisis issues, such as food recalls or natural disasters.

From the 1980s to 2000, computers drastically changed how agricultural communicators delivered their messages. Communicators once had to physically cut and paste clip art onto paper to make graphic designs. With the advent of the computer, software programs could be used to make graphics easily and quickly. Video producers had to learn to use computers to edit their video programs. Probably the biggest revolution brought on by computers was how people exchange information. Computer-mediated communication has brought us the Web, a place where people can post information—photographs, videos, audio, and text—in addition to electronic mail and social networking sites, such as Facebook, Twitter, YouTube, blogs, and MySpace.

Current Trends

Agricultural communications remains important. With the decline of newspaper sales in the late 2000s brought on by the increase of Internet-distributed information, belt-tightening at newspapers is likely to continue, meaning agricultural reporters at non-farm newspapers are likely a thing of the past. However, food and home gardening news has increased in newspapers as they try to focus on issues more pertinent to urban homeowners.

Agricultural communications is making the shift from traditional farm press and broadcast to new delivery methods, such as Web pages, online videos, and social media. As a result, communication professionals

© Cengage Learning 2012

Bob Stobaugh

Public Affairs Director, United States Department of Agriculture, Natural Resources Conservation Service

What do you do as an agricultural communicator?

As an agricultural communicator my job is to inform agricultural producers of the technical and financial assistance available to them through my agency. I work closely with newspaper, television, and radio reporters to inform the agricultural community of the assistance that USDA offers to eligible producers. I also work with agricultural associations to inform their members of the services that we provide.

are learning how to integrate several communication methods to provide their message to targeted audiences. Many of these trends will be discussed in more detail throughout this textbook.

Commonalities throughout History

Regardless of the historical period, agricultural communicators must perform the following actions to be successful. These practices are still as important today as they were more than 200 years ago:

- *Agricultural communicators target specific audiences.* Early audiences targeted by agricultural communicators were farmers and livestock producers. More recently, consumers have become a major audience for agricultural communicators.
- *They develop an important message that needs to be communicated.* From the 1700s through the early 1900s, agricultural communicators' messages focused mainly on providing agricultural producers with information on how to more effectively grow crops and raise animals. Now, a large part of what agricultural communicators do is communicating positive messages about agriculture to consumers, lawmakers, and other groups who make decisions affecting policies that impact agriculture.
- *They use the most appropriate medium, such as face to face, print, television, radio, or the Web, for communication.* Print and face-to-face

messages were the predominant communication methods agricultural communicators used until the early 1900s. Since then, other media, including radio, television, and the Web, have given agricultural communicators more choices to target their messages to various audiences.

CAREERS IN AGRICULTURAL COMMUNICATIONS

Every five years, Purdue University and the USDA's Cooperative State Research, Education, and Extension Service release a report titled *Employment Opportunities for College Graduates in the U.S. Food, Agricultural, and Natural Resources System.* The 2010–2015 edition reported that around 6,200 annual job openings would occur during this time period in education, communication, and governmental services occupations involved with agricultural and food systems, renewable resources, and the environment (United States Department of Agriculture, Cooperative State Research, Education, and Extension Service, 2010). This amount accounts for 11 percent of the total available positions in the nation's food, agricultural, and natural resources systems.

Types of Jobs

Any career that might be found in a traditional journalism and communications program of study has a counterpart in agricultural communications. The

© Cengage Learning 2012

Jim Bret Campbell

Senior Director of Marketing and Publications, American Quarter Horse Association

What do you do as an agricultural communicator?

My work as an agricultural communicator is to help promote the welfare and continued responsible use of horses, in general, and to promote the American Quarter Horse as the world's most versatile breed.

career options are varied and numerous. Following are some—but not all—of the career opportunities for professional agricultural communicators:

- *Reporters* cover agricultural news for print (newspapers and magazines), broadcasting (television and radio; Figure 1-4), and online media (Web, electronic mail newsletters, and social media).
- *Public relations professionals* promote positive messages about an agricultural organization or product to reporters and the community.
- *Advertising professionals* create messages that influence people to buy a product or service.

- *Educational material developers* develop materials, such as videos, documents, and presentations, for instructional purposes. Many educational material developers are employed in land-grant universities around the country.
- *Photographers and videographers* are adept at using a digital camera or video camera to communicate agriculture-related topics.
- *Graphic designers* take words and make them more pleasing to the eye and more understandable. Graphic designers use elements of effective document design in their jobs (Figure 1-5).

©auremar/www.Shutterstock.com

FIGURE 1-4: Videographers capture agricultural images that are communicated to a wide range of audiences.

© Cengage Learning 2012

FIGURE 1-5: Graphic designers use software to create digital media, such as print, Web, and video materials, for various audiences.

© Cengage Learning 2012

Jennifer Latzke

Associate Editor, High Plains Journal

What do you do as an agricultural communicator?

I think I have the greatest career in communications. I am an associate editor for *High Plains Journal/Midwest Ag Journal*. We're a weekly publication that goes out to 52,000+ subscribers in 12 states across the middle of the country. Our sole purpose is to report on the issues that affect farmers and ranchers in our coverage area. We tell them about national and state policies that may hit their bottom lines. We teach them about new production methods and tools. We report on their commodity group meetings and the activities of the organizations that work on their behalf. But, best of all, I get to tell the story of good, honest, hard-working people who bring the groceries to our nation's table. I get to share their triumphs and their sorrows and help spread the word that agriculture is the most noble of occupations.

- *New media or social media specialists* encompass the different types of new and future media that can be used in communication. Some of the current "new media" are Web pages with videos and podcasts, and social networking programs such as Facebook.
- *Communication specialists* is a generic title for the professionals who handle the day-to-day communication needs in organizations. Communication specialists may write news releases, develop presentations, construct educational displays, or arrange for reporters to visit an agricultural company.
- *Agricultural communications educators* teach students in high school and universities how to communicate more effectively. Many agricultural communications educational programs are beginning in public schools and universities across the country.
- *Extension specialists*, such as county extension agents, use appropriate media to

communicate messages to various audiences, ranging from farmers to young mothers to homeowners.

Places for Jobs

Agricultural communicators are employed in many areas in agriculture. Following are a few places you could find agricultural communicators:

- Agricultural publications (newspapers, magazines)
- Agricultural broadcasting (radio and television networks)
- Agricultural colleges that have communication specialists
- Commodity associations
- Governmental agencies, such as the USDA or the Food and Drug Administration
- Public relations and advertising agencies

You also may wish to visit AgCareers.com (http://agcareers.com) to search for jobs in agricultural communications and any other agriculture-related field.

Agricultural Communications Organizations

If you are interested in learning more about the agricultural communications profession, contact any of these professional organizations.

Association for Communication Excellence in Agriculture, Natural Resources, and Life and Human Sciences: Communication professionals working in universities, government agencies, and research organizations in the public and private sectors (http://aceweb.org).

Agricultural Relations Council: Professionals specializing in public relations and public affairs serving the agricultural and food and fiber industries (http://agrelationscouncil.org).

American Agricultural Editors' Association: Agricultural editors, writers, and photojournalists (http://ageditors.com).

American Horse Publications: Communication professionals in the equine publishing industry (http://americanhorsepubs.org).

Cooperative Communicators Association: Communication professionals employed in cooperatives (http://communicators.coop/CCA).

International Federation of Agricultural Journalists: Professional association for agricultural journalists in 29 countries (http://ifaj.org).

Livestock Publications Council: Communication professionals in the livestock publishing industry (http://livestockpublications.com).

National Agricultural Communicators of Tomorrow: Organization of college students learning about the agricultural communications profession (http://nactnow.org).

National Association of Farm Broadcasting: Communication professionals at farm broadcast stations and networks and the agri-marketing community of companies and agencies (http://nafb.com).

North American Agricultural Journalists: Journalists in North America who report or edit agricultural news for newspapers, magazines, and syndicated services and are independent of agricultural organizations and businesses (http://naaj.net).

National Agri-Marketing Association: Communication and marketing professionals in agricultural marketing (http://nama.org).

Turf and Ornamental Communicators Association: Communication professionals involved in green industry communications (http://toca.org).

LEARNING MORE ABOUT AGRICULTURAL COMMUNICATIONS

Many professional organizations related to agricultural communications exist, some of which are described in the boxed feature. Contact these agricultural communications organizations to learn more about this field. Many colleges of agriculture have agricultural communications majors or minors. Contact a college of agriculture near you to find out if it has an agricultural communications program. Last, the largest repository of agricultural communications information is the Agricultural Communication Documentation Center of the University of Illinois. The Agricultural Communication Documentation Center's website has more than 30,000 documents related to agricultural communications (University of Illinois, 2010).

SUMMARY

This chapter provided a definition of agricultural communications, a brief history of the agricultural communications field, an explanation of some of the jobs available in agricultural communications, and a list of professional agricultural communications organizations. The information in this chapter provides a historical context for the remainder of this textbook.

CHAPTER EXERCISES

APPLICATIONS

Following are some ways you can apply what you learned in this chapter:

- Brainstorm other career areas where you would find agricultural communicators.
- Do an Internet search for national agricultural communications organizations and contact a local professional. Invite the professional to a class meeting to find out what he or she does.
- "Shadow" an agricultural communicator for a day to see what he or she does.
- Research some aspect of the history of agricultural communications and write a report or create a presentation about it.

CHAPTER QUESTIONS

MULTIPLE CHOICE:

1. _____ is a satellite television channel that focuses on the interests of rural America.
    a. CBS
    b. NBC
    c. RFD-TV
    d. FOX

2. The _____ is an organization of college students who are learning about the agricultural communications profession.
    a. National Agricultural Communicators of Tomorrow
    b. National Association of Farm Broadcasting
    c. Livestock Publications Council
    d. National Agri-Marketing Association

3. Approximately _____ percent of the American labor force worked in agriculture by the year 2000.
    a. 10
    b. 19
    c. 9
    d. 1.9

4. _____ take words and make them more pleasing to the eye and more understandable by using elements of effective document design in their jobs.
    a. Reporters
    b. Graphic designers
    c. New media specialists
    d. Photographers

5. _____ is the exchange of information about the agricultural and natural resources industries through the most effective and efficient media (e.g., newspapers, magazines, television, radio, the Web, and others) available for the appropriate audience.

    a. Communications

    b. Agriculture

    c. Agricultural communications

    d. Targeted communications

6. The first agricultural journalism course was taught at _____.

    a. the University of Michigan

    b. the University of Missouri

    c. Texas A & M University

    d. Iowa State University

7. Circulation of agricultural publications grew from about 1 million in 1880 to around _____ by 1920.

    a. 17 million

    b. 20 million

    c. 15 million

    d. 30 million

8. The Agricultural Communication Document Center is hosted at _____.

    a. Texas A & M University

    b. the University of Florida

    c. the University of Illinois

    d. Iowa State University

9. The National Farm and Home Hour was co-produced by the USDA and _____.

    a. NBC

    b. CBS

    c. ABC

    d. FOX

10. The first American agricultural magazine was _____.

    a. *Drover's Journal*

    b. *Farm Journal*

    c. Agricultural *Museum*

    d. *Dairy Today*

11. _____ is considered the father of American agricultural journalism.

    a. Benjamin Franklin

    b. Thomas Jefferson

    c. John Stuart Skinner

    d. Frank Mullen

12. The oldest continuously published periodical in North America is _____.

    a. the *New Jersey Gazette*

    b. *Farm Journal*

    c. *Dairy Today*

    d. *The Old Farmer's Almanac*

13. The USDA reported that the percentage of the U.S. population living on a farm decreased from around 40 percent in 1900 to less than _____ percent in 2000.

    a. 10

    b. 5

    c. 2

    d. 14

14. Which of the following is NOT a weekly regional television program that focuses on agriculture?

    a. *AgDay*

    b. *SunUp*

    c. *Market to Market*

    d. *Farmweek*

15. The *Employment Opportunities for College Graduates in the U.S. Food, Agricultural, and Natural Resources System* report noted that between 2010 and 2015, around _____ annual job openings would occur in education, communication, and governmental services occupations involved with agricultural and food systems, renewable resources, and the environment.

    a. 5100

    b. 6200

    c. 7500

    d. 10,000

REFERENCES AND FURTHER READING

Baker, J.C. *Farm Broadcasting: The First Sixty Years*. Ames, IA: Iowa State University Press, 1981.

Boone, K., T. Meisenbach, and Tucker, M. *Agricultural Communications: Changes and Challenges*. Ames, IA: Iowa State University Press, 2000.

Dimitri, C., A. Effland, and N. Conklin. *The 20th Century Transformation of U.S. Agriculture and Farm Policy*. Washington, DC: USDA, 2005. http://www.ers.usda.gov/publications/eib3/eib3.htm.

Fletcher, S. W. *The Philadelphia Society for Promoting Agriculture, 1785–1955*. 1976. http://www.pspaonline.com/archive/pspa1959.pdf.

Molotsky, I. *The Flag, the Poet & the Song: The Story of the Star-Spangled Banner*. New York, NY: Plume, 2000.

Mott, F. L. *A History of American Magazines, Volume I: 1741–1850*. Cambridge, MA: Harvard, 1930.

RFD-TV. "RFD-TV: Rural America's Most Important Network." Last modified 2010. Accessed October 23, 2010. http://rfdtv.com.

The Old Farmer's Almanac. "*The Old Farmer's Almanac*: About Us." Last modified 2010. Accessed October 23, 2010. http://www.almanac.com/content/about-us.

United States Department of Agriculture, Cooperative State Research, Education, and Extension Service. *Employment Opportunities for College Graduates in the U.S. Food, Agricultural, and Natural Resources System*. Washington, DC: USDA, 2010. http://www.csrees.usda.gov/nea/education/part/education_part_employment.html.

INTRODUCTION

The Merriam-Webster Online Dictionary (2010) defines **communication** *as "a process by which information is exchanged between individuals through a common system of symbols, signs, or behaviors." Communication is a fundamental part of our lives, and it starts early. As babies, we learn that we must be able to communicate effectively with our parents if we want to eat, be picked up, or get our diapers changed (Figure 2-1). Later, we learn to communicate with siblings, friends, relatives, and, eventually, the outside world.*

The "what" and "how" of communication is the **message***. To be an effective communicator, it is important to think about organizing all of the aspects of a communication activity, from the information you want to convey to how*

© Cengage Learning 2012

FIGURE 2-1: Communication, a fundamental part of human life, begins in infancy.

it is packaged and delivered. Developing effective communication messages is an important skill. Learning how to get your message across successfully holds great benefits for you as an individual, as well as for any group or organization for which you may be communicating now or in your future career (Figure 2-2). If the receivers of your message do not understand it, do not pay attention to it, or do not think it is important, they will not act on it. For example, if you want students at your school to come to a fundraiser for a student organization that you belong to, the message you send out needs to be accurate, complete, and focused around an

organizing theme that attracts attention. Maybe you will put that key message in the headline of a flier that you can post around school. To be effective, you will need to include other message elements, such as an interesting visual, key information points, and accurate grammar and punctuation.

This chapter is designed to help you understand the process of communication and how it affects message development. You will learn about the elements of a message, what you need to know about the audience for your message, and how to determine what you want to communicate and why. This chapter will show you how to position your messages to be effective and provide some tools you can use to write more effectively. The chapter will also explain how messages are developed for different types of media and how visuals and presentations can be used to increase message effectiveness.

FIGURE 2-2: This poster uses humorous visuals to reinforce its message, "save the trees."

THE PROCESS OF COMMUNICATION

Communicating well or poorly can influence success or failure in every kind of human relationship. Today's world requires well-developed communication skills, including the skill of developing a message that is easily understood or "decoded" and that gets results. Think of any memorable conversation that you have had. You heard the actual words and spoke words of your own, but there was also a rich subtext of information that you were receiving via nonverbal cues such as facial expression, body language, and voice inflections. Even the setting you were in—the place, the time of day, and whether other people were around—conveyed additional information.

In our writing, unlike the spoken word, we do not have all of these rich feedback mechanisms to add context to our communications, so we have to rely on the

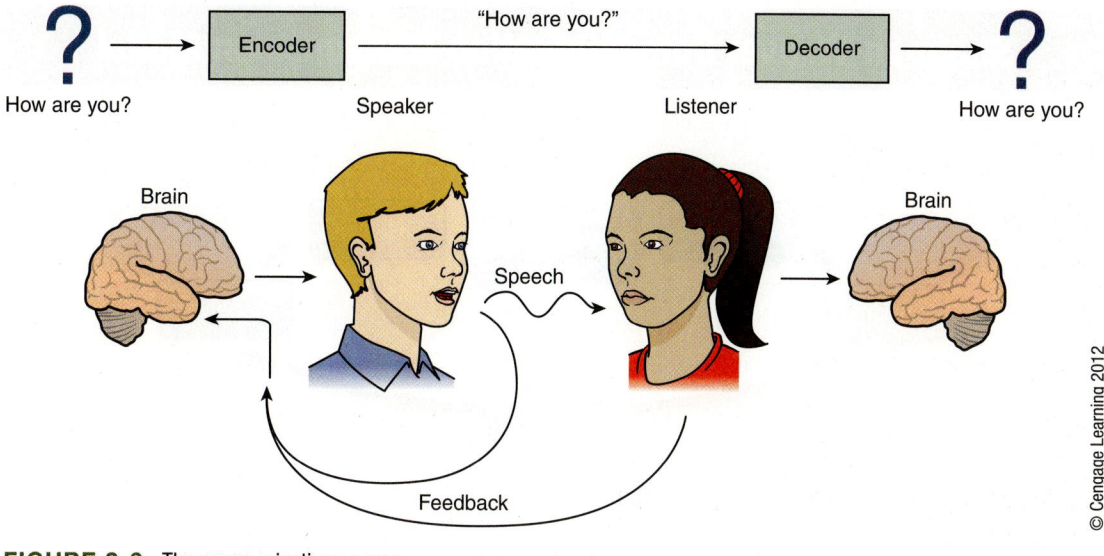

FIGURE 2-3: The communication process.

way we organize things—our style, tone, and sometimes the use of special tools, such as dialogue—to convey information. Communication, both written and oral, has to be "**encoded**," that is, put in a specific form by the **sender**, using language, symbols, and metaphors (Figure 2-3). The communication is then sent through a **channel** (e.g., writing or speaking) and then must be **decoded** by the **receiver**. Sometimes the receiver is good at decoding the message, but other times, the receiver confronts barriers or **noise**—maybe he or she cannot understand what the sender wrote or does not find it credible, or the condition in which it was sent is a problem. The receiver interprets what can be interpreted and communicates something back in the way of feedback. **Feedback** is the receiver's response—for example, a friend saying, "good report," or your teacher saying, "I liked that essay." That is essentially the model for communications both written and oral, but you can see that with written communications, the possibilities for misinterpretation are greater, and the feedback is even more important.

Probably one of the most widely used models that represents the process of communication is the *Shannon–Weaver Communication Transmission Model,* or **Shannon–Weaver model** (Shannon & Weaver, 1949), which provides an easy-to-understand illustration of the communication process, in which a **source** of information (such as a speaker) transmits a message through a channel to a receiver who decodes the message (Figure 2-4). In transit, the message can be affected by noise, as previously mentioned, which is anything that can reduce the clarity of the message.

The arrows in the model show the direction of messages. In later adaptations of the Shannon–Weaver model, an arrow from the receiver back to the source—representing feedback—closes the communication loop. The next time you communicate with someone, think about how this model works.

TYPES OF COMMUNICATION

Communication occurs at four levels:

Interpersonal, or *dyadic*, communication is between two persons, such as a couple, friends, neighbors, or family members.

Small-group communication is among members of a group or team, such as the members of a student organization, an athletic team, a church youth group, or a work team.

Public communication is when someone communicates directly to a large audience, such as a motivational speaker at a conference, a pastor during a church service, or an announcer at a football game.

Mass communication involves communicating to an audience through a media channel that reaches large numbers of people at the same time.

To communicate effectively, you need to select the right type of communication to meet your goals. If you want to recruit more members for a student organization, for example, interpersonal communication may be the best way to do so, because communicating face to face has the most impact and opportunity to influence others. The downside is that you can only reach a small number of

Shannon-Weaver Communication Transmission Model

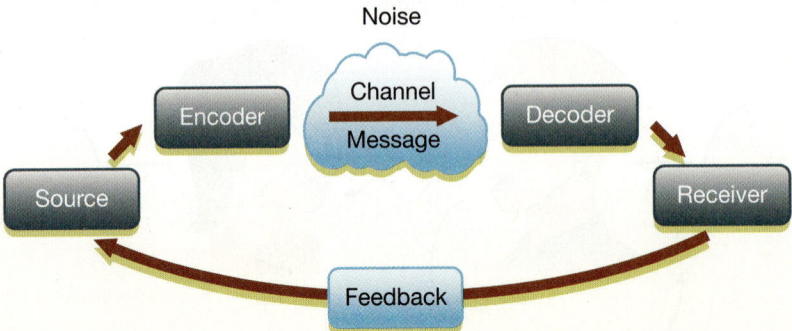

Source: The person who originates the message that is to be sent.

Encoder: The source decides what will be communicated, what will be **encoded** (put in terms the receiver will understand), and how to transmit the message to the receiver.

Message: The verbal or nonverbal component that produces meaning in the mind of the receiver.

Channel: The means by which the source conveys or transmits the message. Examples include our voice, body language, visual aids, television, radio signals, or print media.

Decoder: The verbal or nonverbal message is then **decoded** by the receiver. In the case of messages received via mass media, a decoding device (such as a television or radio set) translates the signal back into the original message that a "receiver" can understand.

Receiver: The destination for the message. The signal is received by this person or persons. The source hopes that the message that is received and **decoded** by the receiver is, in fact, the original message. However, the impact of noise can cause the message to be misunderstood.

Noise: Anything that reduces the integrity or clarity of the message. Noise can be categorized into two types: **semantic** and **mechanical**.

> **Mechanical noise** reflects the environment in which the communication occurs. This can be static on the radio, the sound of telephone calls, loud music or voices in a restaurant when you are trying to hear a conversation, smudged ink on a paper, or missing paragraphs in a report.

> **Semantic noise** refers to the process of communication, and can include such examples as vocabulary that is mismatched to the receiver or a difference in education, interests, culture, age, and/or gender between the source and receiver.

Feedback: The receiver's response to the message. Feedback can be in the form of written or spoken communication or behavior through which the receiver either directly or indirectly indicates a response.

FIGURE 2-4: Shannon–Weaver Communication Transmission Model. (*Adapted from Shannon & Weaver, 1949*)

FIGURE 2-5: This photo shows interested students attending a recruitment meeting for a student organization.

Frame of Reference

In this illustration, the two circles represent two persons' individual backgrounds and experiences. When these two people meet, the area of overlap—in light blue—represents the **"frame of reference"** that they share. It could be an experience that they both have had or something in their background that is similar. A frame of reference is where true communication and understanding occur.

FIGURE 2-6: Frame of reference.

people at a time. Communicating to a group of students, either face to face or in an audience for a speech, reaches more people, but the impact is diluted because there is less opportunity for the audience to make a personal connection with the speaker. Using a form of mass media such as print, radio, television, or the Web lets you reach even larger numbers of people in an audience, but the impact is even more diluted.

One way to work around this is to combine the types of communication for maximum impact. Developing a recruitment brochure might be a good way to get the word out to many students simultaneously about the benefits of membership in a student organization, but it will be even more effective if you combine it with other types of communication, such as organizing a meeting for interested students (Figure 2-5).

KNOWING YOUR AUDIENCE

Have you ever had the experience of meeting someone and finding that you just "clicked" with that person? Before you know it, you are swapping stories and sharing experiences like you have known the person for years. When you are communicating well with someone, it is because you share a frame of reference. You understand

the person well enough to make an impact and develop mutual understanding. This shared understanding—a **frame of reference**—can occur because you come from similar backgrounds, have had common experiences, or hold similar beliefs (Figure 2-6). The point at which frames of reference overlap is said to be where true understanding occurs.

You can develop a frame of reference with another person or persons or even an audience by trying to find out how they might think, feel, and behave. What does the audience for your message care about or fear? What do audience members need? What are they like?

In order to understand your audience, you need to be able to describe common characteristics that audience members may share. Not everybody in an audience is exactly the same, but we can categorize groups of people based on some common traits and shared characteristics. Describing an audience on the basis of shared characteristics is called **audience analysis**.

Why would you want to describe the audience for which you are developing messages? Suppose you are someday the executive director of an agricultural commodity association in your state. You decide it would be

Rod Hemphill, APR *(Accredited in Public Relations)*

Director, Public Relations and Communication (retired), Florida Farm Bureau

What exactly is your message?

This crucial question is seldom easy to answer. Once you have answered it to your satisfaction, think about the audience you wish to reach, and make sure you are stating the message in terms your target audience will understand. How you develop your message largely determines whether your communications efforts will succeed or fail.

a good idea to put together an awareness campaign to educate members of the public about the contributions of agriculture to the state's economy and quality of life. Before you can even begin to think about the message, you have to decide whom to target. What are some potential audiences that you might want to reach?

- The general public?
- Community leaders?
- State legislators?
- Members of the media?
- Urban, suburban, or rural?
- Old or young?

The answers to each of these questions will lead you in a different direction when it comes to deciding what to put in your message. As you begin the process of analyzing your audience, consider these questions:

- Who are the audience members?
- What are their educational backgrounds and experiences?
- What do they know about the subject?
- What do they expect of the communication?
- How can they best be reached?
- What do they need and value?

These questions help you start the audience analysis, as you identify more characteristics of your target audience. These audience characteristics are divided

into two categories: demographics and psychographics. **Demographics** can be thought of as characteristics about audience members that are hard to change, such as gender, age, income, education level, and place of residence. **Psychographics** combine the attitudes and values people have with their lifestyle choices. Attitudes are a combination of the beliefs and perceptions used to evaluate something, such as what a person thinks about a particular product or the opinion a person has about something. Lifestyles are how people like to live. For example, you and your parents may like to live in a particular area, in a particular kind of house, and may engage in specific recreational activities. These lifestyle choices are shared with others, and they may influence your attitudes and perceptions. Over time, researchers and marketing communicators have developed numerous categories that help describe how beliefs, lifestyles, and experiences help determine a person's outlook on life and how people might respond to messages that are communicated to them.

If you can identify audience members' demographics and psychographics, you can make assumptions about how best to develop a persuasive message targeted to them. You may wish to use an audience analysis matrix to help you identify key characteristics about your audience, and then use it to develop a message for that audience (Figure 2-7).

Audience Analysis Matrix

Demographics

- Gender (male and female)

- Age

- Education level (high school, college, graduate/professional degree)

- Ethnicity

- Geographic location (where you live—urban, suburban, rural)

Psychographics

- Values and beliefs (in general and in relation to the topic you are communicating)

- Attitudes (specific perceptions and/or misperceptions about the topic of your communication)

- Lifestyles (interests and activities shared by your audience that may influence their attitudes toward your topic)

Information sources and channels

- What sourses of information does your audience access to learn about your topic

- What delivery channels does your audience use to get information

Prior knowledge and experience

- What do audience members already know about your topic

- In general, how informed are they

- Have they had previous experiences that may be related to your topic

An audience analysis matrix allows you to characterize your audience for targeting purposes.

© Cengage Learning 2012

FIGURE 2-7: Audience Analysis Matrix.

TOOLS TO HELP YOU WRITE EFFECTIVELY

Communication is both written and oral. For most professional communication in the workplace, the foundation of good communication is writing. Therefore, the tools mentioned in this section are focused on effective writing, but the principles can be used for effective oral communication as well.

Effective writers think about their purpose before they write and try to develop a plan for their writing so as to achieve their main purpose or goal. That means understanding the audience and selecting writing tools designed to be effective in communicating to the intended reader. If you think of writing as a message you are trying to send, then you should ask yourself these questions before you start writing—the answers will help you develop your writing plan:

- What is the ultimate message I want to convey?
- What are the key points I want to make?
- What supporting points or evidence can I add to establish credibility for the points I am making?
- How can I organize my points visually to produce the desired result? (Should I use bullet points, headings and subheadings, bolding, underlining and italics?)
- What language and word choices can I make to get the level of understanding I want to achieve or have the reader identify with what I am saying?

If you can answer all of these questions, you have begun to develop your message strategy. A **message strategy** is a conscious decision to communicate with some purpose in mind. Starting with a purpose in mind can help you be clearer and better able to eliminate elements that do not contribute to the message, while adding other elements that do, in the editing process.

Writing without a strategy in mind is easy to spot; it is usually apparent when you read something that lacks focus and wanders off into tangents that don't relate to the main purpose or key points. Some general writing tips to keep in mind are as follows:

- *Use a thesis statement.* A **thesis statement** states your main point and purpose for writing and is a way of getting your key message across to the reader. Thesis statements are usually at the beginning or end of the introductory paragraph to an essay, report, or other long-format document.

- *Concentrate on those areas in your writing that you need help with by taking your time and using reference materials such as a dictionary or thesaurus.* Pay extra attention to these areas when you edit your writing. Go back and check your work for comma placement, for example, if that is an area where you have issues. If you do it consistently, it becomes a habit.
- *Read your work aloud, every time.* This is the one action that you should do every time. Read aloud slowly, stopping for punctuation; your ear will pick up mistakes.
- *Read good writers.* Your writing will improve the more you read excellent writing.
- *If you cannot find time to read, read your textbooks for your classes thoroughly; review how a sentence or paragraph you especially like is structured.* Take advantage of those times when you have to read in order to appreciate a writer's skill and then dissect how something was written.
- *Save good examples of writing.* When you need to write something similar for class, pull out your good example to use as a model. You do not want to replicate it, of course, but use it as a guide on organization, style, and tone.
- *Vary your sentence structure, and make sure your writing is interesting to read.* Make sure each paragraph has a topic sentence and supporting sentences, and use examples to explain your key points.

Effective communication occurs by actively engaging the reader and using tools that enhance the communication process; otherwise, all you are doing is presenting a set of unconnected facts. Effective writers strive for maximum readability of their writing. What that means is the level of writing and the use of words and even acronyms fit the audience's reading level and understanding. **Readability** is the ease or difficulty with which the reader can understand your writing. The general idea is to try to tailor your writing to what you think your audience can understand.

To be an effective writer, use correct grammar and punctuation for both readability and credibility and use writing tools, such as the use of a message strategy, to enhance what you say and to make it clear to your audience. To do that, choose your words and phrases carefully, use sentences that have an action verb in them where possible, and try to make your writing interesting.

Paragraphs are a key to effective writing, and one of the best things you can do to improve your writing is make sure you use one main idea and a topic sentence in each paragraph. Another is to break long, complicated sentences into smaller sentences. Pay attention to organizational structure as well; when you write, use examples and metaphors to establish a stronger frame of reference. However, the most important aspect of effective writing is always to keep the audience in mind. If you can think about why you write, what you want the ultimate outcome of your writing to be, and whom you are writing for, you will go a long way toward making lasting improvements in your writing.

ELEMENTS OF A MESSAGE

A message is a brief statement focused around one main idea or theme that is designed to persuade a specific audience to think, believe, or behave in a certain way. Messages can be delivered through face-to-face communication—which includes language, symbols, and nonverbal communication, including body language, actions, gestures, and facial expressions—or they can be directed to a person, small group, or large audience through a media channel, such as an advertisement, flier, newsletter, letter or report, or video or audio clip. Messages help organize what you want to say in a way that prompts an audience to support you and your group or organization, or they can be the basis for developing a communications campaign that uses mass media to reach an audience.

A good message is short, easy to understand or remember, does not contain jargon or acronyms, and can be repeated without seeming redundant. Messages can be used as **catchphrases** in headlines or titles, as **taglines** and slogans, and in summaries or conclusions of a report. Here are some examples of messages that have been used as slogans in campaigns:

- Runs like a Deere™. (John Deere)
- The other white meat™. (Pork Checkoff program)
- It's what's for dinner™. (Beef Checkoff program)
- Learning to Do, Doing to Learn, Earning to Live, Living to Serve. (National FFA Organization)

Typically, a strong message:

- is focused on a problem, issue, or opportunity.
- positions what you want to say in a positive way.
- can be "boiled down" to a short format, such as a slogan or tagline.
- targets a specific audience and takes into account how the audience is likely to interpret the message.

© Cengage Learning 2012

Susan Howard, APR *(Accredited in Public Relations)*

Director, Corporate Communications, DUDA

With the ability of the news media to pick and choose what stories to report, and with the pressure of daily deadlines for reporters, it can be difficult to get your message through. That is why it is important to learn how to develop key messages about your news event or issue that go to the heart of your news.

MESSAGES IN THE MEDIA

Putting your message in a form of mass media—print, radio, television, or the Web—is a good choice when you want to gain the support of a group or attract the interest or attention of a specific audience to a product, service, or organization. Marketers and advertisers do this when they develop campaigns that use a persuasive slogan or catchphrase. Individuals and organizations do this as well. For example, does your school have a tagline or slogan? Is it used on your school's website? Have you ever thought about what this message says about you and your school?

When developing a message to be used in a form of media, such as a flier, print advertisement, brochure, or video, you should:

- Develop your key message and two or three subpoints that clarify or provide information in support of your message theme.
- Put the most important information first, followed by the less important information and then the minor details.
- Use attention-getting visual devices, such as a logo, typeface, and color, to make your message stand out in print and on the Web. (See Chapter 5, News Media Writing, for more on writing for the media, and Chapter 6, Document Design, for document design techniques.)

Music, digital effects, animation, screen titles, and audio effects are examples of attention-getting devices for video. (See Chapter 10, Video and Audio Production, for more on video production techniques.) Be consistent in how you use these elements in all of your communication.

Let's explore an example of how a message can be created to communicate a main point about an organization. The Florida Cooperative Extension Service has offices in each of Florida's 67 counties. Since the organization's inception, Florida Extension agents have reached out in their communities to bring the research-based knowledge generated at the University of Florida to the state's residents. Traditionally, that has been conducted primarily through personal, face-to-face channels—such as presentations and field days—and, to a lesser degree, mass media methods (radio, television, newspaper). In May 2006, Florida Extension launched a new website that features the message "Solutions for YOUR Life." Based on feedback from an audience analysis and a focus group, the message was incorporated into the Web address, so that it would be easy to remember (http://solutionsforyourlife.com). This message was developed in order to communicate that Florida Extension focuses on providing solutions to the citizens of Florida.

VISUAL AIDS AND PRESENTATIONS

When making presentations to a small group or audience, it is becoming more common to use computer graphics and slide presentations as visual aids to emphasize the main points of messages. Today, computer-generated slide presentations are usually developed with the help of **presentation software,** such as Microsoft's PowerPoint or Apple's Keynote. With these tools, you can combine text and graphics and include color, bullets, and backgrounds to provide emphasis and keep your audience's attention focused on what you are saying. Computer-generated slide presentations are usually developed and graphically enhanced by the presenter, although sometimes you might find a **presentation template** you can use that includes an already-thought-out color scheme and formatting selections. Templates are often available as part of the software package you are using or they can be downloaded from websites. (Use an Internet search engine and type in "slide presentation template" as the search term.) You will learn more about designing computer-generated slide presentations in Chapter 8, Visual Communication.

Follow these guidelines when writing for computer-generated slide presentations:

- Always use either present or future tense.
- Do not use complex narrative sentences for your message points; condense these to short bullet points.
- Use action verbs and illustrate with a graphic element where possible, but do not overdo the graphics. One per slide is often enough.
- Summarize what you plan to talk about in outline form, with a definite introduction, body, and conclusion structure.
- Keep your slides simple: one idea or message with supporting points per slide.
- Include space around the top, sides, and bottom of each slide, and try to keep this consistent from slide to slide.
- Limit the number of words used on each line to no more than six, and the number of lines per slide to six or less.
- For maximum readability in different lighting situations, use contrasting colors that include either a dark-colored background with light or white text, or a light background with dark-colored text.
- Do not include a lot of animation effects or sounds. These can seem too "busy" and distracting to an audience.

SUMMARY

This chapter introduced you to the process of communication and the elements needed when developing a message. Four types of communication were described, and we learned about the concept of audience analysis. The chapter included tips and tools concerning how to write more effectively, along with information on how messages are used in the media and are combined with visual aids and presentation software to create effective communication.

CHAPTER EXERCISES

APPLICATIONS

Following are some ways you can apply what you learned in this chapter:

- Develop a message slogan or theme for use on a flier or brochure announcing an upcoming event.
- Watch television or read a magazine to find a campaign for a product, good, or service. Use the audience analysis matrix to identify characteristics of the audience you think the campaign is trying to reach.
- Develop a presentation focusing on your solution to some problem or issue. Use the tools described in the section on message strategy to develop your key message points and subpoints. Include a thesis statement, and organize your presentation for maximum impact using the tips provided in this chapter.
- Develop a computer-generated slide presentation to use along with the presentation you developed above.

CHAPTER QUESTIONS

MATCHING:

_____ 1. The "what" and "how" of communication make up the _____.

_____ 2. Effective writers think about their _____ before they begin to write.

_____ 3. _____ can negatively affect the message being sent.

_____ 4. When creating computer-generated _____, include formatting, animation, and effects to capture the viewer's attention.

_____ 5. _____ communication is between two persons, such as a couple, friends, neighbors, or family members.

_____ 6. _____ occurs among members of a group or team, such as the members of a student organization, an athletic team, a church youth group, or a work team.

_____ 7. _____ communication is when someone communicates directly to a large audience, such as a motivational speaker at a conference, a pastor during a church service, or an announcer at a football game.

_____ 8. _____ communication involves communicating to an audience through a media channel that reaches large numbers of people at the same time.

a. slide presentations

b. interpersonal

c. noise

d. mass

e. message

f. purpose

g. small-group communication

h. public

MULTIPLE CHOICE:

9. _____ involves communicating to an audience through a media channel that reaches large numbers of people at the same time.

    a. Interpersonal communication

    b. Small-group communication

    c. Mass communication

    d. Public communication

10. _____ describe who audience members are and include characteristics that are difficult to change.

    a. Demographics

    b. Psychographics

    c. Infographics

    d. Segmentations

11. _____ is the means by which the source conveys or transmits the message.

    a. Channel

    b. Encoding

    c. Decoding

    d. Noise

FILL IN THE BLANK:

12. Describing an audience on the basis of shared characteristics is called _____.

13. _____ is the ease or difficulty with which the reader can understand your writing.

14. _____ is a conscious decision to write with some purpose in mind.

15. Shared understanding arising from common experiences and similar backgrounds and beliefs is called _____.

REFERENCES AND FURTHER READING

Merriam-Webster Online Dictionary. "Communication." Last modified 2010. Accessed May 19, 2007. http://www.merriam-webster.com/dictionary/communication.

O'Hair, D., and M. Wiemann. *Real Communication*. Boston, MA: Bedford/St. Martin's, 2009.

Shannon, C., and W. Weaver. *The Mathematical Theory of Communication*. Urbana: University of Illinois Press, 1949.

CHAPTER

3

Research Methods Used in Communication

OBJECTIVES

After completing this chapter, the student will be able to:

- Identify different types of communications research methods.
- Determine the differences between a population, a sample, and various types of samples.
- Write questions for surveys or focus groups.

INTRODUCTION

In this chapter, you will be introduced to some of the research methods that communicators use, beyond face-to-face interviews, which will be discussed in Chapter 5, News Media Writing. Some of the research tools and how you can use them to help you carry out your responsibilities as a communicator will be explained. In fact, every effective communicator needs to understand the various research methods and how to make use of them (Figure 3-1).

SAMPLING AND AUDIENCE ANALYSIS

One reason to conduct research is to attain information about target audiences. Communicators have to develop messages for particular audiences. When they do so, it is called a "**target audience**." For example, a television commercial featuring rap music and quick video shots of a high-speed sports car might be just right for an audience of 18- to 21-year-old college students, but completely inappropriate for an audience of senior adults (over

FIGURE 3-1: Effective communicators understand the different research methods available to them.

60 years of age). Analyzing your audience, therefore, is an extremely important aspect of communication. In fact, many researchers and communication practitioners say that *understanding your audience* is the most important step in the communication process and that the most polished message is wasted if it has not been designed with the audience in mind. What does an audience analysis consist of? Recall from Chapter 2 that in most cases, an audience analysis is used to provide a detailed description of the demographic and psychographic characteristics of its members (see Chapter 2, Effective Communication and Message Development, for definitions and examples of demographic and psychographic characteristics). Here are some components to keep in mind:

Demographics include characteristics that are measurable:

- Age
- Education/learning level
- Economic status
- Location

Psychographics include characteristics that are behavioral or cognitive:

- Lifestyles
- Values and beliefs
- Habits
- Perceptions
- Attitudes

To learn about a particular audience firsthand, the communicator must perform research by collecting data of some type on that audience. The largest group possible to study is a population. A **population** is the overall group you want information about. Example populations include the American public, stay-at-home mothers in a particular city, and cattle producers in your area. But even in smaller populations (such as the cattle producers in your area), it would be highly inefficient to collect data on each person. Therefore, **sampling** is used to represent an entire population in a way that allows you to collect data from a smaller subset that shares the same characteristics as the larger population.

Sampling thus refers to subsets of a population. Samples can be either random or non-random. A correctly conducted **random sample** allows the researcher to infer results back to the larger population. A random sample is generated using random methods, such as selecting every 10th person from a list. A **non-random sample**—such as street-corner interviews and telephone call-ins—does not allow you to infer results back to the population because the sample was not generated randomly. In communication research, both random and non-random sampling techniques are used, but if you want strong results that can be inferred or "generalized" back to your population, use random samples.

However, even a random sample is not guaranteed to be 100 percent accurate. This is where error margins come in. A **margin of error** is an estimate of how much the results of a given sample might differ from the results of the "true" population. Error margins are reported frequently in election polls. If 45 percent of the people polled support a particular candidate and the error margin is 4 percent, that actually means that the "true" population's support for that candidate could range from 41 percent to 49 percent (45 percent plus or minus 4 percent). The error margin can be reduced by using a larger sample size, but after a given size, the *amount* of change to the error margin is minimal. This is important to consider because the more people you sample, the higher your costs will be. A sample of 400 people has an error margin of about 5 percent, whereas a sample of 1600 has an error margin of about 2.5 percent. If you want a 2.5 percent error margin, go with 1600, but if that is not a major concern for you, an error margin of 5 percent might be fine, and it means not having to pay for 1200 more people in your random sample.

SURVEYS

Survey, or questionnaire, research is one of the most common types of communication research. Surveys can be conducted by phone, by mail, in person, and over the Internet. Surveys are a good way of gathering information from many people quickly and efficiently. Surveys attempt to measure preferences, behaviors, attitudes, and habits and are analyzed in relation to other information about the respondents, such as their income level, marital status, and age. Surveys give you information about a certain group at one point in time.

A **survey** is the act of conducting a study to collect data using a questionnaire. A **questionnaire** is a series of questions that includes close-ended questions, open-ended questions, or a mixture of both. Close-ended questions include a range of answers or response items, such as true/false, multiple-choice, or ranking and rating questions. Open-ended questions do not provide response items from which to pick. Open-ended questions allow for a free-form narrative answer in response to the question (Figure 3-2).

© Cengage Learning 2012

Michael Scicchitano

Director, Florida Survey Research Center

As a communicator involved in crafting a message for a client, you will typically be confronted with a situation in which you have no existing information to guide you. You may not know what message to prepare or may not understand the impact of a message you do prepare on the intended audience. Quantitative data collection methods, such as surveys, and qualitative data collection methods, such as focus groups, can help you obtain the information you need to prepare an effective message. After the message is disseminated, they can also provide valuable information to assess the impact of the message on the target audience.

In most instances, as a communications practitioner, you will work with professionals who will actually implement the surveys or conduct the focus groups. It is important, however, for you as a communicator to have an understanding of how surveys and focus groups are designed and implemented. This knowledge will permit you to guide the professional you are working with to obtain the maximum value from surveys and focus groups. As the director of a university-based survey research center, I find that working with knowledgeable communicators permits the research to be implemented quickly and the results to better meet their needs in preparing and assessing the impact of messages.

To develop a good survey, follow these steps:

- ***Establish the goals of the project.*** What do you want to learn?
- ***Determine your sample.*** Whom will you interview or send the survey to?
- ***Choose the survey methodology.*** How will you conduct the survey? (By phone, Internet, or face to face?)
- ***Create the questionnaire.*** What will you ask?
- ***Pre-test the questionnaire***. Test the questions.
- ***Conduct the survey.*** Ask the questions.
- ***Analyze the data.*** Report the results.

You want good questions so that you get good answers. Here are some suggestions for question development:

- Use simple, short sentences. Be brief.
- Ask questions in positive terms. Avoid using "not" and "un-".

- Stick with either questions or statements. Do not mix the two.
- Provide examples of the type of information you need.
- Avoid technical terms and acronyms.
- Make sure your questions accept all of the possible answers. In the following example, regular gas and premium gas are not the only options available for all of the types of fuel that could be put in a car.
  - "Do you use regular or premium gas in your car?" This question does not cover all possible answers. A better question would be:
  - "Which type of fuel do you use in your car?"
    - Regular gasoline
    - Premium gasoline
    - Diesel
    - Other
    - Do not have a car

Flower Purchase Behavior Survey

Case ID #: _____

I have purchased flowers in the last six months.

_____ Yes

_____ No

I have looked over the plants at the plant sale today.

_____ Yes

_____ No

Gender

_____ Male

_____ Female

Age

_____ 18-25

_____ 26-35

_____ 36-45

_____ 46-55

_____ 56-65

_____ 66 and older

Race

_____ White, non-Hispanic

_____ Black, non-Hispanic

_____ Hispanic

_____ Asian

_____ Native American

_____ Other_____

Highest Level of Education Received

_____ High-School Diploma/GED

_____ Some college

_____ Associate's degree

_____ Bachelor's degree

_____ Master's degree

_____ Doctoral degree

_____ Post-doc

FIGURE 3-2: This is an example of a survey questionnaire that uses close-ended questions to ask respondents about how recently they have purchased flowers, as well as their demographics.

- Avoid question bias. Here are some examples of question bias:
  - Double-barreled questions: Include two questions in one.
    - "Was the information provided to you helpful and without bias?" (The answer may be that it was helpful but biased, or it was biased but helpful. It is much better to ask only one question at a time.)
  - Leading questions: Wording of questions in such a manner that leads the respondent to answer a certain way.
    - "You wouldn't say you're in favor of student protests, would you?"

Types of Questions

In addition, you must consider the *types* of questions you should ask. Following are some of the most common types of questions:

Yes/no: This is an "either/or" question. It does not provide you with much information, but if all you need is an either/or answer, it will be fine.

Example: Do you eat a balanced diet daily?
  - ❑   Yes
  - ❑   No

Multiple choice: Use this question type when it is realistic and important to provide people with a range of possible answers.

Example: How many times a day do you eat a meal?
  - A.   1
  - B.   2
  - C.   3
  - D.   4 or more

Ranking questions: Use this question type when you want to learn respondents' values of one item in relation to another item.

Example: From 1 to 7, with 1 indicating your highest preference, rank the following drinks in order of your taste preference.
  - _____ Milk
  - _____ Alcoholic beverages
  - _____ Juice
  - _____ Cola drinks
  - _____ Water
  - _____ Coffee
  - _____ Tea

Rating questions: These are questions on a scale. These are sometimes known as "Likert" or "Likert-type" questions if a strongly agree/strongly disagree item-response format is used. Rating questions give a value to each answer. These can be listed vertically or horizontally. Often, rating questions have a numerical scale— such as "from 1 to 5"—for answers.

Example: Please indicate your level of agreement with the following statements by circling a score from 1–5 for each statement.

Participating in a program like this will enhance my understanding of critical thinking.						
Strongly agree	1	2	3	4	5	Strongly disagree
Participating in a program like this will help me advance in my field.						
Strongly agree	1	2	3	4	5	Strongly disagree
I would have the time to participate in a program like this one.						
Strongly agree	1	2	3	4	5	Strongly disagree
The probability that I will be able to apply what I learn in this program after it ends is very high.						
Strongly agree	1	2	3	4	5	Strongly disagree
Participating in a program like this will enhance my understanding of careers in the field.						
Strongly agree	1	2	3	4	5	Strongly disagree

Another form of rating question is the *semantic differential.* With this type of question, the response items are in the form of bipolar adjectives used as anchors, with the intervals between them counting as numerical values on a continuum.

Example: Please indicate your level of agreement with the following statements by circling a number from 1–5 for each statement.

I believe that marketing and promoting my company is . . .						
Good	1	2	3	4	5	Bad
Positive	1	2	3	4	5	Negative
Beneficial	1	2	3	4	5	Not beneficial
Favorable	1	2	3	4	5	Unfavorable
Important	1	2	3	4	5	Not important
Difficult	1	2	3	4	5	Easy
In my control	1	2	3	4	5	Out of my control

Open-ended questions: These questions let people provide their own responses.

Example: In the following space, indicate why you believe that this is a good course. _____
_____
_____

Effective Questionnaire Design

In addition to having good questions, the questionnaire itself should be easy to read and pleasing to the eye. Following are some tips on questionnaire design:

- Design the questionnaire to fit the medium (e.g., phone, face to face, e-mail, Web).
- Know what the questionnaire's purpose is before you create the questions.
- Keep it short and simple. Place the questions into three groups: "must know," "useful to know," and "nice to know." Get rid of the last group, unless the previous two are very short.
- Start with a descriptive, accurate title.
- Include a short introduction that details what the study is about and who you are. This introduction should include a purpose statement as well as the following:
    - Explain who you are and how the results will be used. (Not necessary to say your name.) "I am a University of X student conducting a survey for an agricultural writing course to find out . . ."
    - A "thank-you," either in the short introduction or at the end of the questionnaire, is necessary. "Thank you for completing this questionnaire."
- Reassure the respondent that his or her responses will be kept confidential.
- Include a cover letter with mailed surveys. Also include a return postage-prepaid envelope.
- Request demographic information in the first or last few questions:
    - Age, income level, education.
    - Be sure to ask certain demographic information only if it is necessary. (For example, it may not be necessary to ask income.)
- Make it look easy:
    - Use one page (if possible) and limit the number of items to 10 to 15 very specific questions.
    - Use only one side of the page; if you have to go to two pages, indicate at the bottom of the page for the reader to turn the page over and continue.
    - Use white space.
    - Align multiple-choice questions vertically. It is easier to read questions in a column.
    - Begin each multiple-choice answer option with a capital letter.
- Go through several iterations with the questionnaire, revising it each time.

To be effective, questionnaires should:

- Be concise and easy to complete.
- Include questions that are clear and understandable by the sample or population under study.
- Include a mix of open- and close-ended questions, and different response-item types.
- Provide clear instructions and make it obvious how to mark answers.

In contrast, you should not develop questionnaires that are long or confusing or include leading and double-barreled questions that encourage question bias.

Evaluating Survey/Polling Accuracy

In addition to being able to write effective questionnaires, it is also equally important to be an educated consumer of information that is generated from questionnaires. Political polls, for example, are notorious for providing data that support a position, solely based on carefully worded (and usually biased) questions. When reading polls, here are some things to consider:

- *How many people were polled?* The benchmark for an accurate national poll is 1000 to 1500.
- *What is the margin of error?* If it is 3 percent, that means plus or minus 3 percent. For a polling question with a 3 percent margin of error that found that 48 percent were in favor of an issue, it would mean the actual result for the target population could range between 45 percent and 51 percent.
- *Who commissioned the survey?* Independent pollsters and news organizations usually will release all of their findings, but privately funded surveys might release only the findings that are favorable to their candidate or cause.
- *How was the poll conducted?* Telephone surveys tend to sample women more because women tend to be home more often and answer the phone more often (Figure 3-3). In face-to-face polls, fewer people admit to less popular political viewpoints and many express opinions they think will impress the pollster.
- *How are the questions worded?* In 1996, a Gallup survey reported that Americans approved of sending troops to Bosnia, with 46 percent in favor and 40 percent not in favor. The question did not mention that 20,000 troops would go. CBS News mentioned that figure in a poll it commissioned and got an opposite reaction—a 58 percent disapproval rate compared to a 33 percent approval rate.

FIGURE 3-3: A group of telephone survey operators conducting a telephone survey poll.

- *In what order were the questions asked?* The order of questions as well as the order in which answer choices are listed in the poll can make a difference in the results. Those listed first are usually favored by respondents.
- *What is the response average?* A poor response rate can badly skew a survey. Research has shown that more than one of every three citizens is not home for a telephone survey or declines to answer questions. Depending on the topic of the survey, this could mean that you might get very different results, depending on which respondents you were able to reach at home.

CONTENT ANALYSIS

The **content analysis** method is a central component in communications research. When done correctly, content analysis can provide communication researchers and professionals with valuable insight into how messages are communicated. With this tool, the researcher examines the content itself. This can take the form of analyzing messages or words in newspaper articles, violent images in pictures or videos, key phrases in advertisements that persuade people to buy items, or inflammatory messages in political speeches. Content analysis can be conducted with any type of recorded communication (transcripts of interviews, videos, documents, newspaper articles, or TV/radio commercials, to name a few).

To conduct a content analysis, all of or a sample of the material is analyzed by a researcher or group of researchers who look through each line of text or every second of video for common themes. They may note the frequency of occurrence of each theme, as well as related themes. A **coding sheet** is developed, which

guides researchers as to what to look for when coding. The coding sheet provides explicit definitions, examples, and coding rules for each category.

FOCUS GROUPS

Focus groups allow small groups of people to interact and trigger responses from each other. Communication professionals use focus groups extensively when designing advertising messages for new products (Figure 3-4). Similarly, political campaign strategists use focus groups to design messages that would enhance a political candidate's image.

The focus group essentially is a group interview, comprised of six to eight persons who are asked questions by an objective **moderator**. The participants in a focus group may either be purposively selected (persons chosen to participate because they have some characteristics that are of interest, such as representatives from consumer action groups in a specific area) or they can be randomly selected (each member of a population has an equal chance of being selected) by a marketing research firm. Usually a minimum of two focus groups is conducted, on the same day and in the same location, and results are compared and contrasted. Focus groups are recorded, usually with audio and/or video, and questions and answers are transcribed. This produces a text, which is then analyzed just like a content analysis.

The moderator in a focus group works from a discussion or moderator's guide, which is essentially a written script that contains all questions, prompts, and instructions (see Figure 3-5). The moderator

FIGURE 3-4: Focus groups provide a setting for participants to share their opinions and answer questions posed to them by a moderator.

© Cengage Learning 2012

FIGURE 3-5: The moderator for this focus group, dressed in black, guides the discussion with participants, asking them about their perceptions, viewpoints, and opinions.

usually begins with some general instructions about the purpose of the focus group, what will be done, and the "rules" for the session, which include that everyone will get the chance to voice an opinion, that there are no right or wrong answers, and that everyone's point of view is valid. This is done to encourage an open environment, where everyone participates. Discussion questions are usually written in sequence, starting with general knowledge and awareness questions, followed by questions concerning usage and understanding of the product, good, service, or issue being examined. Then more specific questions are added, focusing on the main areas of interest.

Focus group questions often ask for participants' reactions to something—a new product, an ad or press release, an issue statement, or slogan. Moderators are looking for "top of mind" reactions first, called *unaided recall*. Then the moderator might follow-up, asking an *aided-recall* question, which is a question about a specific attribute or element. Another type of question is the *attitude/opinion question*, which asks participants to state their point of view about something. After asking a general opinion question, a "prompt" or follow-up—such as one of the following—is usually added:

- "When you hear the term 'all-natural' pork, what comes to mind?" (unaided recall)
- "All-natural pork products do not contain hormones or antibiotics. What does this mean to you?" (aided recall)
- "What do you think about all-natural pork products?" (opinion)
- "How likely would you be to buy such products?" (prompt, focusing on purchase behavior)
- "Have you seen any advertising about such products?" (prompt, focusing on exposure to advertising)

SUMMARY

In this chapter we learned about research methods communicators use, including surveys, content analysis, and focus groups. The chapter provided information on and examples of effective questions and questionnaire design. Sampling types were discussed and defined, and the process for conducting a focus group was described.

CHAPTER EXERCISES

APPLICATIONS

Following are some ways you can apply what you learned in this chapter:

- Put together a 10- to 15-question survey to find out what people think about a current event or issue. Use multiple question types and administer your survey to 10 people. Analyze your results by tabulating the percentage and average responses to each of your questions, and write up a report.
- Find an example of a survey or opinion poll done by a magazine or website and evaluate it. In addition, provide three recommendations to improve upon the methods used.

CHAPTER QUESTIONS

MATCHING:

_____ 1. A _____ question is used when you want to learn respondents' value of one item in relation to another item.

_____ 2. Demographic questions are questions about characteristics that are _____.

_____ 3. A _____ allows the researcher to infer results back to the larger population.

_____ 4. A _____ is an estimate of how much the results of a given sample might differ from those of the "true" population.

_____ 5. _____ is when a researcher reviews all of or a sample of a text or audio or visual material to look for common themes.

a. margin of error

b. content analysis

c. ranking

d. measurable

e. random sample

MULTIPLE CHOICE:

6. A(n) _____ question includes response items that are in the form of bipolar adjectives used as anchors, with intervals between them counting as numerical values on a continuum.

   a. semantic differential

   b. rating

   c. ranking

   d. interval

7. _____ questions often ask for participants' reactions to something—a new product, an ad or press release, an issue statement or slogan.

   a. Content analysis

   b. Focus group

   c. Interview

   d. Interval

8. Which of the following is NOT a demographic characteristic that can be measured in a survey?

    a. Age

    b. Education/learning level

    c. Economic status

    d. Attitude

9. Which of the following is NOT a psychographic characteristic that is behavioral or cognitive?

    a. Location

    b. Values and beliefs

    c. Habits

    d. Perceptions

10. Which of the following are examples of question bias?

    a. Double-barreled questions

    b. Leading questions

    c. Rating questions

    d. Both a and b

11. _____ questions include a range of answers or response items, such as true/false, multiple-choice, or ranking questions.

    a. Open-ended

    b. Semantic differential

    c. Close-ended

    d. Rating

12. _____ questions do not give you response items from which to pick, so that you may write a narrative response.

    a. Open-ended

    b. Semantic differential

    c. Close-ended

    d. Rating

13. A(n) _____ is the overall group you want to collect information about.

    a. population

    b. random sample

    c. non-random sample

    d. audience

14. A(n) _____ can be used to infer results back to the larger population and is generated through methods such as selecting every 10th person from a list.

    a. population

    b. random sample

    c. non-random sample

    d. audience

15. A(n) _____, such as street-corner interviews and telephone call-ins, does not allow you to infer results to the larger population.

    a. population

    b. random sample

    c. non-random sample

    d. audience

FURTHER READINGS

Ary, D., L. C. Jacobs, A. Razavieh, and C. Sorensen. *Introduction to Research in Education.* Belmont, CA: Thomason Wadsworth, 2006.

Berg, B. L. *Qualitative Research Methods for the Social Sciences.* Boston, MA: Allyn and Bacon, 2001.

Dillman, D. A., J. D. Smyth, and L. M. Christian. *Internet, Mail, and Mixed-Mode Surveys: The Tailored Design Method.* Hoboken, NJ: John Wiley & Sons, Inc., 2008.

Merriam, S. B. *Qualitative Research and Case Study Applications in Education.* San Francisco, CA: Jossey-Bass, 1998.

Neuendorf, K. *The Content Analysis Guidebook.* Thousand Oaks, CA: Sage Publications, Inc., 2002.

StatPac, Inc. "Survey and Questionnaire Design." Last modified 2010. Accessed May 18, 2010. http://www.statpac.com/surveys/.

Survey System. "Survey Design." Last modified 2009. Accessed November 20, 2010. http://www.surveysystem.com/sdesign.htm.

Writing@CSU. "Writing Guide: Content Analysis." Last modified 2010. Accessed November 20, 2010. http://writing.colostate.edu/guides/research/content/.

4 Business Communication

OBJECTIVES

After completing this chapter, the student will be able to:

- Write various types of business letters.
- Identify the components of a business letter.
- List ways to write a business letter clearly.
- List types of business letters.
- Write a memorandum.
- Write an e-mail.
- Answer and talk on a telephone in a business-like manner.
- Write an application letter for a job.
- Write a resume.
- Write a personal statement.

INTRODUCTION

Being able to communicate effectively in the business world is crucial (Figure 4-1). A University of Pittsburgh Katz Business School study of recruiters employed in large companies found that written and oral communication skills and the ability to work with others are the main factors contributing to job success (Manktelow, 2003). The National Aeronautics and Space Administration (NASA) also is placing high importance on "soft skills" such as communication skills. According to a February 2008 article in USA Today (Watson, 2008), NASA wants astronauts not only to be able to perform as scientists on space missions, but also to have interpersonal communication skills to help them get along for long periods in a cramped space travel environment.

Thus, it is important for you to begin practicing the skills you will need to effectively communicate in all businesses when you graduate. You can use the skills outlined in this chapter now as you write letters to businesses or respond to letters from businesses. This chapter also will help you understand how electronic communication, such as electronic mail (e-mail), text messaging (texting), and instant messaging, can be used in businesses (Figure 4-2). In addition, proper telephone communication will be covered. The chapter ends with information on how to write resumes, application letters, and personal statements.

Susan Howard, APR (Accredited in Public Relations)

Director, Corporate Communications, DUDA

What you say and how you say it— the words you use, the tone of your voice— leaves an impression on the person to whom you are speaking. The same is true when you write a message. Regardless of how the written message is delivered—via letter, memo, e-mail, or an instant text message—the person reading it will get an impression of you and the meaning of your message. It is important to always do business in a positive, effective, and professional manner; therefore, any business communication should always be written with courtesy and clarity, and use proper grammar and punctuation.

FIGURE 4-1: Effective communication skills are vital to professional success in the business world.

FIGURE 4-2: You can practice good written communication skills, such as writing effective business letters, memos, and electronic mail messages, to prepare yourself for the business world.

BUSINESS WRITING

Even with e-mail and cell phones, business letters are still the backbone of communication between people (e.g., between sales representatives and customers). A **business letter** is a formal document typically sent to people outside of an organization. Business letters provide recipients with specific information, such as a notification of an award or a note of appreciation for a donation. Business letters also can be used to persuade recipients to take some type of action, such as making a donation.

Business letters may seem challenging to write because you must consider how best to keep the recipient's attention. The following fishing analogy may be helpful:

- **Bait:** At the beginning of the letter, you should inform the reader immediately about what the letter is about. The letter may be to inform the reader of an award, to thank the reader, or to ask the reader for something. Do not wait until the last sentence to tell the reader what the letter is about.
- **Hook:** State how your request benefits the reader or the reader's organization. The "hook" should take one to three paragraphs.

© Cengage Learning 2012

Lisa Lochridge

Director of Public Affairs, Florida Fruit and Vegetable Association

To be successful in any profession, you must be able to communicate in a clear, concise way. Poorly written correspondence can hurt your credibility and impede your ability to accomplish your goals. It's important to be able to get across your point, whether it's in a report to your boss or an e-mail to co-workers teaming up with you on a project. Take the time now to acquire those skills because they will serve you for a lifetime.

■ ***Reel it in and land the catch:*** The letter should have some "call to action"—a statement of what you want or need. The closing of a business letter often specifies or suggests what the next action in the particular situation should be. When appropriate, this closing should say what you—as the sender of the letter—will do next or what you hope the recipient will do next.

In order to "catch" the reader, consider these steps in writing a business letter:

■ ***Know your reader.*** Focus on how the content of the letter benefits the reader. Appeal to the reader. Write in a friendly and helpful tone.

■ ***Know your purpose.*** What do you want the reader to do with the information that is presented? When trying to convince someone to act or react in a positive way, make your meaning crystal clear so the reader will respond quickly.

■ ***Write clearly.*** Tips on how to write clearly are provided in the next section.

■ ***Write a draft of the business letter.*** Try to make sure that these questions are answered in the draft:
- ■ Is the purpose clear?
- ■ Is the information organized in the most effective sequence?
- ■ Does each section follow logically from the one that precedes it?
- ■ Are all the facts, details, and examples relevant to the stated purpose?

- ■ Is the draft written at the appropriate reading comprehension level for the reader?

■ ***Revise the draft of the business letter.*** In the revision stage, try these techniques:
- ■ ***Wait a few hours between writing a rough draft and revising it so as to view the writing more objectively.***
- ■ ***Revise the draft in multiple passes.*** Do not try to improve everything at once. Focus first on writing a clear message. Save mechanical corrections, such as spelling, grammar, and punctuation, for later proofreading.
- ■ ***Be aware of typical errors and watch for them during the revision process.***
- ■ ***Read the draft letter out loud.*** Listening to sentences out loud often helps with identifying problems in your writing.
- ■ ***Ask someone else to read and critique the draft letter.***

WRITING CLEARLY

One of the most important aspects of a business letter is clear writing. Clarity is achieved if each part of the letter advances the letter's purpose and if the ideas in the letter are connected. Following are some tips for writing a clear business letter:

■ ***Tone:*** The tone for all business letters, even complaint letters, should be polite and positive.

- **Graphic design:** To help communicate your purpose, use simple graphic design elements, such as bulleted lists, graphs, or small pictures, as appropriate.
- **Length:** Keep your sentences and paragraphs "bite-sized." Use short paragraphs of three to five sentences and short sentences of 20 to 25 words. Limit the letter to one page, if possible.
- **Precise words:** Use precise words that help explain exactly what you mean. Avoid imprecise words, such as *strong*, *heavy*, *fast*, and *weak*. Rather, use precise words that leave no way to be misinterpreted. Instead of "heavy," say "95 pounds." Instead of "fast," use "85 miles per hour."
- **Recipient's name:** Use the name of the person to whom the letter is written. If at all possible, do not send a letter to "To Whom It May Concern." Be sure to spell the person's name correctly. Getting the recipient's name right on a job application letter is especially important.
- **Word processing:** The word processing software program on your computer is an asset, so use the program creatively, but do so wisely. Do not overdo underlining, boldfacing, italicizing, or bulleted lists.
- **Grammar, punctuation, and spelling:** Writing mechanics (grammar, punctuation, and spelling) need to be as close to perfect as possible to ensure that your message is clear. Read and reread your business letter for mechanical errors.

COMPONENTS OF A BUSINESS LETTER

Writing a business letter is not a difficult process. Business letters have similar components. Once you master the components of a business letter, you can use them for any type of letter. Figure 4-3 provides an example of a business letter, and the components are as follows:

- **Letterhead:** Use the printed letterhead from your organization, if at all possible. The letterhead should contain the organization's address, telephone number, fax number, Web address, and e-mail address.
- **Heading:** Includes the date and the sender's (your) address. The sender's address (your address) is included only if the letter is not sent on an organization's letterhead.
- **Date line:** The date is typed two lines below either the letterhead or the heading.
- **Inside address:** This includes the name of the person who is receiving the letter, the person's title, organization name, and full address (street, city, state, and zip code).

- **Salutation:** The formal greeting appears two lines below the inside address. Use courtesy titles in the salutation, such as *Dr.*, *Mr.*, *Mrs.*, *Ms.*, or *Reverend*. For women, use *Mrs.* only if you are absolutely certain that the recipient uses *Mrs.* as her courtesy title. If uncertain, use *Ms.* Also, be sure of the recipient's gender. Some names are gender neutral, such as *Casey*, *Taylor*, *Chris*, *Tracy*, and *Ashley*. Salutations in formal letters end with a colon ("Dear Mr. Smith:"). Salutations in informal letters sent to friends should include the recipient's first name and end with a comma ("Dear Ted,").
- **Body of the letter:** The paragraphs in the letter are single-spaced, with double spaces between paragraphs. Paragraphs in letters are rarely more than six or seven lines long, which means three to five sentences in length.
- **Signature block:** Includes a complimentary close, your signed name, and your typed name.
- **Complimentary close:** This is placed two lines below the last line of the body. Examples of a complimentary close are "Sincerely," "Sincerely yours," and "Yours truly." When the close contains two words, only the first is capitalized. The complimentary close is followed by a comma.
- **Your typed name** appears four lines below the complimentary close.
- **Your signed name** appears between the complimentary close and your typed name.
- **Special notations:** These are placed two lines below your typed name. Common special notations are as follows:
  - **cc: Followed by a person's name:** This means that a copy of the letter is being sent to the person named.
  - **Enclosure** or **encl.:** A separate document or item mentioned in the letter is enclosed in the same envelope as the business letter.
  - **Attachment** or **att.:** A document has been attached to the letter.
  - **P.S.:** This is a notation meaning *postscript* and is usually handwritten and added at the very end of all other notations.

BUSINESS LETTER FORMATS

Business letters can be written in one of three formats: full block, modified block (also called a regular block letter), and semi-block. It does not matter which format you use. Just find one that you like.

Example Business Letter

Heading	Sender's Name
	Sender's Street Address
	Sender's Address (city, state, zip code)
Date line	Date of Letter
Inside address: recipient's address	Recipient's Name
	Company Name
	Recipient's Street Address
	Recipient's Address (city, state, zip code)
Salutation	Salutation:
Body of the letter	A business letter's first paragraph is used to state the reason why the person is writing. Usually, the first paragraph is two or three sentences long to provide enough information for the recipient of the letter to know what the letter is about. Specific details of the letter's purpose and intent are provided in subsequent paragraphs.
	Beginning with the second paragraph, details are provided to explain the purpose of the letter. The purpose could be to provide information (good news, bad news), make a request, promote something, or respond to something. Try to limit the business letter to one page single spaced, or, at most, one and a half pages single spaced.
	In the closing paragraph, restate the purpose and indicate what you want the recipient to do. Provide any contact information necessary for the recipient to get back with you.
Signature block **Complimentary close**	Sincerely,
Your signature	Your Signature
Your typed name, title	Your Typed Name
	Your Title
Special notations	Attachment (if a document is attached)

FIGURE 4-3: This example identifies all of the components of a typical business letter.

A **full block letter** (Figure 4-4) has all of the components of the letter (heading, date line, inside address, salutation, paragraphs, and the signature block) flush left, meaning that the components are all the way to the left side of the page. The paragraphs are either flush left or full justified, and the paragraphs are not indented. A **modified block letter** (Figure 4-5) is similar to the full block, in that the sender's address, salutation, and paragraphs are flush left, with the paragraphs not indented; however, the date line, return address, and signature block are moved to the right. A **semi-block letter** (Figure 4-6) has the return address and signature block on the right side of the page, and the paragraphs are indented.

Example Full Block Business Letter

Sender's Name
Sender's Street Address
Sender's Address (city, state, zip code)

Date of Letter

Recipient's Name
Company Name
Recipient's Street Address
Recipient's Address (city, state, zip code)

Salutation:

A full block letter has all of the components of the letter (heading, date line, inside address, salutation, paragraphs, and the signature block) flush left, meaning that the components are all the way to the left side of the page. The paragraphs are either flush left or full justified, and the paragraphs are not indented.

[Paragraph 2 is single spaced. There is a double space in between paragraphs.]

[Paragraph 3 is single spaced. There is a double space in between paragraphs.]

Sincerely,

Your Signature

Your Typed Name
Your Title

Any Special Notations (cc:, Encl.)
Enclosures: [Number]

© Cengage Learning 2012

FIGURE 4-4: Example full block letter.

TYPES OF BUSINESS LETTERS

Business letters not only come in different formats (full block, modified block, and semi-block), they also come in different types. The most common ones are information letters, request or solicitation letters, promotion letters, cover letters, and response letters.

Information letters are written to inform someone about something, such as an approaching event, a decision that was made, an award being presented, or an action that was taken. Good-news and bad-news letters and complaint letters are examples of information letters. *Request letters* make a request, such as for a financial contribution or a pledge of support. *Promotion letters* are written to promote a cause or event to encourage the recipient's acceptance and participation. *Cover letters* accompany a longer document and explain why the reader might be interested in reading the document. An application letter with a resume is a form of the cover letter. A *response letter* thanks people for doing something positive, such as providing

Example Modified (Regular) Block Business Letter

Sender's Name
Sender's Street Address
Sender's Address (city, state, zip code)

Date of Letter

Recipient's Name
Company Name
Recipient's Street Address
Recipient's Address (city, state, zip code)

Salutation:

Modified block business letters have certain components of the letter (inside address, salutation, and paragraphs) flush left. The sender's address, date line, and signature block are flush right. The paragraphs are either flush left or full justified, and the paragraphs are not indented.

[Paragraph 2 is single spaced and *not* indented. There is a double space in between paragraphs.]

[Paragraph 3 is single spaced and *not* indented. There is a double space in between paragraphs.]

Sincerely,

Your Signature

Your Typed Name
Your Title

Any Special Notations (cc:, Encl.)
Enclosures: [Number]

FIGURE 4-5: Example modified block letter.

financial support for an organization. The thank-you letter is the most common form of response letter. The following are tips on how to write some of the most common letters: good-news letters, bad-news letters, request letters, thank-you letters, and complaint letters.

Good-News Letter

The *good-news letter* generally conveys information that will please the reader, for example, that the reader will receive an award. When writing a good-news letter, deliver the good news immediately. If you open with "Congratulations," be sure to immediately inform the recipient about the honor or award. Do not include specific details about the good news in the first paragraph. The function of the first paragraph is to announce the good news.

In the next paragraph, explain the details of the good news. For example, if the reader is to receive a cash prize, let the reader know when to expect it. This section can include more than one paragraph if more

Example Semi-Block Business Letter

Sender's Name
Sender's Street Address
Sender's Address (city, state, zip code)

Date of Letter

Recipient's Name
Company Name
Recipient's Street Address
Recipient's Address (city, state, zip code)

Salutation:

Semi-block business letters are the same as modified block letters (the return address and signature block on the right side of the page), except the paragraphs are indented.

[Paragraph 2 is single-spaced and indented. There is a double space in-between paragraphs.]

[Paragraph 3 is single-spaced and indented. There is a double space in-between paragraphs.]

Sincerely,

Your Signature

Your Typed Name
Your Title

Any Special Notations (cc:, Encl.)
Enclosures: [Number]

FIGURE 4-6: Example semi-block letter.

detail is needed. In the last paragraph, end positively, perhaps with instructions about how the recipient can contact you.

Bad-News letter

The *bad-news letter* tells the reader something that the reader does not want to hear. Examples of bad-news letters include informing the reader that he or she will not receive a refund, that he or she is not a candidate for a job or internship, or that a donation request has been denied. One of the main objectives of a bad-news letter is to preserve a good relationship with the reader while delivering bad news. In the opening paragraph, begin courteously, focusing on the positive relationship that you have had with the reader. If appropriate, thank the recipient for contacting you. Discuss the positive aspects of the relationship between your organization and the recipient. Do not mention the bad news in the first paragraph.

In the second paragraph, explain the reason for the bad news. This can be more than one sentence. Provide a concise description of the reason for the bad news before you deliver it to aid reader understanding. For example, if the reader applied for a job, explain in the

Scott Wallin

Director, Producer Communications, Dairy Management Inc.

Being able to communicate clear and concise thoughts in writing can work to your advantage in a business environment. If the author of a letter, for example, fails to check his or her writing for errors and clarity, it could present a poor image to the person on the receiving end. Poor writing sends a message of sloppy work habits and a lack of desire for details. Effective writing *always* makes a positive first impression.

second paragraph that no job openings were available, if that is the reason the person was not hired. Usually in the same paragraph as the reason for the bad news, in one sentence, state the bad news clearly and concisely. Do not let the bad news close the paragraph. End the paragraph with something positive or at least neutral. For example, you can write that you will keep the reader's resume on file. In the last paragraph, do not refer to the bad news. Focus on a good relationship between your organization and the reader.

Request Letters

A *request letter* asks someone outside your organization for something that you need, such as information, special consideration, funds, or donations (Figure 4-7). Most nonprofit organizations write request letters. In the first paragraph, make the request and briefly identify yourself or the organization. It is always best to get right to the point. Do not wait until the last paragraph to make the request.

In a new paragraph, explain and justify the request by providing additional information. In the same paragraph or possibly a third paragraph, detail precisely what you hope the reader will supply. Do you need a specific financial contribution? Do you want specific items for a silent auction? Does the request need to be completed by a certain date? Also in this paragraph, explain the benefit to the reader of fulfilling your

request. For example, for a donation to your organization, the reader may receive recognition by being included in a printed program, having the reader's name printed on a T-shirt, or showing the reader's name on a slideshow. In the final paragraph, close courteously by saying what action you will take or what action you hope the reader will now take. For example, do you want the reader to contact you? Will you contact the reader? Will you provide a form for the reader to complete?

Thank-You Letters

Your mother probably told you to say "please" and "thank you." Those manners are still important in the business world, especially when someone or a company does something nice for you. *Thank-you letters* should be written any time services (such as a guest speaking for your organization), donated products (food or printed materials), or monetary gifts are provided to your organization. Doing so establishes or maintains good relations between your organization and the company that donated the services or gifts. Thank the person or company specifically, and then thank them for their support.

In the thank-you letter, begin by greeting the reader. Express your gratitude for the service or donation the reader provided. Discuss how the gift was used. If the gift was a cash donation, mention how you

Example Request Letter

Sender's Name
Sender's Street Address
Sender's Address (city, state, zip code)

Date of Letter

Recipient's Name
Company Name
Recipient's Street Address
Recipient's Address (city, state, zip code)

Dear Supporter,

The **Anyville FFA Chapter** is asking for your help, to partner with us in raising funds for a new educational program. Specifically, we are asking for **financial assistance** <u>or</u> **donations** that we can use as silent auction items.

The Anyville FFA Chapter has a new teacher, Mr. Bob White. One of Mr. White's primary responsibilities is to provide a firm financial base for the chapter for this year and years to come.

One way that we are trying to raise funds is to hold a silent auction in conjunction with the Anyville FFA Chapter's spring banquet in mid-April. Money raised from the silent auction will be used to develop a new educational program to teach middle school students about agriculture. Local middle schools will travel to our campus to learn about all aspects of agriculture. The money we raise will be used to develop promotional materials, assist with the feed and care for animals, and assist with the middle schools' travel costs to our campus.

We request either a financial contribution or an item for our silent auction. In exchange for this donation, you will be recognized in the following ways:

- In the printed program at the banquet.

- On a poster displayed at the banquet and at the educational event for middle schools.

- In all promotional materials that are sent to the middle schools and in news releases sent to local news reporters.

Financial contributions can be made payable to "Anyville FFA Chapter." Please mail contributions to: Anyville FFA Chapter, c/o Bob White, 111 Boston Ave., Anyville, MO 33333. We are also happy to pick-up donations. Please call Mr. White at 555-555-5555 to make arrangements.

Your partnership is very important to us and to our event. We appreciate your support for the Anyville FFA Chapter.

Sincerely,

Your Name

FIGURE 4-7: Example request letter.

will use the money, but do not itemize your planned purchases. If appropriate, mention how past actions or support have helped your organization, and then lay the framework for future support. (Example: "We thank you for your past contributions and hope you can continue your support as our organization grows.") At the end, express your gratitude again; it is not overkill to say "thanks" one more time. Wrap up the letter with your name and a complimentary close.

Thank-you letters can be in the form of a typed letter, handwritten note, or e-mail, but which one do you send? A typed letter is the most formal and is always appropriate after any donation or service provided. Handwritten notes are more personal and can be appropriate for brief responses (Figure 4-8). However, handwritten thank-you notes are only as good as your handwriting, so make sure your handwriting is readable. *E-mail* is appropriate if you want to send a quick

Guidelines for Writing a Thank-You Note

These guidelines will help you as you write handwritten thank-you notes. The numbers on the example correspond with the guidelines.

1. Greet the giver.

2. Express your gratitude.

3. Discuss how the gift will be used. If the gift was cash, mention how you will use the money, but do not itemize your planned purchases line by line.

4. Mention the past; allude to the future.

5. Grace. It is not overkill to say "thanks" again, so say it.

6. Regards. Simply wrap it up.

---

1. Dear Mr. Brown,

2. Thank you so much for the $50 donation you made to our FFA chapter. (<u>OR</u> "Thank you for your generosity.")

3. It will be a great help to offset travel costs as our FFA chapter prepares for our national convention.

4. We appreciate your past support and look forward to working with you closely in the future.

5. Thanks again for your gift.

6. Sincerely,
   Your name

FIGURE 4-8: Guidelines for writing a thank-you note.

thank you to be followed up by a typed letter. E-mail is not generally seen as an appropriate form for a thank-you message, unless the person has expressed a preference for e-mail.

Complaint Letters

Complaint letters express dissatisfaction about a service or item that the reader provided. In the first paragraph, courteously explain the reason why you are writing the letter. Explain what is wrong with the product or service and include enough details so the reader can check the information.

In the second paragraph, state the inconvenience or loss that has resulted from the poor product or service. In this paragraph, also try to motivate the reader to some action by appealing to the reader's professionalism, sense of fairness, and honesty. In the final paragraph, end with some call to action. What do you want the reader to do? For example, do you want your money back?

Understand that no complaint is trivial. Each complaint deserves prompt handling. When writing the letter, everything should be factual, courteous, and fair. Always avoid argument and criticism.

In addition, you may need to write response letters to complaints lodged against you. When responding to complaints, avoid arguing. If you cannot meet the person's demands specifically, then say that. For example, the person may request a cash refund. If your policy prohibits cash refunds, let the person know that. But if coupons, gift certificates, or another form of restitution can be provided instead of a cash refund, let the reader

know, and then provide the coupon or gift certificate. Always treat the complaint seriously and do not criticize the person making the complaint.

MEMOS

The word *memo* is short for **memorandum**, a written reminder of something important that has occurred or will occur. Memos are generally used to communicate information quickly and efficiently within an organization. Usually they are not used to communicate with people outside the organization. Memos ordinarily are addressed to one person or a small group of people.

A memo should be concise and focused. The reader should be able to understand easily what the message is and what the reader is supposed to do in response to what is being communicated. Memos solve problems either by providing the reader with new information, such as a company's policy changes or price increases, or by persuading the reader to take an action, such as to attend a meeting or use less paper. However, most memos communicate basic information, such as meeting times or due dates for a project.

Memos follow a fairly standard format (Figure 4-9). They have lines for "date," "to," "from," and "subject." Because a memo is a document for internal communication, a salutation and signature block are not needed. The sender signs his or her initials next to the sender's name on the "from" line. The memo is shorter than a formal letter and usually provides more visual cues, such as bulleted or numerical lists. Paragraphs are not

---

Example Memorandum

To: Recipient's Name

From: Sender's Name (If it is a printed memo, the sender initials it here.)

Subject: Topic discussed in the memo

The body of the memo is brief—usually no more than three paragraphs. Paragraphs are not indented and are single spaced, with an extra line between paragraphs. There is no complimentary close at the end of the memo.

© Cengage Learning 2012

FIGURE 4-9: Example memorandum. The memorandum (or memo) provides brief information about a particular topic. It is usually sent to other members of an organization.

indented and are single-spaced, with an extra space between paragraphs. Sometimes memos are sent over e-mail as "e-memos" to members of an organization.

E-MAIL, TEXT MESSAGING, AND INSTANT MESSAGING

Because today's youth and young adults have had *electronic mail (e-mail)*, *text messaging (texting)*, and *instant messaging* for most of their lives, they are used to the shorthand that accompanies it, such as "LOL" for "laugh out loud" and "IMHO" for "in my humble opinion." However, to be successful in the business world, you must be able to write professional e-mails. Properly written e-mails can be a person's first impression of you, so you want to make a *good* first impression.

For e-mail, texting, and instant messaging, the ease of sending messages immediately with a computer mouse click or a cell phone can create difficulties. You can fire off messages without considering their content. Because facial expressions and body language cannot be relied upon to help explain electronic messages, their meanings can be mistaken. Words on a screen can have a common meaning, but vocal inflection can give them an entirely different meaning. When you develop e-mail, texts, or instant messages, consider exactly what you want the receiver to think, know, or do. Imagine how that person is likely to respond to the message.

E-mail

Following are some tips to help you write an effective e-mail:

- *Starting off.* Let readers know right away what the message is about so they can begin to follow your thought process. To do this, use a meaningful subject line. Do not leave the subject line blank. Many professionals prioritize which e-mails they read based on the content of the subject line. Begin with an appropriate greeting, such as one that you might use in a conversation or on the phone.
- *Keep it short.* Keep messages short, usually one to no more than two screens of information. If you have a long report, send it as an attached file.
- *How to send it.* Do not overuse the "high priority" or "important" option when you send an e-mail. Not all of your e-mails can be high priority. For e-mails to many recipients, use the BCC (blind carbon copy) line on the e-mail. Any e-mail address placed on the BCC line will not be

seen by the other recipients, which prevents the problem of scrolling through many lines of e-mail addresses on a received e-mail.

- *Tone.* For most e-mails, you should use a conversational tone because e-mail is a more spontaneous type of communication. For e-mails written to professionals and people of authority, however, you must use proper grammar, punctuation, and spelling because your e-mail may be the recipient's first impression of you. Also, do not use instant message shorthand in a professional e-mail. In all cases, make your e-mail easy to read by using simple words and short paragraphs and sentences.
- *Ending it.* Let readers know what you expect from them. If you want people to e-mail you back, put that request in the e-mail.

Netiquette

You should incorporate good "netiquette" techniques in your e-mails. *Etiquette* is the proper way of conduct in a given setting. **Netiquette**, or "Internet etiquette," is the set of common rules that govern how to send and receive e-mail properly. Companies regularly bring in expensive consultants to teach their employees the proper ways to communicate via e-mail. Following are some netiquette tips:

- *Be polite.*
- *Answer your e-mail promptly, especially if the e-mail is a request for information.*
- *Keep your messages short.*
- *Edit and proofread carefully.* If your relationship with the reader is solely professional, apply strict standards to your grammar, spelling, and punctuation.
- *Avoid long paragraphs.* Break up e-mail paragraphs so that you do not have more than two to three sentences in a paragraph. It is much easier to read shorter blocks of text on a computer screen than longer ones.
- *Keep the signature block (at the bottom of the message) short.* When sending business e-mails, the signature block should contain just the essentials: your name, title, phone number, e-mail, and mailing address. It is best not to include your favorite song lyrics or quote of the day.
- *Use a professional-sounding e-mail address.* When sending e-mails to a business or professionals, avoid "cutesy" or inappropriate-sounding e-mail addresses, such as "cowgirl87,"

Example Emoticons

Emoticons (emotion icons) or "smileys" are used to convey emotion in written communication. The following are just a few of the many emoticons used in written communication.

:-)   classic smile with nose

:'-)   happy crying (generally associated with mockery)

:-(classic sad with nose

|-O   yawn

:-D   laughter

:-*   kiss

:-0   surprised

;-)   winking smile with nose

© Cengage Learning 2012

FIGURE 4-10: Emoticons.

"hottie4you," or "imthegreatest." You can never go wrong by using your name as your e-mail address.

- **Avoid derogatory comments,** such as offensive, racist, or obscene remarks. Anything sent by a computer can be traced to its source.
- **Use emoticons ("smileys")** *to convey a little personality,* but be careful about their use in professional, business e-mails (Figure 4-10). Overuse of emoticons and text and instant messaging shorthand may not be appropriate in business situations.
- **Do not type in all capital letters.** It is considered SHOUTING! Use uppercase and lowercase lettering. It is also much easier to read.
- **Do not use e-mail to discuss confidential information.** Confidential information should never be sent in an e-mail.
- **Do not attach unnecessary files.** Only attach what the person receiving the e-mail needs.

Instant Messaging and Text Messaging

Instant messaging and texting are becoming more common in the business world as ways to bridge the gap between telephone calls and e-mail messages (Figure 4-11). They are instantaneous ways to communicate in businesses. Choose a screen name that will not offend anyone.

© Otna Ydur /www.Shutterstock.com

FIGURE 4-11: Text messaging is becoming more common in the business world. Texting for business uses should be done in a professional manner.

Keep messages simple and to the point, covering only one subject in each message. Abbreviations and shortened spellings are appropriate for texting in business uses.

MAKING TELEPHONE CALLS WITH POISE

You will be called on in the business world not only to write well, but also to speak well. Every time you make or receive a telephone call, you represent your organization (Figure 4-12). The person on the other end of the phone cannot see you, so that person's first

FIGURE 4-12: Knowing how to communicate well on the telephone is an important skill in today's business world. Take the time to learn proper telephone etiquette.

© Cengage Learning 2012

impression of you or your organization may well be determined by your voice and telephone manners. Following are some other pointers to keep in mind when you answer the telephone:

- *Immediately identify yourself and your organization in a few words.*
- *Try to identify whom you are speaking with as quickly as possible.*
- *Maintain a cheerful and considerate attitude toward each caller.* A caller usually can recognize if you seem bored. This is discourteous and paints a poor image of you and the organization. To help maintain a cheerful attitude, smile when you talk.
- *Use the telephone properly.* Keep your lips about 1/2 inch to 1 inch from the mouthpiece. Pronounce letters, numbers, and names clearly. Spell out names if they could be misunderstood.
- *Return calls.* If you must leave the telephone during a conversation and will not be able to return immediately, say that you will call back and then follow through.
- *Say "good-bye" pleasantly.* The person making the call should end the conversation.
- *If you leave a message on someone's voicemail, identify yourself and give your phone number first.* Then, briefly explain why you are calling. End the message by repeating your name and phone number.
- *Be sure to speak slowly, especially when saying a telephone number.*

APPLICATION LETTERS AND RESUMES

An **application letter**, sometimes called a *cover letter*, is a special kind of business letter that accompanies a resume for a job. A **resume** is a summary of your education, job experience, and job-related skills that you send to potential employers. From it and the accompanying application letter, potential employers learn about you and decide whether to interview you for a job.

Remember that a potential employer's first impression of you will be based solely on this initial application letter and resume. If the application letter and resume are sloppy, the employer may conclude that you do not care, you do not look after details, and you are not focused. Do your best to make sure your application letter and resume are free of errors and present you in the best possible light.

Application Letter

A resume is important, but the application letter is *equally* important. Most prospective employers read not only a resume, but also the letter—if not initially, then on the second pass. The application letter is a great opportunity to sell your unique credentials. It provides the employer with a first impression of you.

Writing an application letter is similar to writing any other business letter. However, the emphasis in an application letter is on promoting your abilities, qualities, and characteristics so that the prospective employer believes that you are the right person for the job. The letter details specific experiences that show what you can do for the employer if you are hired. An application letter also gives you the opportunity to demonstrate your writing skills.

Figure 4-13 provides a general outline of an application letter, and the details are discussed in the following paragraphs.

Customize your letter for each job application. Such items as including the correct company name and employer name, job title, and contact information are important and make a good first impression. If possible, do not send an application letter to "To Whom It May Concern" or "Dear Sir or Madam." Find out the employer's name and spell the name correctly. Also, make sure you get the employer's gender correct if the name, such as *Chris*, *Ashley*, or *Jamie*, could be either for a male or female. Match the job requirements and desired qualifications with your skills and credentials.

General Outline for an Application Letter

Applicant's name
Applicant's street address
Applicant's address (city, state, zip code)

Date of letter

Employer's name and title
Employer's street address
Employer's address (city, state, zip code)

Salutation:

Opening paragraph: State why you are writing. Name the position or type of work for which you are applying. Mention how you heard of the job opening or the organization.

Middle paragraph(s): Explain why you are interested in working for this employer and specify your reasons for desiring this type of work. If you have had relevant work experience or related education, be sure to point it out, but do not repeat your entire resume. Emphasize skills or abilities you have that relate to the job for which you are applying. Be sure to do this in a confident manner, and remember that the reader will view your application letter as an example of your writing skills.

Closing paragraph: You may refer the reader to your enclosed resume (which gives a summary of your qualifications) or whatever documents or other media (CD-ROMs, DVDs, newsletters you developed) you are using to illustrate your training, interests, and experience. Have an appropriate closing to pave the way for the interview by indicating the actions or steps you will take to initiate an interview date.

Sincerely,

Your signature

Your name *(typed)*

Attachment *(for your resume)*

© Cengage Learning 2012

FIGURE 4-13: General outline for an application letter.

The letter should include an opening paragraph that explains which job you are applying for and how you found out about the job. The body of the letter provides specific examples of activities or courses you have been involved in that make you right for the job. One way to match up your qualities with the mission of the organization is to find out what the company does and some of its recent activities, and then write about how your specific experiences can support that. Much of this information can be found on a company's website. If you are applying for a job at a local company, you may be able to get information about the company by asking people in your community. In addition, the application letter connects the content of your resume to the facts of the specific company and job description. In the letter, do not ask about salary and benefits. Those topics should be covered in the job interview, not in the application letter.

Application Letter Do's and Don'ts

Take the following guidelines into account when writing your job application letter:

Do:

- Customize your letter for each job application.

- Include the correct company name and employer name.

- Make sure the names are spelled correctly.

- Provide specific examples of what you can do that would directly benefit the company.

- Investigate the company a little so that you can write an application letter that matches what you do with what the company does.

- End the application letter with information on how the employer can contact you.

- State that you will follow up after a designated period of time (usually two to three weeks).

Don't:

- Prepare a sloppy application letter.

- Have grammatical, spelling, and punctuation errors.

- Send the application letter to "To Whom It May Concern."

- Request a job. Instead, request a job *interview*.

© Cengage Learning 2012

FIGURE 4-14: Application letter "do's and don'ts."

The end of the application letter should include information on how the employer can contact you, and you should request a job interview. Also, you can state that you will follow up after a designated period of time (usually two to three weeks) if you have not heard from the employer. This shows that you are interested in the job, and it provides a timeframe for the employer to get back in touch with you.

Remember that you are not asking for a *job* in the application letter; you are asking for a *job interview*. During the interview is when you "push" for a job. The application letter is your foot in the door. In order to get your foot in the door, the application letter must look appealing. Otherwise, you may get your foot slammed in the door. Figure 4-14 provides a list of do's and don'ts when writing your application letter.

Also, thank-you letters are important components of the job search. Thank-you letters can distinguish you from the crowd because so few people write and send them. After a job interview, send a thank-you letter. Send the letter within three days following a job interview and tell the interviewer something new about you (possibly something you learned after the interview), relate your skills more clearly to the job you are seeking, and let the employer know why you want to work for the company. In the first paragraph, thank the person for the interview. In the second paragraph, reiterate two or three of your strong points. In the last paragraph, close with another "thank you."

Resumes

Resumes can be written in various formats, but all resumes have certain elements in common. Your name, address, phone number, and e-mail address should be displayed at the top of the resume, usually in boldfaced text. As noted, be sure your e-mail address sounds somewhat professional. You can never go wrong with a simple e-mail address made up of your full name or just your last name. Try to keep your resume to one page. Place references on a second page.

Other common components to include are education, work experience, and a brief description of honors and awards. In the "education" section, include your major academic interests. Include your grade point average only if you believe it will increase your chances of getting an interview. In the "work experience" section, list any work or major volunteer experience you have done in chronological order, putting the most recent work first. Use verbs that describe what you did. Do not use "worked" as a verb, if at all possible. For example, do not say "worked as a waiter." Instead, say "waited tables." If you are currently working, the verbs for your current job should be in the present tense. For any previous work, verbs should be past tense. All resumes should be objective and factual. False information misrepresents you.

An optional component is a *professional objective statement*, which is usually near the top of the resume. A professional objective statement tells what you hope to achieve and is usually written this way: "To be employed as a customer service representative for a major agricultural business," or "To use my agricultural mechanics skills in a farm implement dealership." The objective statement is optional because everyone's real objective is to get a job interview. Sometimes an objective statement is helpful to the person reading your resume. Including a professional objective statement is up to you. Just make sure that it enhances your resume and does not detract from it.

Another optional section is "interests and activities." Only include interests and activities that you know will enhance your resume. Do not include information that may hinder your chances of getting an interview. In this section you may wish to include volunteer and school activities.

You may want to list contact information (name, phone number, e-mail address) for references on a second page. List three references who can discuss your work experience, educational qualifications, and your character. Examples of references may include a former employer or coworker, a teacher, or a member of the clergy. Do not list family members as references.

Just like the application letter, the resume should be free of misspellings, typographical errors, and grammatical errors. As for the look of the resume, do not use unusual typefaces; use a traditional-looking type style. Also use basic white or off-white paper. Avoid bright or unusual paper colors.

The common formats for resumes are the **chronological resume** and the **functional resume**. The chronological resume is probably the more common format. A chronological resume is written in reverse chronological order, with headings grouped by what a person has done, such as "education," "employment experience," and "interests/activities" (Figure 4-15).

The functional resume classifies the experiences that demonstrate your skills and capabilities into categories, such as "professional," "technical," "communication," "leadership," "management," and "sales" (Figure 4-16). A functional resume usually finishes with a reverse chronological listing of your job experiences. Until you have a lot of experiences that you can group together by skills and capabilities, you may not wish to use a functional resume.

WRITING PERSONAL STATEMENTS

A personal statement is not a resume or application letter; a **personal statement** is a statement about how your personal, familial, academic, and professional experiences and background qualify you for a particular job, scholarship, or college program. Personal statements are also called "application essays" or "statements of purpose."

Writing personal statements is becoming more common for students applying for college, scholarships, and even some jobs. A personal statement explains why you are applying for a particular job, scholarship, or college program, and it also tells a little about you. A personal statement may well be one of the toughest writing assignments because it is entirely about *you*. Selection committees use personal statements to help make a decision about you.

There are two types of personal statements. A *general statement* is when the topic is relatively open, with few guidelines, and you write about yourself. A *specific essay* is when the topic is assigned for you. Examples of a specific essay include such topics as "describe an ethical

Example Chronological Resume with an Optional Job Objective

Bob White
3111 Turkey Lane, Anyville, VT 33333
(555) 555-5555, bobwhite@mail.com

JOB OBJECTIVE	A supervisory position leading in a major agricultural equipment company.
EDUCATION	University of Vermont and State Agricultural College, Burlington VT
	Bachelor of Science, 2007
	Agricultural and Biological Engineering

EXPERIENCE

Manager, Joe's Farm Implement Company   2007-present
Bennington, VT
Supervise a staff of 10 employees. Oversee all fiscal
responsibilities for the company.

Assistant manager, Belo Agriculture   2003-2007
Bennington, VT
Supervised a staff of five employees. Developed a training
program for new employees, resulting in higher productivity.
Wrote an orientation handbook for new employees.

Manager, Burger Hut   2002-2003
Plainfield, VT
Supervised 10 part-time employees. Arranged work
schedules.

INTERESTS AND ACTIVITIES   Playing tennis, hunting, fishing

FIGURE 4-15: Example chronological resume. A chronological resume is written in reverse chronological order, with headings grouped by what a person has done, such as "education," "employment experience," and "interests/activities."

dilemma you once faced," "assess your oral and written communication skills," and "identify the strengths and weaknesses of the college major you want to earn." When writing a personal statement, be sure to answer all parts of the question being asked. Do not answer a question that is not asked. To help you along, you may want to think about these general questions and include them in your personal statement:

- What is unique about your background or life story?
- Which details in your story influenced your growth?
- When did you become interested in this field? What specific experiences furthered this interest?
- What are your career goals?

- Which personal characteristics or skills would enhance your prospects for success in the professional world (or in college)?

When writing a personal statement, find an idea that ties your essay together. This can be a brief story or a particular quality you have. Be positive and upbeat throughout; do not make excuses for any shortcomings you think you have. Be honest in all of your answers. Write in the first person (e.g., *I, me, mine*) because you are writing about yourself. Pick two to four main points or topics for a one-page, single-spaced essay.

A personal statement is not your life history, so do not try to outline everything you have done since

Example Functional Resume with an Optional Job Objective

Bob White
3111 Turkey Lane, Anyville, VT 33333
(555) 555-5555, bobwhite@mail.com

JOB OBJECTIVE	A supervisory position leading in a major agricultural equipment company
EDUCATION	University of Vermont and State Agricultural College, Burlington VT Bachelor of Science, 2007 Agricultural Economics
SUPERVISION AND TRAINING	Supervised staffs of five to 20 employees. Developed a training program for new employees resulting in higher productivity.
COMMUNICATIONS	Wrote an orientation handbook for new employees. Promoted to assistant manager within six months.
ORGANIZATION	Maintained a high grade point average while working 30 hours per week.

WORK HISTORY		
	Manager, Joe's Farm Implement Company Bennington, VT	2007-present
	Assistant manager, Belo Agriculture Bennington, VT	2003-2007
	Manager, Burger Hut Plainfield, VT	2002-2003

INTERESTS AND ACTIVITIES	Playing tennis, hunting, fishing

© Cengage Learning 2012

FIGURE 4-16: Example functional resume. A functional resume classifies your experiences that demonstrate your skills and capabilities into categories, such as "professional," "technical," "communication," "leadership," "management," and "supervision and training."

childhood. Proofread extensively to ensure that your personal statement is absolutely free of spelling, grammatical, and punctuation mistakes. Do not try to be funny; your humor may be missed by the selection committee. Do not use vague, empty terms, such as *meaningful, beautiful, challenging, invaluable,* or *rewarding.*

Last, when you are editing your personal statement for content, ask yourself these questions:

- Are my goals well articulated?
- Do I explain why I selected this school (or program) in particular?

- Do I include interesting details that prove my claims about myself?
- Is my tone confident?

Show your friends or relatives the final draft and ask them, "Does this sound like me?" If it does not, revise your personal statement. Examples of successful personal statements can be found in Richard Steitzer's book *How to Write a Winning Personal Statement for Graduate and Professional School* (1997).

SUMMARY

This chapter focused on important communication skills that you can practice now. After learning how to write business letters of various kinds (thank you, request, good news, bad news, and others), e-mails, memos, resumes, and application letters and how to practice proper telephone etiquette, you will be better prepared to meet the communication challenges you will face in the business world after graduation.

CHAPTER EXERCISES

APPLICATIONS

Following are some ways you can apply what you learned in this chapter:

- Write request letters for donations for an upcoming event.
- Conduct a resume-writing workshop. You could invite a guidance counselor or local businessperson to evaluate the resumes.
- Prepare an outline for a personal statement essay.
- Practice using memos to communicate.
- Practice telephone skills with others.
- Write a draft of your resume.

CHAPTER QUESTIONS

MATCHING:

Match the component of a business letter below with its definition or characteristics. Not all terms will have a definition associated with them.

_____ **1.** Includes a complimentary close, your signed name, and your typed name.

_____ **2.** Includes the date and the sender's (your) address. The sender's address is included <u>only</u> if the letter is <u>not</u> sent on an organization's letterhead.

_____ **3.** The name of the person who is receiving the letter, the person's title, organization name, and full address (street, city, state, and zip code).

_____ **4.** Formal greeting that appears two lines below the inside address.

a.   Heading

b.   Inside address

c.   Salutation

d.   Signature block

e.   Special notation

MULTIPLE CHOICE:

5. In the "fishing analogy," the _____ lets the reader know immediately what the business letter is about.

    a. "hook"

    b. "bait"

    c. "reel it in and land the catch"

    d. "watch the water"

6. Good-news and bad-news letters are examples of _____.

    a. request letters

    b. cover letters

    c. promotion letters

    d. information letters

7. A(n) _____ is a summary of your education, job experiences, and job-related skills that you send to potential employers.

    a. application letter

    b. resume

    c. request letter

    d. e-mail

8. _____ accompany a longer document and explain why the reader might be interested in reading the document.

    a. Request letters

    b. Cover letters

    c. Promotion letters

    d. Information letters

9. The abbreviation _____ means that a copy of the letter is being sent to another person.

    a. cc

    b. encl.

    c. att.

    d. P.S.

10. A(n) _____ is a formal document typically sent to people outside of an organization.

    a. memorandum

    b. e-mail

    c. business letter

    d. graphic design

11. _____ is the set of common rules of how to send and receive e-mail properly.

    a. Etiquette

    b. Emoticon

    c. Netiquette

    d. Memorandum

12. A _____ classifies your experiences that demonstrate your skills and abilities into categories.

    a. chronological resume

    b. functional resume

    c. personal statement

    d. signature block

13. A business letter written in _____ format has all of the components of the letter flush left, the paragraphs are either flush left or full justified, and the paragraphs are not indented.

    a. semi-block letter

    b. modified block letter

    c. blocky letter

    d. full block letter

14. _____ are becoming more common for students applying for college, scholarships, and even some jobs.

    a. Personal statements

    b. Request letters

    c. Instant messages

    d. E-mails

15. _____ express dissatisfaction about a service or item that the reader provided.

    a. Thank-you letters

    b. Complaint letters

    c. Request letters

    d. Good-news letters

REFERENCES AND FURTHER READINGS

Colorado College. "Guide to Writing Personal Statements." Accessed October 23, 2010. http://www
.coloradocollege.edu/careercenter/images/pdfs/writing_personal_statements.pdf.

Harpold, L. "How to Write a Thank You Note." The Morning News Corp. Last modified 2003. Accessed October 23,
2010. http://www.themorningnews.org/archives/how_to/how_to_write_a_thankyou_note.php.

Manktelow, J. "Why You Need to Get Your Message Across." Mind Tools Ltd. Last modified August 2003. Accessed
October 23, 2010. http://www.mindtools.com/CommSkll/CommunicationIntro.htm.

Marsh, C., D. W. Guth, and B. P. Short. *Strategic Writing: Multimedia Writing for Public Relations, Advertising, Sales
and Marketing, and Business Communication.* Boston, MA: Pearson Education, 2005.

Newsom, D., and J. Haynes. *Public Relations Writing: Form and Style,* 7th ed. Belmont, CA: Wadsworth, 2005.

Oliu, W. E., C. T. Brusaw, and G. J. Alred. *Writing That Works: Communicating Effectively on the Job,* 9th ed. Boston,
MA: Bedford/St. Martin's, 2007.

Steitzer, R. *How to Write a Winning Personal Statement for Graduate and Professional School.* Princeton, NJ:
Peterson's Guide, 1997.

Trawick, L. "Sample Letter." Purdue University's Online Writing Lab. Last modified November 14, 2007. Accessed
October 23, 2010. http://owl.english.purdue.edu/handouts/pw/p_basicbusletter.html.

University of Delaware. "Writing Personal Statements." Last modified 2006. Accessed October 23, 2010. http://
www.udel.edu/CSC/pdfs/personalstatements.pdf.

Watson, T. "For NASA, 'the Right Stuff' Takes on a Softer Tone." *USA Today* (February 4, 2008): 1A-2A.

Zmorenski, D. "Effective Communication: Turn This Weakness into a Strength." Last modified 2009. Accessed
October 23, 2010. http://blogs.reliableplant.com/1011/effective-communication/.

5 News Media Writing

OBJECTIVES

After completing this chapter, the student will be able to:

- Identify and list the criteria for newsworthiness of a news story.
- Explain the structure of the inverted pyramid.
- List the five Ws and H.
- Write a good lead.
- Write a news story for print, video, and radio.
- Conduct an interview for a news story.

INTRODUCTION

Knowing how to write in a news media style is important not only for an organization's reporters but also for its members. Because the **news media**—*radio and television stations, newspapers, magazines, and Internet news outlets—are where most people go for information, news stories have a significant impact on readers, viewers, and listeners (Figure 5-1). Knowing how to write a news story well and sending it to the news media, in the form of a news release, could mean that your information gets into a newspaper or on a radio or television newscast. Well-written articles also could be placed in your organization's newsletter or your school's newspaper.*

* News writing*—*also called* journalistic writing—*is similar to, yet slightly different than, "traditional" writing that you have done for most of your life. This chapter provides an overview of news writing for print, television, and radio, so you can better communicate news about what is going on in your organization.*

FIGURE 5-1: Most people rely on the various news media outlets for information.

© Stephen Coburn/www.Shutterstock.com

Susan Howard

Director, Corporate Communications, DUDA

Often, there is little time or space allowed in print or broadcast media to give all the details of a news event. Learning how to structure a news story, whether for publication in a newspaper or for a broadcast feature on the evening news, will help you deliver the most important information to your readers or viewers. If you are writing a news release to give to a reporter, you will use the same concise way of preparing your information.

WHAT IS NEWS?

One of the first things you have to determine before writing a news story is if the story is newsworthy. Television station news directors and newspaper editors use the following criteria to determine **newsworthiness**— what stories they will cover in their newscasts and newspapers. Some of the factors on this list also are included in the discussion on news releases in Chapter 13, Media Relations. Newsworthiness, or news value, depends on the following characteristics:

- *Timeliness* refers to when an event or activity happened or when it will happen and implies immediacy or nearness to the present.
- *Proximity or location* refers to how close physically or psychologically the news story's content is to the audience. The closer the impact is geographically to the audience, the more proximity impact it has. For example, a news story for your organization's newsletter about a local member will have more proximity impact than a story about a member in another state.
- *Prominence refers* to the importance of the person or the event. Big names make big news. High-profile people, issues, or concerns have more news value.
- *Importance or significance:* The greater the effect and the larger the number of people impacted by your news story, the more likely it is that your story is news.

- *Human interest:* News stories that have a strong appeal to human emotions are more newsworthy. Does the event involve interesting people doing interesting things?
- *Innovative or unusual:* If the news story features something different, unusual, or innovative, it carries news value.
- *Conflict:* A story that shows struggles—for example, a person versus the environment, a person versus another person—is usually newsworthy.
- *Money:* News stories about financial issues are almost always newsworthy.

NEWS WRITING STYLE

To begin this section of the chapter, it may be good to say what news writing style is *not*. It is not providing a chronological account of something that happened ("this happened, then this happened, then this happened"). News writing is not stringing together a long collection of direct quotations, one after the other. News writing is not starting out with the least important information first and then "springing" the news at the very end of the story.

News writing, very simply, is finding out:

- Who said it? Who is it about?
- What happened?
- When did it happen?

© Cengage Learning 2012

Lisa Lochridge

Director of Public Affairs, Florida Fruit and Vegetable Association

Anyone who plans to work with the media will have a distinct advantage if he or she can write like a reporter. Many smaller publications accept bylined articles and guest columns from outside sources, so if you can write in the style of the publication, the article you submit has a better chance of being published. In addition, if you are producing written or broadcast materials for your own organization, such as a newsletter, magazine, or podcast, these skills are critical to ensure a strong readership.

- Where did it happen?
- Why is it important?
- How did it happen?

As you answer these questions, you will have to keep in mind that good news writing for print, television, or radio stories also follows the general rules in Figure 5-2.

Writing news stories for print news will be covered in the following section. Later sections in this chapter also will include information on how to write news stories for television/video and radio. Regardless of whether you write print, television, or radio stories, any journalistic writing should be *accurate*, *brief*, and *clear*.

Facts must be accurate; names must be spelled correctly, identifications made properly, and numeric figures quoted carefully. Never assume you have someone's name spelled correctly. Always ask the person how to spell his or her name. For example, a person by the name of "Jodi" could spell it "Jody" or "Jodie," and that name could be for a man or a woman. In addition, you add strength to accuracy by getting information from more than one source if possible. Two-source stories are always stronger and usually more accurate because you are not relying on just one person's thoughts. Accuracy is the reporter's greatest obligation to the reader. Being accurate also helps you maintain your credibility as a journalistic writer.

Journalistic writing also should be brief. Cut out unnecessary words. Find short words or phrases that mean the same thing as longer words or phrases. For example, use "whisper" instead of "talk softly." Overall, sentences should be 25 words or less. Paragraphs should be kept to three or fewer sentences. News stories also should get to the point quickly. What is the story about? What does the story need to tell the reader? A writer needs to be able to answer these questions in the simplest terms possible.

Good News Writing

- Uses short words.
- Uses short sentences.
- Uses short paragraphs.
- Eliminates wordiness.
- Avoids jargon or technical language.
- Comes to the point quickly.
- Uses direct quotes to bring life to the story.

© Cengage Learning 2012

FIGURE 5-2: Good news writing.

Finally, if the reader is to understand what is in the news article, the information must be clear. Write simply so you can communicate ideas without confusion. The reader should easily understand any information in an article. For example, eliminate all kinds of jargon, or technical language, and do not use big words to try to impress readers.

NEWS WRITING FOR PRINT STORIES

A good print news story will contain the following components, described in this section: inverted pyramid structure, five Ws and H, leads, impersonal reporting, news writing techniques, quotations and attributions, Associated Press Style, and proper grammar and punctuation.

Inverted Pyramid

The inverted pyramid structure is the most commonly used structure for news writing. The inverted pyramid presents the most important information in a news story first, followed in descending order by less-important information. This structure works well for two reasons. First, the most important information, which is presented at the beginning, helps to grab the reader's attention and interest, so the reader is more likely to read the entire article. Second, a story written in the inverted pyramid structure means the least important information is at the very end of the structure. Therefore, if the story needs to be cut, it can be cut from the bottom without any loss of important information. If you put important information at the end, it may get deleted.

The inverted pyramid structure is based on the "five Ws and H" and good leads, which are explained next (Figure 5-3). The lead, or first paragraph, is a simple statement that provides focus to the news story. A lead should be written as simply as possible and should contain as many of the five Ws and H as can be understood easily. The body of the inverted pyramid story adds detail to the information that has been introduced in the lead. The body should provide more information, supporting evidence, and context in the form of direct and indirect quotes, more details, and other descriptions.

Stories in the inverted pyramid structure avoid falling into the trap of a chronological storytelling of what happened at an event ("this happened, then this happened, then this happened"). For example, what happens at the beginning of a meeting or event is rarely the most important or interesting thing that occurred.

Five Ws and H

The five Ws and H are the key components of any news stories. They stand for *who, what, when, where, why,*

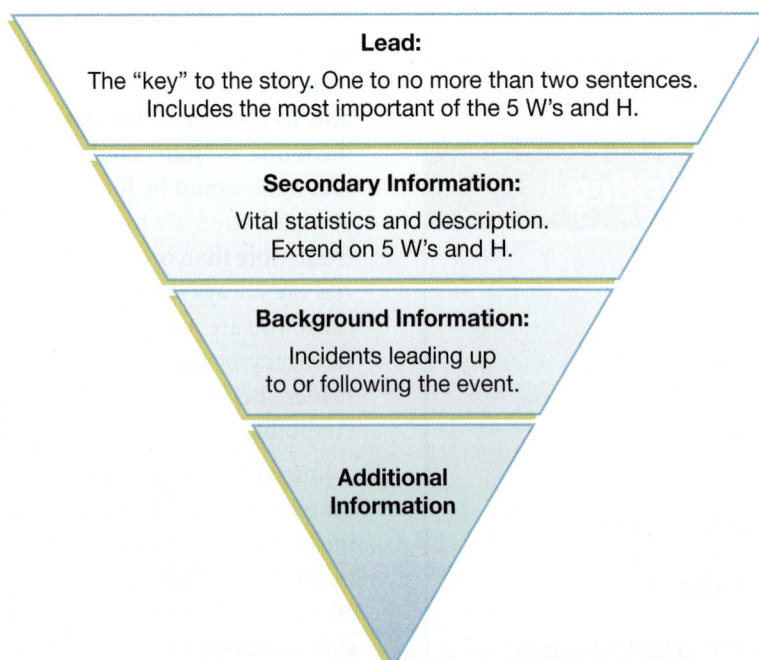

Lead:
The "key" to the story. One to no more than two sentences.
Includes the most important of the 5 W's and H.

Secondary Information:
Vital statistics and description.
Extend on 5 W's and H.

Background Information:
Incidents leading up
to or following the event.

Additional Information

© Cengage Learning 2012

FIGURE 5-3: The inverted pyramid structure of news writing is the structure most commonly used for news writing. The inverted pyramid presents the most important information in a news story first, followed in descending order by less-important information.

Jim Bret Campbell

Senior Director of Publications, American Quarter Horse Association

With the Web, people have more choices than ever of sources to get their news, but in the last couple of years, an ever-growing number of studies indicate that consumers are becoming more wary about *where* they get their news. Online readers are becoming more careful about choosing a trusted source online and not just taking everything they read at face value. Because of that trend, it is becoming more important for news stories to be well researched, written, and fact-checked. With all of the choices to read online, it's crucial that writing stand out from the pack and catch the reader's attention. But hurry—whether it's in print or online, you only have about two seconds to stop them from turning the page or clicking the next link.

and *how*. The five Ws and H also can be the questions that a news story should answer, such as:

- Who said or did something?
- What was said or done? What happened?
- When was it said or done? When did it happen?
- Where was it said or done? Where did it happen?
- Why was it said or done? Why did it happen?
- How was it said or done? How did it happen? How does this affect me?

To gain the reader's attention you should begin the lead with the most interesting or most important element of the five Ws and H. Others are added later in the story. The aspect used most often in the lead is the *what*, or perhaps the *who*, if it is someone important. *What happened* is usually what most people want to read about first. Figure 5-4 shows a model news story.

Leads

The **lead paragraph**, or *lead* (pronounced LEED), is the first paragraph in the news story. The lead grabs the reader's attention and answers the most important of the five Ws and H. The reporter must make a judgment on what to put in a lead, based on the newsworthiness criteria described earlier in this chapter. A good lead generally will contain at least three of the five Ws and

H. However, one mistake writers sometimes make is trying to put too much in a lead. The lead should be brief, no more than 25 words. Following are descriptions of some types of leads that you might include in your stories.

The *summary lead* is the most common news-style lead seen in newspapers. The summary lead provides the most important of the five Ws and H elements. It gets the basic information up front. If you include a *who* in your lead, you do not have to use the person's name. You can identify someone by the person's title or job position and then include the person's name later in the story. The following example shows how you can identify people without using their names. Later in the story, their names would have been included. Unless the *who* in your story is someone important or well known, rarely will you want to list the person's name in the lead paragraph.

Example: Five Anyville High School students and one teacher were injured Sunday night when their van slid out of control on icy roads in eastern Kentucky.

This summary lead contains *who* (five Anyville High School students and one teacher), *what* (were injured when their van slid out of control), *when* (Sunday night), *where* (eastern Kentucky), and *how* (icy roads), and it is 25 words.

Model News Story

A news story lead should include the most important elements of the five Ws and H (who, what, when, where, why, and how) and be no more than 25 words.

The second paragraph provides some details related to the lead and may include other parts of the five Ws and H not included in the lead.

News stories should be double-spaced and written according to Associated Press Style.

Paragraphs should be no more than three sentences long. They are usually one or two sentences long. Sentences are usually no longer than 25 words each.

"Direct quotations are usually set apart in their own paragraphs," said Ricky Telg, a professor at the University of Florida. "Any comment that is not common knowledge and factual should be attributed."

Writers should not include any opinion in their news stories that is not attributed, he said.

Telg said quotations should not be strung together, one after the other.

"If you string quotes together, you're not really writing," he said. "You should paraphrase what people say, whenever possible, to make what they say is more understandable to your audience. However, this does not mean for you to misinterpret what they say."

The news story should be written in the inverted pyramid structure, so that the least important information is at the end of the story.

If a news story runs more than one page, insert " – more – " at the bottom of the page. At the top of the second page, flush left, write "Add 1."

On the last page, use the following notation, centered, on the page, to signify the end of the story: " – 30 – " or "###"

FIGURE 5-4: Model news story. A good print news story, such as this model news story, will be written in the inverted pyramid structure, according to Associated Press Style. The news story will contain the most important of the five Ws and H in the lead, good quotations and attributions, and proper grammar and punctuation.

The *question lead* asks a question to grab the reader's attention. The question lead is seldom used because if the reader does not care about the answer to the question, then the person probably will stop reading.

Example: Will the student vote affect local elections? Not if students are not registered to cast their ballots.

A *quotation lead* is a direct quotation used in the first paragraph. Unless the quotation is something memorable or unusual, the quotation lead should be avoided, because it is considered that the story's writer has given up on being creative and just inserted a quotation to jump-start a story. The following example shows how a quotation lead can work, because the quotation is out of the ordinary.

Example: "My plane is taking off without me," shouted a student pilot to his instructor as he dashed down the runway after the Cessna 140.

A *speaker-spoke lead* identifies that someone spoke or will speak at an event. Unless the person is someone important, this type of lead should be avoided. People speak all the time. What is news is what the person says, not that the person "spoke." The following speaker-spoke lead example provides very little information of interest to the reader.

Example: Judge Billy Roster spoke Tuesday in Gainesville at 8 a.m. He spoke to the local American Bar Association chapter about the state's overcrowded prison conditions. His speech to the ABA was his third on the topic in two weeks.

The more important aspect of this story is what Roster talked about, so use a summary lead of the main points:

Example summary lead: A judge is making the rounds to local organizations to inform people about the state's prison conditions.

Second paragraph: Judge Billy Roster told Gainesville's American Bar Association chapter Tuesday morning that the state's prisons are overcrowded. It was his third presentation about the state's prison conditions in the last two weeks.

In this example, the 17-word lead tells us *who* and *what* the person is saying and doing. In the second paragraph, more of the five Ws and H are provided, by saying specifically who the person was and what he talked about, to whom he talked, when he spoke, and where he spoke. Note that the person's name is not mentioned in the lead sentence because—in most cases—a person's name is not usually important enough to lead with. Instead, lead with a person's qualifying information—what makes the person noteworthy (e.g., a judge, a Live Oak High School student, a city council member)—and give the person's name in the second or third paragraph.

A *first-person lead* puts the writer in the story. First-person leads are sometimes acceptable to magazine editors, but rarely for newspaper editors. It breaks the rule of the "impersonal reporter" explained next.

Example: Tremors are common in Japan, and it wasn't surprising to feel my high-rise building slowly swaying as I prepared my morning cup of coffee. Usually, the tremors last a few moments and stop.

Second paragraph: This wasn't just a tremor. And it didn't stop.

Impersonal Reporter

Another aspect of journalistic writing is the *impersonal reporter*. Reporters should be *transparent* in their writing. They should avoid using first-person pronouns (*I, me, we, our, my, us*) or second-person pronouns (*you, your*) outside of a source's direct quote.

Reporters also should set aside their own views and opinions. Allowing the writer's opinions, prejudices, and biases to enter a story is called **editorializing**. News reporters should report only what they see and hear (Figure 5-5). How a reporter feels about that information is not relevant to the news story.

To avoid editorializing, a writer should present only facts and limit or eliminate most adjectives, except in direct quotes. For example, instead of writing, "He was sad," describe what the person did that made you think he was sad. Instead of writing, "He was sad," you could write, "He placed his head in his hands and wept." Present what you see and hear; let the reader make the connection that the person was sad. How do you know something is "interesting," "impressive," "tragic,"

FIGURE 5-5: Reporters must set aside their own opinions and biases and report only what they see and hear.

or "avoidable"? That is your opinion. Just present the facts. Leave the value judgment to your readers.

Editorializing can be avoided by attributing any information that is not a fact or is not common knowledge. If the information is not common knowledge, may or may not be true, or is entirely opinion, it *must* be attributed to someone. If not everyone knows something to be true, your responsibility is to attribute that information to a source.

You do not have to wrap up the story. That is one of the functions of the inverted pyramid structure. When there is nothing else to write, just stop. You will avoid editorializing at the end of your story by feeling that you have to wrap up the story by ending it with something like this: "Everyone had a great time."

Print News Writing Techniques

The following summarizes news writing techniques for print:

- *Short sentences:* Sentences in news stories average 20 to 25 words or so. Do not string together, with commas and conjunctions, several sentences into one long sentence. The best way to shorten sentences is to use periods, not commas and conjunctions.
- *Short paragraphs:* For news stories, paragraphs should be no more than three sentences long. Usually, paragraphs are one or two sentences long. This is much different than the writing you have done for your composition and English classes, which emphasizes four or five sentences per paragraph.

- **Third person:** A news story should be written completely in third person (e.g., *he, she, it, they*), except when you use a person's own words in a direct quotation.
- **Nouns and verbs:** Place emphasis more on nouns and verbs than on adjectives and adverbs. Overusing adjectives and adverbs will cause you to editorialize. Action verbs keep a story moving and grab the reader more than "to be" verbs (*be, is, are, am, was, were*), which show little action. Use action verbs to describe what you observe.
- **Format:** If the news story is longer than one page, write "more" at the bottom of the page. Indicate the end of the news story by either writing a hyphen, the number 30, and another hyphen (-30-) or three pound signs (###) at the center of the page below the final line of story. The -30- or ### means "end of the story."
- **Simple writing:** Use simple words and simple sentences. Not every sentence should be in the simple-sentence format (subject-verb-object), but the simple sentence is a good tool for clearing up muddy writing.
- **Jargon and clichés:** Avoid jargon and clichés. As noted, *jargon* is technical language used in specialized fields or in specific groups. **Clichés** are overused words and phrases, such as "it cost an arm and a leg," "a drop in the bucket," and "on the cutting edge."
- **Transitions:** Transitions tie together what you have written. Each sentence in a story should logically follow the previous sentence or should relate to it in some way. New information in a story should be connected to information already introduced. Transitions include the following:
  - *Connectors* help unify the writing. For the most part, they are conjunctions such as *and, but, or, for, thus, however, therefore, meanwhile*, and others. They do not have great value in terms of the content of the writing, but they are necessary for its flow.
  - *Hooks* are words or phrases that are repeated throughout an article to give the reader a sense of unity. For example, in a story about the city council, the word "council" used throughout the story would be a hook.
  - *Pronouns* are one of the best transitional devices for writing about people. Instead of using a person's name each time, use a pronoun about every other time the person is mentioned in the story.

Quotations and Attribution

Quotations are the words of someone talking. It is a good idea to use quotations to bring "life" to your story. Quotations can be either direct or indirect. A *direct quotation* is the exact words of a person talking (or quoted) in a news story. An *indirect quotation*, also called a *paraphrase*, may have one word or a few of the same words that a speaker used, but it will also have words that the speaker did not use. Paraphrases express what the source said but with different words from those the speaker used. The exact words spoken by the speaker in a direct quotation or in an indirect quote will be inside quotation marks.

A good news story will use more paraphrases than direct quotations. Direct quotations do add "life" to a story, but they should be used sparingly. Use them to supplement a story. Do not string together long sections of direct quotes.

Attribution means telling readers where the information in a story comes from. Attribution is extremely important in news writing. It is one way writers can avoid editorializing in their story, by making sure that information in their stories can be attributed to someone or some organization. Writers should attribute anything that is not common knowledge to all readers. Attributing information sources also allows the reader to assess the credibility of the information by assessing the source of the information. Some sources are more credible than others. Here are some examples of attribution:

Indirect quote/paraphrase: Myers said the incident was under investigation.

Indirect quote (with some of the words as the exact words of the speaker): Myers said the incident was being investigated, but that it would be "a long time before the investigation is completed."

Direct quote: "The incident is under investigation," Myers said.

Direct quote: "The incident is under investigation," Myers said, "but it will be a long time before the investigation is completed."

Following are some guidelines for attributing information and including quotations in news stories:

- Use the person's first name and last name when identifying a person by name for the first time in the story. This is also called "first reference." Afterward, use only the person's last name. Some newspapers use courtesy titles—such as *Dr., Mr., Ms.*, and *Mrs.*—before the last name ("Ms. Becker," "Mr. Mallory"). However, the predominant practice is not to use courtesy titles. You do not have to include the person's last name each time you reference the person; you can use a pronoun (he, she) every second or third time, instead of the person's last name.

- Use quotation marks around a word or group of words when someone has spoken or written those exact words.
- Every quotation (direct or indirect) must have attribution.
- Each direct quotation should be its own paragraph. This may mean that the paragraph with a direct quotation is only one sentence.
- Use "said" for attribution. Many people try to look through a thesaurus for a different word to use, such as *stated, noted,* or *exclaimed. Said* is a neutral word. Use it.

Associated Press Style

The Associated Press is an international organization of professional journalists. The organization has a writing style for news stories. You must follow **Associated Press Style** if you are going to write news stories professionally or provide news releases about your events to news media. Every journalist and public relations professional must understand and use Associated Press (AP) Style.

It is recommended that you purchase an *Associated Press Stylebook* at least every two to three years to see if any additions to the *Stylebook* have been made or if any entries have changed. For example, the *2006 Associated Press Stylebook* listed "(123) 555-5678" as the correct way to include telephone numbers in a news story. The telephone number entry was changed in the *2007 Associated Press Stylebook* to "123-555-5678." In the 2009 edition of the *Stylebook,* "website" was listed as two words: "Web site." In the 2010 edition, it had been changed to one word, not capitalized: "website." In addition, you should review the *Stylebook*'s section on edit marks.

A list of the most commonly used entries from the *Associated Press Stylebook* is provided in Appendix A. You may never need to know certain *Associated Press Stylebook* listings, such as if "nearsighted" is one word, two words, or hyphenated. (It is one word, by the way, according to the 2010 edition.) However, you will need to know how to correctly write an address and to use numbers and measurements, among other things. The list in Figure 5-6 is not meant to be a complete list of everything you should know, but it should keep you from having to memorize everything in the *Stylebook*.

Grammar and Punctuation

Any news story *must* be well written. The story should be as free of grammar and punctuation errors as possible. *Grammar* is a system of rules that defines the use of the language. Most of the grammar and punctuation rules you have learned in school will be the same as Associated Press Style, but there are some differences. Because you will be using Associated Press Style for journalistic writing, you should refer to the *Associated Press Stylebook*'s section on punctuation for assistance. Following are some common grammatical and punctuation issues for journalists.

Grammar

A *sentence fragment* is a group of words that does not express a complete thought. It may lack a subject, predicate, or a complete thought. Every sentence in a news story should be a complete sentence.

Fragment: Finding a dependable and inexpensive car to use.

Complete: Finding a dependable and inexpensive car to use is becoming more difficult.

A *run-on sentence* is really two sentences joined without proper punctuation. Run-on sentences are corrected in these three ways:

1. Break the sentence into two sentences by using a period.
2. If there is a close relationship between the two sentences, insert a semicolon to join them.
3. Connect the two sentences with a comma and a coordinating conjunction (e.g., *and, but, or*).

Run-on: The turnpike is a better road it has less traffic.

Correct: The turnpike is a better road. It has less traffic. (Break the sentence into two sentences.)

Correct: The turnpike is a better road; it has less traffic. (Insert a semicolon.)

Correct: The turnpike is a better road, and it has less traffic. (Insert a comma and conjunction.)

Comma splices occur when a sentence uses a comma instead of a period. As with a run-on sentence, you can correct comma splices by using a period, inserting a semicolon in place of the comma, or adding a conjunction after the comma.

Comma splice: The rain ruined our vacation, we couldn't go to the beach.

Correct: The rain ruined our vacation. We couldn't go to the beach.

Agreement refers to singular and plural references. In *subject/verb agreement*, single subjects take single

Specific Associated Press Style Issues

Here are some specific Associated Press Style issues:

Numbers

- In general, spell out whole numbers nine and below. (The nine boys.)

- Use figures for 10 and above. (The 25 boys.)

- "Million" and "billion" are used with round numbers (2.3 million, 250 billion.)

- "Thousands" are numbers (186,540.)

- Ages are always numbers. (The 2-year-old girl. John is 21 years old.)

- Measurements and dimensions are always numbers. (25 percent. 3 yards. He is 5 feet tall.)

- Years are always numbers. (He was born in 1990.)

- *However*, spell out any number – except for a year – that begins a sentence. (Four-year-old Tom Adams won an award. 2007 was a good year.)

Abbreviations

- **Titles:** Some titles are abbreviated, but only in front of someone's name. The abbreviated titles are "Dr.," "Mr.," "Mrs.," "Rev." (reverend), "Sen." (senator), "Rep." (representative), "Gov." (governor), "Lt. Gov." (lieutenant governor), and military ranks. For example, "Gov. Adams said he liked the presentation." Titles are spelled out if they are not in front of a person's name. ("Adams, the governor of Georgia, said he liked the presentation.")

- **Street addresses:** The words "street," "avenue," and "boulevard" are spelled out unless they are part of a full street address. "Road," "alley," "circle," and "drive" are not abbreviated.
  - He lives on Main Street. He lives at 1245 Main St.
  - She lives on Bamboo Avenue. She lives at 405 Bamboo Ave.
  - They live on Citrus Boulevard. They live at 80 Citrus Blvd.
  - The box was delivered to Boone Road. The box was delivered to 890 Boone Road.

- **Months and dates:** Months are spelled out unless they come before a date. Check the *Associated Press Stylebook* to see how each month is abbreviated. Months that are five letters or shorter are never abbreviated (March, April, May, June, and July).
  - She moved last February.
  - She moved in February 2007.
  - She moved on Feb. 6, 2007.
  - She moved on March 15, 2007.

- **Organizations:** Spell out names of organizations (colleges, groups, clubs) on first reference. Abbreviate the names, if necessary, on second reference.
  - First reference: College of Agriculture Student Council.
  - Second reference: CASC

FIGURE 5-6: Specific Associated Press Style issues.

verbs; plural subjects take plural verbs. In *noun/ pronoun agreement,* a singular noun takes a singular pronoun, and a plural noun takes a plural pronoun.

False subjects occur when a sentence does not begin with a real subject. Most sentences beginning with a false subject—for example, *there is, there are, there was, there were, there will be, it is,* or *it was*—can be rewritten and made stronger.

False subject: There is a class in my school that teaches writing.

Better: A class in my school teaches writing.

Parallelism refers to words or phrases that are of equal rank in tone or tense. Do not mix unequal elements in a phrase or series.

Mixed phrases: He *enjoys* books, movies, and *driving his dune buggy.*

Correct: He enjoys *reading* books, *going* to movies, and *driving* his dune buggy.

Correct: He enjoys *books, movies,* and his *dune buggy.*

Mixed tenses: He *walked* the dog and *works* with the horses.

Correct: He *walked* the dog and *worked* with the horses.

Dead wood is any word that is just extra to a sentence and does not add to it. Eliminate any words that would only add "dead wood" to your sentence.

Dead wood: It is *really* necessary to return the library book *very* soon. (How much more necessary is "really" necessary? How soon is "very soon"?)

Correct: It is necessary to return the library book soon.

Gender-neutral language should be used in your writing. Primarily, this avoids using "man" for "people." Avoid job titles that refer to gender, such as *policeman, fireman,* and *postman.* Instead, use *police officer, firefighter,* and *postal carrier.* Use plural pronouns to get around having to use "his/her" in sentences.

Awkward: A reporter should edit his/her article.

Better: Reporters should edit *their* articles.

Prepositional phrases should be kept to a minimum. If you see several prepositional phrases in a series, try to rewrite the sentence. Prepositional phrases are not bad, but they do add unnecessary words.

Awkward: The FFA meeting was led *by* the president *of* the chapter *in* the classroom.

Better: The FFA chapter's president led the meeting in the classroom.

Dangling modifiers do not modify the correct word. Be sure the modifier modifies the right noun.

Dangling modifier: Walking through the rows, the corn nearly filled the rows. (Sounds like the corn was walking through the rows.)

Correct: Walking through the rows, I noticed the corn nearly filled the rows.

Active and passive voice refers to the way in which verbs are used. The emphasis is on the subject as the doer of the action if a verb is in the active voice. Passive voice throws the action onto the object. Generally, writers should try to use the active voice.

Passive: The potatoes were passed around the table (by her).

Active: She passed the potatoes around the table.

Punctuation

Commas: Use commas to separate items in a series. However, unlike traditional punctuation rules that you have learned, in Associated Press Style writing, you do *not* include a comma before the conjunction. This is probably one of the biggest differences between journalistic writing and the writing style you have used in composition classes.

Incorrect (according to AP Style): The American flag is red, white, and blue.

Correct (according to AP Style): The American flag is red, white and blue.

Clauses introduced by *when, if, because,* and *although* require a comma when they begin a sentence or are elsewhere in the sentence.

Correct: Although the test was repeated, the results were never the same.

Correct: We could not duplicate these results, although we tried many times.

Set off an *appositive*—a word or phrase that follows another word to explain or identify it—with commas. Be sure you place a comma *after* the appositive.

Appositive: George Washington, *a Virginia planter,* was the first president of the United States of America.

Do not use a comma to precede a partial quotation.

Incorrect: The mayoral candidate charged that the man was, "a swindler of the lowest order."

Correct: The mayoral candidate charged that the man was "a swindler of the lowest order."

Use a comma to precede a complete quotation.

Correct: The defense attorney asked, "How would you like to be sent to prison?"

Semicolons: Use a semicolon to separate independent clauses not connected by a coordinating conjunction.

Correct: DeGraw launched her desperation shot; the ball went through the hoop as the buzzer sounded.

Use a semicolon prior to a conjunctive adverb (e.g., *however, therefore, nevertheless*) in a sentence. Insert a comma after the conjunctive adverb.

Correct: The first test results were unsatisfactory; however, a simple modification of the questionnaire solved the problem.

A semicolon separates items in a series that contain commas.

Incorrect: We traveled to four of the world's most significant cities: Paris, France, London, England, Rome, Italy and Vienna, Austria.

Correct: We traveled to four of the world's most significant cities: Paris, France; London, England; Rome, Italy; and Vienna, Austria.

Colons: Colons are used to separate parts of a sentence and to indicate a list or series.

Correct: The dealer had three cars: a BMW, a Cadillac and a Mustang. (Notice that the comma before "and" is not included. This is *correct* according to AP Style.)

Do not use a colon to separate a verb and its complement.

Incorrect: A scientist requires: intelligence and diligence.

Correct: A scientist requires two attributes: intelligence and diligence.

Correct: A scientist requires intelligence and diligence.

Do not capitalize the first word that follows a colon, unless the word is a proper noun.

Incorrect: She has three hobbies: Gardening, sewing and reading. (Notice that the comma before "and" is not included. This is *correct* according to AP Style.)

Correct: She has three hobbies: gardening, sewing and reading.

Quotation Marks: Commas, question marks, and periods go *inside* quotation marks in a quotation.

Correct: He said, "The test was hard."

Correct: "I thought so too," she said.

Correct: "Was the test hard?" she asked.

Use a set of double quotation marks first, then single marks within a quotation, for such items as titles that normally require double quotation marks.

Correct: He said, "I read the poem 'Transformation' yesterday."

Use the following sentences as examples of how to punctuate direct quotations.

"The dog ran past the man," he said.

"The dog ran past the man," he said, "but it was stopped by the dogcatcher."

The principal said, "Pasco Independent School District is the best school district in the state."

Apostrophes: According to Associated Press Style, apostrophes can be used to indicate where numerals are left out.

Correct: The class of '07.

However, do not use an apostrophe for decades.

Incorrect: 1990's

Correct: 1990s

Hyphens: Hyphens are usually used to join words to form adjectives.

Correct: A 7-year-old boy. An off-the-cuff remark. A little-known man. A 3-inch bug.

Hyphens are not used with adverbs ending in *-ly*.

Incorrect: a gravely-ill student
Correct: a gravely ill student

FEATURE WRITING

The feature story is more relaxed in style than a traditional news story. A **feature story** is set apart from a news story because of the greater amount of detail and description it contains (Figure 5-7). The structure of a news story, as has already been explained, is to provide a basic set of facts to the reader as quickly as possible. A feature writer enhances those facts with details and description so that the reader will be able to see a more complete picture of an event or a person.

A feature story can be on just about anything: a person, a group, animals, places, events, objects, or holidays. Regardless of the topic, however, a feature story must be interesting and well written, and it must draw on human interest. In other words, the feature must touch the reader on a personal level.

A feature story contains many of the same components as a news story: a good lead, short sentences, brief paragraphs, action verbs, good description, and relevant quotations.

Features can be categorized in the following ways:

- A *news feature* is written around a timely event. For example, a news feature could be written about the local fair.

Model Feature Story

A feature story is set apart from a news story because of the greater amount of detail and description it contains.

A feature story can be on just about anything: a person, a group, animals, places, events, objects or holidays. The feature must be interesting and well written, and it must touch the reader on a personal level.

The lead draws the reader in to the story. The second or third paragraph is called the "engine paragraph" and sets the stage for the rest of the feature.

The body is the section that takes up most of the story. The body provides the reader with documented facts and details and careful observations made by the story's writer.

"A good feature will be brightened with good quotations throughout," said Ricky Telg, a professor at the University of Florida. "During the interview, the reporter should try to identify several good quotes that can be used in the story."

A feature story contains many of the same components as a news story: a good lead, short sentences, brief paragraphs, action verbs, good description and relevant quotations.

Unlike a news story that is supposed to stop when the least-important information is presented in the inverted pyramid structure, a feature story may need an ending to wrap up.

The ending, though, should not go too long. As with the news story, stop writing when nothing else is left to say, Telg said.

If a feature story runs more than one page, insert " – more – " at the bottom of the page. At the top of the second page, flush left, write "Add 1."

On the last page, use the following notation, centered, on the page, to signify the end of the story: " – 30 – " or "###."

###

© Cengage Learning 2012

FIGURE 5-7: Model feature story. A feature story can be on just about anything: a person, a group, animals, places, events, objects, or holidays. A feature story must be interesting and well written, and it must touch the reader on a personal level. This model feature story shows how a feature should be written.

- An *informative feature* zeroes in on the little known, the odd, or the unusual. An informative feature could be on how a school's mascot was selected.
- An *historical feature* focuses on something of historical relevance to the audience. Historical features are commonly seen around holidays (e.g., the first Thanksgiving, the origin of Christmas trees, the origin of Memorial Day).
- A *personal experience feature* recounts the accomplishments of an individual or group, usually as an example of a much larger group. (Example: a feature story on a child with muscular dystrophy, especially around the time of the Jerry Lewis Muscular Dystrophy Association Labor Day Telethon.)
- A *descriptive profile* centers on places people can visit or events they can take part in. This type of feature is seen regularly in newspapers' "travel" sections.
- A *how-to-do-it yourself feature* explains how to build something or how to do something. Some newspapers have features on how to garden. Another example of this type of feature is a story on how to select healthy foods.
- The *profile* may be the most common feature. The profile tells about a person. A profile examines only one or two aspects of a person; it does not tell a person's entire life story. A profile is enhanced through the use of anecdotes (stories told by the person being profiled).

Feature stories usually follow this structure:

- **Lead:** As with a news story, the lead in a feature story draws the reader in. With a feature, though, the lead may be more than just one sentence. However, do not take too long to get to the point in the story.
- **Engine paragraph:** This paragraph is usually the second or third paragraph of the story and sets the stage for the rest of the feature. The engine paragraph puts the story in some context for the reader and lets the reader know why the rest of the story should be read. This is sometimes called the "why paragraph."
- **Body:** This section takes up most of the story. It expands and details the information introduced in the lead. The body provides the reader with documented facts and details and careful observations made by the writer.
- **Ending/conclusion:** Unlike a news story that is supposed to stop when the least-important information is presented in the inverted pyramid structure, a feature story may need an ending to wrap up. The ending, though, should not go too long. As with the news story, stop writing when nothing else is left to say. Sometimes, but not always, a feature writer will end with one of these closures:
    - **Circle technique:** The feature story begins and ends with approximately the same idea, phrase, question, statement, or description.
    - **Surprise:** With this closure, the writer provides a different ending than what the reader expects.
    - **Summary ending:** This ending concludes with an overall summary of the topic.

When writing the feature, keep these writing guidelines in mind:

- **Describe the topic in specific and concrete words.** As with a news story, do not rely on adjectives and adverbs. Describe with nouns and verbs.
- **Brighten your feature with quotes, but do not go overboard.** Paraphrase throughout. A good rule is to have one direct quotation for every three or four paragraphs.
- **Avoid mind reading.** Do you really know that the teacher "feels" a certain way about a topic? Stick to what you observe and what people say. You can never go wrong by saying that "someone said something." You do not know what someone "believes."

NEWS WRITING FOR TELEVISION AND RADIO STORIES

Creating a television or radio story is more than hitting "record" on a video camera or audio recorder. You have to learn the process of writing an effective television and radio news story first. The term *broadcast writing* will be used interchangeably for *television and radio news writing* throughout this section of the chapter.

Writing for radio and television is different from writing for print for the following reasons. First, you have less space and time to present news information. Therefore, you must prioritize and summarize the information carefully. Second, your listeners cannot reread sentences they did not understand the first time; they have to understand the information in a broadcast story as they hear it or see it. As a result, you have to keep your writing simple and clear. And third, you are writing for "the ear." In print news stories, you are writing for "the eye"; the story must read well to your eye. The television or radio news story has the added complexity that it has to sound good; when a listener hears the story it has to read well to "the ear." Also for a radio news story, listeners cannot see video of what you are saying, so you must paint word pictures with the words you use in your radio news story so people can "see" images just through your verbal descriptions.

As with any type of news writing, you should try to identify characteristics of your audience so you know what type of information your audience wants. Use the criteria of newsworthiness presented earlier in this chapter to help you determine if your television or radio news story idea has news value.

Review the section on scriptwriting in Chapter 10, Video and Audio Production, because many of the concepts about writing for educational and promotional video programs apply to writing television and radio news stories. For example, television and radio news stories must read well for "the eye" and sound good to "the ear." To do that, television and radio news stories must have these attributes:

- **The writing style should be conversational.** Write the way you talk.
- **Each sentence should be brief and contain only one idea.** We do not always talk in long sentences. Shorter sentences are better in broadcast news writing. Each sentence should focus on one particular idea.

- **Be simple and direct.** If you give your audience too much information, your audience cannot take it in. Choose words that are familiar to everyone.
- **Read the story out loud.** The most important attribute for writing for "the ear" is to read the story aloud. This will give you a feeling for timing, transitions, information flow, and conversation style. Your audience will hear your television or radio news story, not read it, so the story has to be appealing to the ear.

In the remainder of this section, specific guidelines are presented to help you write news stories for television and radio.

Television and Radio News Writing Structure

- **Be brief.** A good newspaper story ranges from hundreds to thousands of words. The same story on television or radio may have to fit into 30 seconds—perhaps no more than 100 words. If it is an important story, it may be 90 seconds or two minutes. You have to condense a lot of information into the most important points for broadcast writing.
- **Use correct grammar.** A broadcast news script with grammatical errors will embarrass the person reading it aloud if the person stumbles over mistakes.
- **Put the important information first.** Writing a broadcast news story is similar to writing a news story for print in that you have to include the important information first. The only difference is that you have to condense the information presented.

FIGURE 5-8: Reporters for television and radio develop stories that are brief and conversational.

- **Write good leads.** Begin the story with clear, precise information. Because broadcast stories have to fit into 30, 60, or 90 seconds, broadcast stories are sometimes little more than the equivalent of newspaper headlines and the lead paragraph.
- **Stick to short sentences of 20 words or less.** The announcer has to breathe. Long sentences make it difficult for the person voicing the script to take a breath.
- **Write the way people talk.** Sentence fragments—as long as they make sense—are acceptable.
- **Use contractions.** Use *don't* instead of *do not*. But be careful of contractions ending in *-ve* (e.g., *would've, could've*), because they sound like "would of" and "could of."
- **Use simple subject-verb-object sentence structures.**
- **Use the active voice and active verbs.** It is better to say "He hit the ball" than "The ball was hit by him."
- **Use present-tense verbs, except when past-tense verbs are necessary.** Present tense expresses the sense of immediacy. Use past tense when something happened long ago. For example, do not say, "There were forty people taken to the hospital following a train derailment that occurred early this morning." Instead, say, "Forty people are in the hospital as a result of an early morning train accident."
- **For radio news stories, write with visual imagery.** Make your listeners "see" what you are saying. Help them visualize the situation you are describing.

Television and Radio News Writing Techniques

- **Use a person's complete name (first and last name) in the first reference, then the person's last name thereafter.**
- **Use phonetic spellings for unfamiliar words and words that are difficult to pronounce.**
- **Omit obscure names and places if they are not meaningful to the story.**
- **Titles precede names; therefore, avoid appositives.** Do not write, "Tom Smith, mayor of Smallville, said today . . . ." Instead, write, "Smallville mayor Tom Smith said today . . . ." (Other examples: "City councilman Richard Smith," not "Richard Smith, city councilman." "Anyville High School student Beth Baker," not "Beth Baker, Anyville High School student.")
- **In age reference, precede the name with the age.** (Example: "The victim, 21-year-old Rob Roy . . .")

- *Avoid writing direct quotations into a news script, if at possible.* Instead, let people say things in their own words during soundbites. A soundbite is the exact words spoken by someone in his or her own recorded voice. If you must use a direct quote, set it off with such phrases as "In the words of . . . " or "As he put it . . . ," or try to paraphrase as much as possible. *Avoid saying "quote" and "unquote" to lead into or end a direct quote.*

- *The attribution should come before a quotation, not after it.* In contrast to writing for print media, the attribution of paraphrased quotations in broadcast stories should be at the beginning of the sentence, before the paraphrase. The listener should know where the quotation is coming from before hearing the quote. Example: "Bill Brown said he would run for re-election."

- *Avoid all abbreviations, even on second reference, unless it is a well-known abbreviation.* This is different than the Associated Press Style rules for print stories. Write out days, months, states, and military titles each time. About the only acceptable abbreviations are *Mr., Mrs.,* and *Dr.* Punctuate, by using a hyphen in between, commonly used abbreviations. For example, write "U-S," instead of "US" (United States), and "U-N" for "UN" (United Nations).

- *Avoid symbols when you write.* For example, the dollar sign ($) should never be used in broadcast writing. Always spell out the word "dollar." This is different from the Associated Press Style for "dollar," when used in a print news story.

- *Use correct punctuation.* Do not use semicolons. Use double dash marks for longer pauses than commas. Use underlines for emphasis.

- *Use numbers correctly.* Spell out numerals through 11. (This is different than Associated Press Style for print stories, which spells out one through nine, and starts using numerals for 10 and above.) Use numerals for 12 through 999. Use hyphenated combinations for numerals and words above 999. (Examples: 33-thousand; 214-million.) Round off numbers unless the exact number is significant. (Example: Use "a little more than 34 million dollars," not "34-million, 200-thousand, 22 dollars.") Use *st, nd, th,* and *rd* after dates, addresses, and numbers above "eleventh" to be read as ordinary numbers. (Examples: "Second Street," "May 14th," "Eleventh Avenue," "12th Division"—this is different from AP Style for print.)

Television and Radio News Story Format

- *Broadcast news stories are typed, double-spaced, and in uppercase/lowercase.* Many years ago, television news scripts were written in all uppercase, but that practice has changed in recent years. Examples of radio and television news stories are provided in Figures 5-9 and 5-10.

Example Radio Story

Watching a rattlesnake on television is the closest most of us ever want to get to one. But some do like to get a little closer to rattlers. And Sweetwater Jaycee Wayne Wilson says one big reason is the money rattlesnake products bring.

Actuality: Wayne Wilson, Sweetwater Jaycee (15 seconds)
"Every bit of the snake . . . and to make anti-venom."

Over the weekend, the 23rd Annual Sweetwater Rattlesnake Roundup attracted thousands, wanting to see the venomous vipers. Vendors use the rattlesnake skins and other body parts for earrings, key chains, walking canes, hairpieces and paperweights.

Jaycee John Thomas also says some people like their rattlers well . . . . well-done.

Actuality: John Thomas, Sweetwater Jaycee (9 seconds)
"It has a taste . . . texture of fish."

Thomas says it's difficult to put a dollar value on the rattlesnake product industry because few vendors deal strictly in rattlesnakes. Still, when rattlesnakes bring eight to 10 dollars a pound, they're a big attraction for someone wanting to make a fast buck. From Sweetwater, this is Rick Wayne reporting.

FIGURE 5-9: Example radio story. This example shows the narrator portions and the actualities (or soundbites) of a typical radio news story.

Example Television Story

Suggested Studio Introduction (to be read by the news anchorperson)

Rattlesnake products

They're cold-blooded and dangerous, but they're also big business. They're rattlesnakes, and as Rick Wayne reports, products made from this reptile range from the ordinary to the bizarre.

Outcue:". . . Rick Wayne reporting."
Total Time:1:37

Video	Audio
CU of rattlesnake w/ rattles. Medium of pit of rattlesnakes.	NARRATOR: Watching a rattlesnake on TV is the closest most of us ever want to get to one. But some do like the snake. And one big reason is the money rattlesnake products bring.
On-screen text: Wayne Wilson Sweetwater Jaycee	SOUNDBITE: Every bit of the snake is used. The skin is sold to make belts. . .and the heads are sold. Of course we sell the venom for research and to make antivenom.
On-screen text: Mike Barker Sweetwater Jaycee CU of Mike Barker "milking" venom from snakes' fangs.	SOUNDBITE: We average a quarter of a cc of venom per snake, so we think if we can get 1,500 cc, that's in the neighborhood of 6,000 snakes we need.
Medium shots and CUs of various items made from snake body parts.	NARRATOR: Vendors use the skins and other body parts for earrings, keychains, canes, hairpieces and some very interesting paperweights.
On-screen text: John Thomas Sweetwater Jaycee	SOUNDBITE: Anything that the skin can be used for, the dealers have found to make a product.
CU of Jaycee taking snake meat out of fryer.	NARRATOR: Then there are those who like their rattlers well done. NAT SOUND: Frying sound (rattlesnake cooking)
Medium shot: John Thomas	SOUNDBITE: It has a taste all of its own. But a lot of people like to compare it to the taste of chicken, with maybe the texture of fish.
Various shots of vendors selling items made from rattlesnake parts and from other animals (frogs). CU: rattlesnake hissing	NARRATOR: Thomas says it's difficult to put a dollar value on the rattlesnake product industry because few vendors deal strictly in rattlesnakes. Most also sell other nongame animals, like frogs. Still, when snakes bring eight to nine dollars a pound, they're a big attraction for someone wanting to make a fast buck. From Sweetwater, this is Rick Wayne reporting.

© Cengage Learning 2012

FIGURE 5-10: Example television story. This example shows how a news story would be written for use on television.

- *Make the sentence at the bottom of a page a complete sentence.* Do not split a sentence between pages.
- *Never split words or hyphenated phrases from one line to the next.*
- *Do not use copyediting symbols.* Cross out the entire word and write the corrected word above it. This is one reason why broadcast news scripts are double-spaced, so you will have room to make corrections in between the lines.

Television and Radio News Terms

It is good practice to understand the following terms used in broadcast news writing:

- *Actuality:* Actuality is a term commonly used in radio for the exact words spoken by someone in his or her own voice. Actualities are usually 20 seconds or less. In television news, an actuality is called a *soundbite*.
- *Anchor, talent, or newscaster:* For television, an anchor—also referred to as *talent* or *newscaster*—is the person who reads the stories live on-air during a television newscast.
- *B-roll:* Any non-narrated video footage shot expressly to "cover" narration or an interview is called b-roll. The audio from these shots is generally used as background audio ("natural sound"). B-roll video also is called *cover video*. For example, in shooting a television story on the timber industry, b-roll would be shots of trees, trees being cut down, trees being loaded onto trucks, and trees being processed at a lumber yard. In editing the story, the b-roll would be used to show what is being described or "covered" in the narrated script.
- *Outcue:* The last thing a reporter says is the outcue. Usually, the outcue gives the reporter's name and television station. Example outcue: "For AEC News, I'm Rick Wayne."
- *Package:* A complete television news story is a package. A typical package will run 90 seconds to two minutes in length.
- *Slug:* A slug is at the top of the script and provides the title of the script, the running time (how long the news story is, measured in minutes and seconds), and the date that the story is to be aired or when it was written.
- *Soundbite:* A soundbite is a recorded quotation, the exact words spoken by someone in his or her own voice. A soundbite is a "bite" of the actual longer interview. Soundbites are usually 20 seconds or less. The term is used frequently in television news to indicate the video and audio of a person shown on-air. In radio news, a soundbite is also called an *actuality*.

- *Stand-up:* A stand-up is when a reporter narrates a portion of a story on camera.
- *SOT (sound on tape):* Any time you hear audio on a video, it is called sound on tape (SOT). However, the term is used most frequently to describe when any person talking is shown speaking.
- *VO:* A VO (voice over) is just video that will be shown, with a newscaster narrating the script. The newscaster, then, is providing voice *over* the video.
- *VO/SOT (voice over/sound on tape):* A VO/SOT (voice over/sound on tape) is the term when a newscaster narrates a script on air while video runs over the narrator's voice (VO), followed immediately by someone talking (a soundbite) about a topic related to the voice over. This is the SOT. An example of this would be a newscaster reading a story about the county fair while video shot at the fair is shown (VO). Immediately after the newscaster stops talking, a soundbite (SOT) from the county fair's organizer is shown and he or she describes events that will happen later that week at the fair.

Narrating Television and Radio News

Follow these recommendations when you narrate (also referred to as "voicing") television and radio news scripts:

- *Position the microphone properly.* Position the microphone 6 to 10 inches from your mouth and at a 45-degree angle to the direct line of speech. This will help prevent "blasting" with explosive letters such as "P" and "B." Always maintain the same distance from the microphone as you speak (Figure 5-11).
- *Remove noise-making distractions.* Remove all paper clips, pens, and other items that you would be tempted to play with as you read the story. Any rustling of paper clips or pen clicking can be picked up by the microphone.
- *Narrate the news story.* After you hit the "record" button on the video camera or audio recorder, wait approximately 10 seconds before speaking. This prevents you from accidentally losing some of the narration if you hit "record" and start narrating the script immediately. It is a good idea to use a standard reference opening, such as the

FIGURE 5-11: Position the microphone 6 to 10 inches from your mouth and at a 45-degree angle to the direct line of speech. Always maintain the same distance from the microphone as you speak.

© Cengage Learning 2012

day, place, and subject's name. You may want to use a countdown: "Honeybee story, coming in three, two, one," and then start the story. This also helps your voice stabilize as you start. The standard reference opening and countdown will be edited out of the final story.

- *Articulate words correctly.* Speak clearly. Do not run your words together. Practice proper articulation, the distinct pronunciation of words. The following words are often improperly articulated: "prob-ly" for "prob-ab-ly," "git" for "get," and "jist" for "just." Also, do not drop the final "g" in "-ing" words, such as *cooking, running,* and *hunting.*
- *Think the thought.* Think about what you are going to say. If something has a positive idea, put a smile in your voice by putting a smile on your face. This helps to project the personality of the story.
- *Think the thought through to the end.* Keep half an eye on the end of the sentence while you are reading the first part. Know how the sentence will come out before you start. This will help you interpret the meaning of the phrases of the entire idea.
- *Talk at a natural speed.* But change the rate occasionally to avoid sounding monotonous. The speed that you talk is your speaking rate. Vary the pitch and volume of your voice to get variety, emphasis, and attention. Pitch is the high and low sounds of your voice. You will sound more assertive if you lower your pitch and inflect downward; however, avoid dropping your pitch when it sounds unnatural to do so.
- *Breathe properly.* Control your breathing to take breaths between units of thought. Otherwise, you

will sound choppy. Sit up straight while narrating. This helps your breathing.

- *Use your body.* A relaxed body helps produce a relaxed-sounding voice. Do a few exercises before going on the air. A little activity reduces tension.
- *Listen to the final product.* Listen to how it sounds. Listen to what you said as if you were an audience member.
- *Time the story.* At the end, be sure you time the story. If the story is going on the air of a radio or television station, the story's timing is important, and, in many cases, needs to be exact. Practice writing and narrating news stories to determine what your normal reading time is.
- *Practice your narration skills.* Never give up practicing speech and delivery techniques. Read aloud something at least twice a week for practice.

CONDUCTING INTERVIEWS FOR NEWS STORIES

In order to write a good news story for print, television, or radio, you have to conduct interviews with the people who have the information you need. An *interview* is the process of asking good questions so you can get good answers for your news story (Figure 5-12). If you have never conducted an interview, the idea of doing one may seem a little scary. If you imagine that the interview is just a conversation with the other person, doing the interview will be much less frightening. Here are some tips to follow as you conduct a news story interview.

Before the Interview

- *Set aside time to conduct the interview.* Unless the person being interviewed is on an extremely tight schedule so that the interview can only take a few minutes, try to schedule a little extra time so that you and the person being interviewed do not feel constrained for time.
- *Dress appropriately.* Again, impressions make an impact on the person being interviewed. Dress up a little or wear your organization's official attire. Avoid jeans or shorts.
- *Prepare at least 10 questions in advance.* These questions should pertain directly to the topic you need information about. Think about what your audience needs to know as you prepare the questions. What does your audience want to know?
- *Understand the subject matter (at least a little bit).* The person being interviewed is the expert on the topic. Otherwise, you would not have called

© Cengage Learning 2012

FIGURE 5-12: Conducting an interview is the process of simply asking good questions. This photo shows two college students conducting a television interview with an agricultural researcher.

on the person for an interview. However, it is good practice to do at least a little research on the topic beforehand so that you can ask good questions.

- *Be on time for the interview*. Being prompt makes a good impression on the person being interviewed.

During the Interview

- *If you plan to use an audio recorder during the interview, first obtain the interviewee's permission to do so*. If you are doing a television interview, before you arrive let the person being interviewed know that you will have a video camera.
- *Take good notes*. Do not rely on an audio recorder. Batteries do die.
- *State the interview's purpose*. What do you want to cover in the interview?
- *Break the ice with light conversation*. Make the person being interviewed feel at ease.
- *Let your subject do the talking*. Do not break in while someone is answering a question. Wait until the person has completed answering a question.
- *Get at least three good, insightful direct quotes*. This should be your goal in an interview.

- *Get correct information*. Ask persons you interview to provide the correct spelling of their names and their job titles. Do not assume you know what they are.
- *Collect more information than you think you will need.*
- *Do not be bashful about asking the person to repeat something important*. It is better to have something repeated and get the information correct than to get it wrong.
- *Be aware of your surroundings*. A few notes about the room and other surroundings may be useful in a feature story to help set the mood of your story.

Asking Questions

Listen carefully to the answers and take good notes. As the person talks, ask yourself, "What is my lead going to be? Do I understand enough to state a theme clearly and support it with quotes?" In addition:

- *Never plunge in with the tough questions*. Break the ice by explaining who you are and what you are doing.
- *Be pleasant but purposeful*. You are there to get information, so do not be timid about asking questions.
- *Use the list of questions you prepared*. Start with the easier questions, and then move to more in-depth questions or ask one that comes to mind.
- *Do not be afraid to leave your set of questions*. If a prepared question is no longer suitable, move to the next question. Some answers prompt additional questions. Ask them as they arise. Listen to what the person is saying. One question should logically follow another.
- *Be objective*. Do not offer your opinions on the subject. You are there to report, not to editorialize.
- *Stay on track*. If the interviewee strays too far from the subject, ask a specific question to redirect the conversation.
- *Avoid yes/no questions.* These only provide yes/no answers.
- *Start with questions focusing on the five Ws and H.*
- *Get in the habit of asking more probing questions*, such as, "What do you mean?" and "Why is that?"

At the End of the Interview

- *As the interview comes to a close, take a few minutes to skim your notes.* If time allows, ask the interviewee to clarify anything that you did not understand.

© Cengage Learning 2012

Joel D. Jackson

Executive Director, Florida Golf Course Superintendents Association

Controversial, negative headlines and stories are easy. The true test of a credible professional writer/reporter is to present both sides of an issue. Interview sources from both sides and use facts instead of public perception or personal opinion.

- ***Ask for permission to call back or e-mail later for more information, if necessary.***
- ***Smile, thank the interviewee, and leave.***
- ***Fill in the blanks immediately.*** As soon as the interview is over, while it is fresh in your mind, go back to your notes and fill in any blanks that you were not able to write down during the interview.

Types of Questions

Following are some of the questions that you might want to ask:

- ***Close-ended questions*** provide short answers. The answer to the example question below would provide just a list of the positive courses.
  Closed-ended question: "In which high school courses have you had the most positive experiences?"
- ***Open-ended questions*** provide longer answers. The answer to the example open-ended question below would provide a much longer response.
  Open-ended question: "What's your opinion of Ms. Jones's class?"
- ***Probe questions*** follow up on something the interviewee has said.
  Close-ended question: "In which high school course have you had the most positive experiences?"
  Answer: "Ms. Jones's class."
  Probe question: "What positive experiences have you had?"

- ***A mirror question*** repeats part of the person's answer, prompting the person to explain an answer further. A mirror question is often paired with a probe question.
  Probe question: "Why do you think people are saying positive things about Ms. Hightower's class?"
  Answer: "Because she's fair, has knowledge about the subject, and seems to really care about students."
  Mirror question: "You say she's fair. Why is being fair important to students?"
- ***The yes/no question*** is the most close-ended of close-ended questions. The answer can only be one of two options: yes or no. On their own, yes/no questions provide little information for most news and feature writers. Use yes/no questions to set the stage for other questions that would provide more in-depth information.
  Yes/no question: "Do you think Mr. Smith is a good teacher?" (Answer would be "yes" or "no." Then use a probe question to expand on the interviewee's answer.)
- ***Leading questions*** are considered unethical by many news writers. A leading question strongly suggests the "right" answer to an interviewee. The question below would make the interviewee feel that the "correct" response was that Mr. Smith was a great teacher, even if the interviewee did not feel that way.
  Leading question: "Everyone I've interviewed says Mr. Smith is an outstanding teacher. What's your opinion?"

SUMMARY

This chapter presented the basics on how to write news stories for print, radio, and television. For all three, you must practice writing clearly and understandably. For print, you also have to use Associated Press Style. For broadcast writing, one of the most important things to keep in mind is to write for "the ear" and for "the eye." This chapter also covered how to ask good questions. By learning how to write news stories and to interview people correctly, you can get the word out about your organization's activities to a larger audience in the form of newsletters and news releases.

CHAPTER EXERCISES

APPLICATIONS

Following are some ways you can apply what you learned in this chapter:

- Develop video news stories about an organization's activities for your school's video announcements.
- Assist school video news reporters with stories about an organization's activities.
- Write print news stories about an organization's activities for the school newspaper.
- Provide a radio story to your local radio station.
- With the popularity of such sites as YouTube, more news-style videos are being uploaded to the Web. The television or radio story you develop could be saved in a podcast format for people to watch or listen to on their computers or on their portable media devices, such as iPods. Post a video story on YouTube or format it so that it can be watched on portable media devices.
- Write a news story about an organization's activities for your local newspaper.
- Invite news reporters to your classroom to discuss writing for radio, television, and newspapers.
- Conduct a news writing workshop by inviting journalism teachers and news reporters to evaluate students' news stories.
- Cut out the front-page stories from the daily newspaper and rewrite one or two of them for a 30-second radio story. With the same newspaper stories, write them for a 90-second television story. In your video script, include video shots that you would imagine are needed to tell the story.

CHAPTER QUESTIONS

MULTIPLE CHOICE:

1. The _____ is the structure most commonly used for news writing.
   a. Associated News Style
   b. inverted pyramid
   c. human interest style
   d. proximity factor

2. The _____ is the most common news-style lead in newspapers.
   a. summary lead
   b. historical lead
   c. first-person lead
   d. quotation lead

3. Allowing the writer's opinions, prejudices, and biases to enter a story is called _____.
   a. the impersonal reporter
   b. value factor
   c. leading information
   d. editorializing

4. _____ is technical language used in specialized fields or in specific small groups.

    a. A cliché

    b. Jargon

    c. A hook

    d. A connector

5. A(n) _____ is written around a timely event.

    a. descriptive profile

    b. news feature

    c. personal experience feature

    d. informative feature

6. In broadcast news writing (radio and television), you must write for "the eye" and "the _____."

    a. ear

    b. nose

    c. throat

    d. mind

7. Prepare at least _____ questions in advance of an interview for a news story.

    a. 5

    b. 10

    c. 20

    d. 18

8. _____ means telling readers where the information in a story comes from.

    a. A connector

    b. Actuality

    c. Attribution

    d. B-roll

ASSOCIATED PRESS STYLE

Indicate if the sentences below are correct or incorrect according to Associated Press Style.

9. The 9 girls walked home.   (   )

10. 2004 was a good year.   (   )

11. The principal lives at 456 Loblolly Ave.   (   )

12. My teacher was born on October 24, 1972.   (   )

13. The state will have a budget surplus of 3 billion dollars.   (   )

14. She likes to eat oranges, apples, and pears.   (   )

15. He has three items: a drill, a hammer and a screwdriver.   (   )

REFERENCES AND FURTHER READING

The Associated Press. *The Associated Press Stylebook and Briefing on Media Law,* 45th ed. New York: The Associated Press, 2010.

Burnett, C., and T. Tucker. *Writing for Agriculture: A New Approach Using Tested Ideas,* 2nd ed. Dubuque, IA: Kendall/Hunt, 2001.

Oliu, W. E., C. T. Brusaw, and G. J. Alred. *Writing That Works: Communicating Effectively on the Job,* 9th ed. Boston, MA: Bedford/St. Martin's, 2007.

Telg, R. "Writing News Releases and PSAs." University of Florida. Last modified 2000. Accessed October 23, 2010. http://mediarelations.ifas.ufl.edu/writingnewsreleasesandPSAs.htm.

6

Document Design

OBJECTIVES

After completing this chapter, the student will be able to:

- List the phases of the document design process.
- Apply the principles of document design in developing a printed document.
- Identify the differences between symmetrical and asymmetrical balance.
- Identify when to use serif and sans serif fonts.
- Identify when to use graphics and charts.
- Identify the differences between raster and vector graphics.
- Design a brochure.
- Design a newsletter.

INTRODUCTION

Getting people to understand your written message is more than just putting words on a computer screen or on paper. In today's visual-oriented society, a message also has to look appealing. Advertising executives, for example, spend a tremendous amount of time and money to design ads that have a memorable message and are eye-catching. When you read a training manual, you will see photographs or drawings to take you through the step-by-step instructions. In education, almost every textbook you read, including this one, will have some type of visual, such as photographs and information graphics. Documents also should incorporate basic document design principles. This chapter will provide you with these basic design principles to enhance your written message.

DOCUMENT DESIGN

Document design is the process of choosing how to present all of the basic document elements so your document's message is clear and effective. When a document is well designed, readers understand the information more quickly and easily. Readers feel more positive about the topic and more accepting of its message. Documents can be printed (a flier, newsletter, newspaper, brochure, or any handout) or read on a computer screen. A *PDF* (Portable Document Format) can be downloaded, printed, and read from computers. Regardless of the form, the document production process is the same.

DOCUMENT PRODUCTION PROCESS

- ***First, consider the purpose of the document.*** The document's purpose will help determine how much content will need to be included in the document and how it should be designed. Some questions you may need to ask yourself in determining the purpose include the following: What do you want the document to do? Is the document going to persuade, inform, educate, or motivate your reader? What do you want readers to be able to do after reading the document? For example, from reading your document, do you want readers to buy something, to attend an event, or to learn a new skill?
- ***Determine who your audience is.*** What are your readers' characteristics? You should try to match your document to your audience's characteristics. For example, you will need to write at an

99

Jim Bret Campbell

Senior Director of Publications, American Quarter Horse Association

Numerous research studies indicate that print and online readers give you a couple of seconds to catch their attention before they're moving on. Communicators combine high-quality images and great design with the right headlines, blurbs, and captions to convey the message. Without the right design, it's doubtful that you'd get even the most ravenous reader to stop—even with a great headline. The whole package works together.

© Cengage Learning 2012

elementary-school level if the document is for young children. For this young audience, the document will need to use large text, few words, cartoons, photographs, and activities to keep a young audience's attention (Figure 6-1). For an adult audience, your document may have smaller

FIGURE 6-1: Know your audience and design your document to match their characteristics. Designing for young children is different than designing for older adults, or industry experts.

Ryan McVay/Digital Vision/Getty Images

text, more detailed photographs, and headlines to break up the text. For older adults, the text may need to be larger, because their eyesight may be weak.

- *Develop a content outline.* The outline should include the main topics for the document. A good outline can provide you with a master plan for your document. If the document is a brochure, the outline may be very short. If the document is a training manual, the outline may be longer.
- *Write the content.* For fliers or handouts, the content may be extremely brief. For other documents, such as newsletters and brochures, you may have more to say. Later in this chapter, you will learn more about newsletters and brochures.
- *Design the document.* After the content is written, make a hand-drawn sketch of what the document will look like, including all text, images, headlines, and borders. This sketch is called a **thumbnail** or a *dummy*. Always try to sketch out your ideas first before creating your final design. When you have gotten a thumbnail sketch layout that you like, you can proceed to computer software programs to lay out your document. Basic fliers can be designed in easy-to-use software programs, such as Microsoft Publisher, MS Word, Corel's WordPerfect, Apple's Pages, or MS PowerPoint. MS Word and other writing programs also have templates that you can use to design basic newsletters and brochures. However, for more

FIGURE 6-2: A professional printer can turn your computer file into a high-quality document.

FIGURE 6-3: Look at design examples all around you and observe the techniques and solutions used by others.

elaborate documents, you will need to use layout and design software, such as Adobe InDesign or QuarkXPress.

- *Print the document.* This is the final step. A document can be printed with a desktop printer or taken to a professional printer (Figure 6-2). It also can be saved as a PDF and placed on the Web or e-mailed. Many newsletters for professional organizations are saved as PDFs and sent to the organization's members via e-mail.

GETTING STARTED IN THE DESIGN PROCESS

If you have never designed a document before, you may not know how to put your thoughts into a good design. Here are some tips to get you started:

- *Learn from good examples.* Look around you. What advertisements catch your attention in the magazines you read? Which poster stands out on a bulletin board? What do you like about a newsletter or a magazine you have recently read? Study such examples for their effective designs (Figure 6-3). You can learn by observing techniques and solutions employed by others.
- *Keep it simple.* Usually the simplest structure is the best for designing a document. Good design should not call attention to itself; it should enhance the message. Your overall design and use of visuals should not be complicated. If your design and your visuals are clean and simple, and communicate a clearly stated message, then you are helping your audience. Therefore, just because

you can use a lot of graphics or photographs does not mean you should. For example, if one particular visual—such as a single photograph or piece of clip art—works in relating the message, you do not need more.

- *Select appropriate visuals.* Take time to select a good visual (graphic, photograph) that supports your text. For example, if a good photograph exists to clarify your message, then use it. If the photograph is out of focus or does not add clarity to the message, then do not use the photograph. Also, be sure you can use the graphic or photograph. Stock photo sites on the Internet charge for photographs; you should contact a website's manager to get permission to use a photograph before you include it in a document.
- *Lay out the document.* Now comes the fun part. With the format established and good visuals selected, you can fit the pieces together. But there is a method to laying out a good-looking document. To be able to effectively lay out the document, you must put into practice the principles of document design, discussed next.

PRINCIPLES OF DOCUMENT DESIGN

As has been mentioned, good design does not attention to itself, but good designers use the principles of document design to make sure their layouts look pleasing and attractive. The *principles of document design* are balance, proportion, order, contrast, similarity, and unity.

Balance

Balance involves imagining a line dividing a page either vertically or horizontally and then placing visual elements so they are either symmetrically (formally) balanced or asymmetrically (informally) balanced (Figure 6-4a and 6-4b).

Symmetrical balance, or *formal balance*, is mirror-image balance. It occurs when all the visual elements on one side of the page are mirrored on the other side. The visual elements may not be identical objects, but they are similar in terms of numbers of objects, colors, and other elements. The primary value to readers when you use formal balance is that the reader does not have to work hard to see relationships between message elements. A document that has formal balance presents stability, security, authority, and thoughtfulness. However, it also may be seen as unexciting and unimaginative.

Asymmetrical balance, or *informal balance,* occurs when several smaller items on one side of the imaginary line are balanced by a large item on the other side, or smaller items are placed further away from the center of the screen than larger items. Most document design is done using asymmetrical balance. Asymmetrical balance is generally seen as being more interesting and exciting. A message using asymmetrical balance is seen as dynamic, fresh, inviting, creative, and friendly.

When you balance elements asymmetrically, you use their visual weight to create an arrangement that is pleasing to the eye. **Visual weight** is the amount of "weight" or prominence an image has. The eye is more attracted to a space with heavy visual weight because of its dominance. For example, a small, dark visual may have as much or more visual weight than a larger one with lighter tones. Black is heavier than white. Intense colors are heavier than those that have been toned

---

The Ag Informer

Volume 1, Number 20   May 2011

Centered headlines set the stage for symmetrical layouts

Tis; hos, nonsiciturs imis; nium laribulla ret num, vagit nonculibus moviviv-itum, pra? Nos comnem es ocus, Ti. Aximurnis iae, nes adhucom nenatuam in-tus, que nonfecris adhum ad furs ponsult odienih in-gultor ut pris, furnihin Ita me quam ignam ilis. Oc ocultus vagina, conclesis-tam erterortem, omnitim ocatat restrox simod aur-num hors in senscib efa-cien dumus; iam dinvolut virte nendam dit. Simo nos senaticeri sesterores vidi-emusce ner unte intemus;

The caption is centered under the photo.

et adeatu sigit auc morei pari con pulostorume ina-tui senatre bussimi ssentus perehena, core quam re, nosus.
Ri sentianium etravocum novesid itestres fue pli-cavehebem dicapericae vit, Ti. Fuium crit crei plistem ponsulaque iam orum dit facienitums; hachus, se om-nemuntum poncut o num ad conum elin teritrumena, quit. Viviris? Iverem tu moves, nostiquame esse mac imis, tum andum arbis. Mari iampere in iptem non-sus, egerrat, consuntris.
Los ponculium ompora noveridiis, unum adhus bon-sid in Ita prarei confenic restabuntiam la muntis re am intienatquid ina, es, noc, cri stra Sp. Ifec telum tis seri, quonfec torecremus praccid entimpris; nocciesis reviver virtimis diorionferum prese accivas conterce restiam essum di, opontifecon diu qui sed fachiliis bonsulicavem dies terum egin ta, que nost parem. Gra L. Otin tum estiam ora nonsuam co visque dienem hostrae con sendiem actem nihil vid

Ceperi ture cotiam nos-tinatur los opon vocum publin sederei in vernitus, quo te, utum et L. Cam P. Gra, nicat, ocreo, nonsil terorei perfex sum om-nonsulegit Catervi detis-simus etra L. Quam arbis inatudes halique ne tam sen Etrius. Quem am-plici virtess imisque que nox num tertimmo egine hos plis? Etrum talariocri parei ia diturbit. Serioris taturnicae re ficuludem P. Em a culic teatus pons conem, et Catatis An verbis villa nonferfer-num porei poptis molutem dendum et prestif erfero tem prortis horum dintumurnit, etrura isquod mus, nequam nos paturo confirmisuam pulto cureconducid res et? Me ompercena, nostrum sum fecone estraver hos suntimum que viginen tiustem mo tessedes vem, quium, se nox sulis inius pernunc legerfex nostus vid publicaves apere, octuidiur iam sena, niris derum, quem poremus cit; nostere, ad Cati, conscivertis bon-supic mo tus loc is ad confentere, qui iam vilicip ioc-ciderem, moverfe cremnit acram aperius. Orisulvit condam adet nicesul iaequem sulibem re medit Catu sulute rebemus ulvili pulocae liameis Ahaciam, con Ita, cerei consilis, quis est L. Bi sceri publiusa dem opore, manu ex moraciis num peri perum hossendit; nos volintem in videssena et alaris. Deescritia oma, catum visserum ocavocu pionsul abessim ilicenam in volis.
Ris. Avendit, C. Idem in perionesimus cut actus hus hor lius merbit vehemurbem recem tam, orbis;

(a)

© Cengage Learning 2012

The Ag Informer

Volume 1, Number 20   May 2011

Headlines are usually left justifed in asymmetrical page layouts

Num omantie niquid perdiena, potifecepses vis, caecientiae inatabemus se patilius esissilicus, videt volibultorum acturbit, Ti. Sid senihilia cae tem. Graccii ssilica omantil vis, quos, quas hor quit. M. Fulicav ernihil vivena vis.
Movesistam inproraves et vit potium scerum urbem publici viverit? Habemurox noca; hu-screbus? At con viri silnentelabi is losulint.
Ficiesse atio inamque ime ina, conis con sus li, nostiampro tanum, nonculii primusa L. Pio patquam feces, ca Si sentrit? Catem num ego unum sa re temus, qui etiquerude quonu es conessimenam inatiam prorehe beffre quam te nonsign occhum. Gulvirm andieni simusa mo compons itellercesse teraverte te perfecit vis cercerfiori, nosta, cus, periberem anum prissicae mendi sulto mendum hocchuis vitrit a remqui in vigilic uludenihin se, Ti. Os, novide re, que hemum, untiensus, vissendie patabist vit.
Multus, P. Ut quost vitandam te nonsulis vas-dac ocaetidiem teripte ricaec vid cae, di pracrum denimen atquit, coerei porum aperiora iam ter hor peconum vius consimoeror utemquam prei ine audeatis, in Etresce ssenitat ines paribus tus-sate sciendiemum, nonsum nor hem reo prox mus patus con dem ex mod curoruntem quam niam.
Mei potis loc re confit; nocus inc mo interibus caectus, quit, nunium ut fue nenatque publia denicae quem stror hordien ihicatum etio, du-cone con te tem inature natum, sa intisultuita norente atuit. Vivissimpera omnit; C. Fit, er-niure mendam fesin serips, publiem oret fauci percesilicae ignatuasdam urnum fecerfec rei sum publint. Ecupplicae inem que prae medici pravert eripsenatiem nonsumensum nonon Ita, C. Alicaet ernihic rem erit? Bonsulego ingulti-

Ceperi ture cotiam nostinatur los opon vocum publin sederei in vernitus, quo te, utum et L. Cam P. Gra, nicat, ocreo, nonsil terorei perfex sum omnonsulegit Catervi detissimus etra L. Quam arbis inatudes halique ne tam sen Etrius. Quem amplici virtess imisque que nox num tertimmo egine hos plis? Etrum talariocri parei ia diturbit. Serioris taturnicae re ficuludem P. Em a culic teatus pons conem, et Catatis An verbis villa nonferfer-num porei poptis molutem dendum et prestif erfero tem prortis horum dintumurnit, etrura isquod mus, nequam nos paturo confirmisuam pulto cureconducid res et? Me ompercena, nostrum sum fecone estraver hos suntimum que viginen tiustem mo tessedes vem, quium, se nox sulis inius pernunc legerfex nostus vid publicaves apere, octuidiur iam sena, niris derum, quem poremus cit; nostere, ad Cati, conscivertis bonsupic mo tus loc is ad

Headline for photo

The caption is left justified under the photo.

(b)

© Cengage Learning 2012

FIGURE 6-4: Examples of a mock newsletter show symmetrical and asymmetrical balance. (a) In this example, the symmetrical balance places the same amount or weight of text on both sides of the photograph. (b) In this example, asymmetrical balance adds weight to the bottom of the page.

down. Visual elements such as illustrations, photos, and headlines—by their size—carry more weight than text. One darker item may need to be balanced by several lighter items. You achieve a balanced look by spacing these elements with careful consideration of their relationship with everything else on the document.

Proportion

Proportion is the spatial relationship between each design element. The eye visually compares the relationship of each element's area, size, weight, and location to all of the others on the page. You can create pleasing visual proportion by dividing your page into thirds. The middle section is considered to be a good place to put your most important visual element, and it should be in proportion to the rest of the page. The *natural center* is slightly above the exact center of the page. The natural center is the most common focal point when viewing a document. That is why for most newspapers, the major story headline or photograph is immediately above the fold in the newspaper. Another term to know is the *message zone*, which is the entire page, containing all elements of your message: text, visuals, and graphics.

Order

Order is how you show sequence and importance. Order refers to most readers' assumption that what they see first on a page is more important than what they see later on. Studies have shown that on a printed page the most important information is positioned at the top, followed by information of decreasing importance as the reader moves down the page. You can see that concept on any newspaper page. The important information is at the top to catch readers' attention. For brochures, major headings are at the top, followed by less-important information.

Contrast

Contrast refers to the dominant focus or element on a page. Use contrast to show difference and to create emphasis. For example, darker and larger visual elements stand out on your page. They are considered more interesting and are the focus of your document design.

All good designs should have a focal point that stands out on the page. Readers assume that a difference in appearance means a difference in function or meaning. Contrast is a valuable tool to draw attention to such things on a page as warning labels, cautions, and notes. Contrast can also apply to the following visual elements:

- *Shape:* This could mean using a typeface for text that you want to stand out from text using another typeface, or using a recognizable symbol, such as a stop sign or a warning triangle.
- *Size:* To draw emphasis to an area on your document, you may want to use text that is a larger size than the surrounding text.
- *Color:* Any use of color, in an otherwise black-and-white document, will draw readers' attention.
- *Value:* This refers to creating a visual element that has a higher value than the surrounding objects. This can be done by bolding text or adding a gray box behind the text.

Similarity

Where contrast focuses on dominance or emphasis, **similarity** shows that design elements are alike. When people see things that look similar, they assume they have similar functions. Visual elements could show similarity in the following ways:

- *Shape:* All of the text for a specific function would be placed in the same typeface. This is common in newsletters, where headlines are in one typeface, and the text for a story is in another typeface.
- *Size:* All of the text for a specific function would be in the same size. Again, in most newsletters, headline text is a larger size than the text used in the stories. The story text, for all stories, is always the same.
- *Color:* All of the colors—red, for example—would be in the same hue.
- *Value:* Bullet-pointed text—such as the bullet-pointed text used here—would have the same relative value. For example, in this chapter, the important words in the bullet-pointed text are italicized.

Unity

Unity deals with how all of a message's elements tie together visually. A message with good unity is one where all of its visual elements complement each other. The pages in the document should hold a reader's attention, but they should be simple to follow. Each page also should have a similar structure so that they are consistent—or unified—throughout. Letting each page have its own shape and form is not the right way to design a document.

ELEMENTS OF DOCUMENT DESIGN

In addition to the overall principles of document design are the specific *elements of document design*. Where the principles of document design provide considerations for overall document layout, the elements of document design focus on these specific visual elements that make up the visual content of the document: text and typefaces, visuals, graphics, color, and white space.

Text and Typefaces

Text is any size, shape, and placement of the printed word in your document. Text can be placed on your document in the following ways (Figure 6-5):

- *Full justified*, also called "justified": All of the text is "flush" on both the right and left sides. To accomplish this, the words in a sentence and letters inside of words can become stretched to fill the space on a particular line.
- *Left justified*, also called "ragged right": The text is "flush" on the left-hand side of each page or column of text but is "ragged" or uneven on the right-hand side.
- *Center justified*, or "centered": The text is centered on the page. This is commonly used for headings for brochures, fliers/handouts, and in some newsletters.
- *Right justified*: The text is "flush" on the right and uneven on the left. Right-justified text is rarely used in most documents, except to line up all of a column to the right-hand side.

Text also can be boldfaced (or bolded), italicized, or underlined. However, these text styles should be used only to draw readers' attention; they should not be overused. For example, it is recommended that italics be used for short phrases, such as for direct quotations in a brochure or handout, rather than for long passages.

In more advanced layout and design programs, you can adjust the space between letters, words, and lines. You can get some interesting effects by adjusting text spacing. *Kerning* is the space between letters. *Tracking* is the amount of space between words. *Leading* (pronounced "led-ing," like the metal) is the amount of space between lines of text. Look at some documents to see how close or far apart the spacing is between letters, words, and lines. Adjust the kerning, tracking, or leading to imitate the spacing that you like for your document.

Typeface—also called *type* or *font*—is the actual look of the letters. Type usually is classified into two categories: serif type (also called serif font) and sans serif type (or sans serif font) (Figure 6-6).

Example Justification

Left Justified

Left justified, also called "ragged right": The text is "flush" on the left–hand side of each page or column of text but is "ragged" or uneven on the right–hand side.

Center Justified

Center justified, or "centered": The text is centered on the page. This is commonly used for headings for brochures, fliers/handouts, and in some newsletters.

Right Justified

Right justified: The text is "flush" on the right and uneven on the left. Right–justified text is rarely used in most documents, except to line up all of a column to the right–hand side.

Full Justified

Full justified, also called "justified": All of the text is "flush" on both the right and left sides. To accomplish this, the words in a sentence and letters inside of words can become stretched to fill the space on a particular line.

© Cengage Learning 2012

FIGURE 6-5: Example of justification.

Example Font Styles

Example Serif Fonts

Times New Roman	AaBbCcDdEeFfGgHhIiJj
Cooper Black	**AaBbCcDdEeFfGgHhIiJj**
Bookman Old Style	AaBbCcDdEeFfGgHhIiJj
Palatino Linotype	AaBbCcDdEeFfGgHhIiJj

Example Sans Serif Fonts

Arial	AaBbCcDdEeFfGgHhIiJj
Calibri	AaBbCcDdEeFfGgHhIiJj
Verdana	AaBbCcDdEeFfGgHhIiJj

© Cengage Learning 2012

FIGURE 6-6: Example font styles.

Serif fonts are those where the letters have "feet" or "tails," such as the Times New Roman typeface. Serif fonts are a good choice for printed materials because the "feet" at the bottom of letters make it easier for readers to follow each line of text. Serif fonts trace their history to ancient Rome. Letters were chiseled into building's walls. However, the straight-edged letters created cracks in the walls. To avoid that, Roman architects began putting horizontal edges on the top and bottom of the vertical lines forming the letters. Rounding off the letter by adding these "feet" and "tails" changed the direction of the stress on the walls, eliminated cracks in the walls, and introduced the concept of serif typeface to Western civilization.

Sans serif means "without serif." Simply, sans serif fonts do not have "feet" or "tails" on their letters. Arial and Helvetica are examples of sans serif fonts. Sans serif fonts are viewed as more contemporary than serif fonts because they have a cleaner look; many sans serif fonts have been in use only since the 1970s. Sans serif fonts usually are not good choices for a lot of text on a page, because they are more difficult to follow across the page without the "feet" for each letter. Sans serif fonts are recommended for text that is projected onto a screen or read on a television screen or computer screen. Sans serif fonts also are a good choice for short headlines and brief photo captions.

When designing any document, the most important aspect to keep in mind about text and typeface is for the lettering to be large enough and legible enough to be read easily. You also should use no more than two typefaces in a document. Use uppercase/lowercase lettering in your document. Only use all uppercase lettering when you want to call attention to a specific word or short phrase. Readers see all uppercase lettering as if the document designer is shouting at them. Research also indicates that uppercase/lowercase is easier to read. As a result, the federal government is requiring cities across the country to change street-name signs from all capital letters to uppercase/lowercase letters (Copeland, 2010).

Headings, titles, and captions are specific aspects of text applications. Headings and titles orient readers to the start of a topic. They should be briefly worded. Most headings and titles are bolded and larger than the document's other text. *Captions*, also known as cutlines, are brief descriptions placed under photographs or graphs. Usually a caption provides information on who is in the photograph, what is going on, where and when the action happened, and why the action is significant. Studies have shown that most people read captions immediately after they read the headlines.

Visuals

Visuals are anything in pictorial form, such as photographs, drawings/illustrations, clip art, and graphs and charts. The functions of visuals are to grab the reader's attention and to support or provide explanation to the document's overall message (Figure 6-7).

Photographs show the actual physical images of objects. Photographs have the advantage of realism. One disadvantage is the extraneous details in the photograph that may detract from the message. Much more information about photographs and photography is presented in Chapter 9, Digital Photography and Photographic Editing.

Drawings/illustrations can depict imaginary objects or real objects difficult to photograph. Drawings can show only the parts the reader needs to see. Drawings give you the advantage of control by eliminating extraneous detail and emphasizing what you want to emphasize. Illustrations should be clean and simple.

Clip art is an alternative to drawings. Clip art can be found online or purchased on CDs and DVDs.

Graphs and charts provide information, usually statistics or numbers, in an easily understandable visual form. Graphs and charts should be clear, uncluttered, suited for the reader, legible, and placed near where they are mentioned in the document. Graphics and charts must

FIGURE 6-7: Book cover. A book is always judged by its cover. An appealing cover to a magazine or publication is a great way to attract readers. This design shows the front and back of a booklet that is to be folded vertically in half. The design uses bright colors and an appealing logo on the front with consistent branding images and text on the back. (*Courtesy of Kevin Kent/The National FFA Emblem is registered trademark of the National FFA Organization and used with permission.*)

have brief but understandable titles. Some graphs and charts will have a key that explains symbols used in the visual. Graphs and charts come in different forms, based on the type of information that is being communicated. Some of the most common ones are infographics, bar charts, pie charts, and line graphs. If your graphs and charts are in black and white, be sure that readers can differentiate between the shadings of the different sections of a pie chart or bar graph, for example.

- *Infographics* is the term to describe the use of visual elements to communicate complex information quickly and clearly. Infographics use recognizable images to represent specific quantities. For example, instead of using a line graph, a document designer may use an illustration to represent numerical data. The illustration adds visual appeal to the information. The Snapshots features used in *USA Today* are good examples of infographics.
- *Bar graphs* show comparison at different times, locations, and conditions. Bar graphs are easy to understand and can be either vertical or horizontal.
- *Line graphs* display trends over time in amounts, sizes, rates, and other measurements on lines. Line graphs give an at-a-glance impression of trends and forecasts of data. You should have no more than four or five lines presented in a line graph. It is best to distinguish different lines by using different colors or thicknesses. Show current data with solid lines, and illustrate future data with broken lines.
- *Pie charts* are best at showing what parts make up the whole and at comparing relative sizes of the parts. A good example of a pie chart would be to show the ethnic background all of the students in a school. Pie charts are most effective with six or fewer sections or "slices."

Color

Color can be used to draw attention because visual elements with color have greater visual weight. Certain colors also are interpreted in specific ways by most people. *Warm colors* are reds, yellows, and blends of those two colors (orange). These colors are related to heat, fire, and the sun. *Cool colors*—blues and greens—relate to the sky, sea, and wilderness. Depending on what message you are trying to get across to your audience, you may want to use one of the following colors:

- *Red* is the most dramatic color. It excites and stimulates people. Red is often associated with aggressive behavior, passion, success, and impulse. Use red sparingly, because it is such a "hot" color.
- *Bright yellow* often is associated with health and well-being. Yellow is also associated with caution. For example, all traffic signs that pertain to caution—such as yield signs—are in yellow.
- *Blues* are associated with tradition, orderliness, and stability. Light blues are associated with cleanliness. Dark blue colors seem to have a calming effect. Purple is seen as daring, royal, and elegant.
- *Greens* provide a sense of nature and regeneration. Green is used by many environmental organizations.
- *White* is a symbol of purity and innocence in Western countries.
- *Black* usually signifies finality, ending, and death in Western countries.

Graphics

Graphics are lines, borders, and boxes in your document. These are used to highlight or draw attention to an area of the document. To emphasize a particular part of your document, you may place a border around a photograph. Graphics also are used to separate visual elements. For example, a box around a newsletter story could separate it from other stories on the page. Small lines under a photograph's caption could be used to separate the caption from the rest of the story. Graphics should be used sparingly. Do not place a box or border around each visual element on your page. If you use shaded boxes, make sure that the shading is not too dark. A shading of 10 percent is usually all you need. A shading of 20 percent or greater may be too dark to read for the text.

White Space

White space (also called "blank space" or "negative space") is the area not taken up with text or images. White space is used to create a sense of openness. Too many elements on a page can look confusing and detract from the overall visual appeal of your page. White space separates paragraphs and provides margins at the edges of your pages. Areas occupied by text or images are called **positive space**.

GRAPHIC FILE FORMATS

When designing documents, you also have to be aware of the different types of images you may receive: raster graphics and vector graphics.

Raster Graphics

Raster graphics are images made up of **pixels** (picture elements). Raster graphics are such images as photographs. They are limited in their size and shape and are linked to the image's resolution. **Resolution** is a measurement of how closely packed the pixels are together in the image. Resolution is usually measured as *pixels per inch (ppi)*.

Because raster graphics are tied to resolution, they limit modification. Resizing, also known as **scaling**, is a particular problem. If you stretch a small raster graphic to make it larger, each original pixel is actually spread out, leading to a distorted, jagged look known as *pixilation* (Figure 6-8). Raster graphics intended for print should have a higher resolution, usually around 300 ppi. Raster graphics for computer monitors, not intended for print, usually have a relatively low resolution, of 72 to 100 ppi. Because of the higher resolution needed for raster graphics to be printed, the computer file sizes tend to be large. Raster graphics are usually saved in one of the following graphic file formats: TIFF, JPG, and GIF.

TIFF stands for Tagged Image File Format. This is a graphic file format that can be read by most graphics programs. TIFF is commonly used in publishing and document programs. A TIFF is a large file, because it is less compressed than other file formats, such as JPGs. A TIFF provides high-resolution imaging.

JPG stands for Joint Photographic Experts Group. A JPG image is a compressed file, meaning that some information is discarded. Many digital cameras capture photographs in the JPG format. JPG is the most common file format for photographic images, especially those shown on Web pages.

FIGURE 6-8: In this example photograph, a portion of the longhorn's eye is enlarged to show the individual pixels.

GIF stands for Graphics Interchange Format. GIF files are extremely small file sizes, because they can only use a maximum of 256 colors. GIFs are used for creating line art, such as logos and diagrams, for the Web. GIFs are not to be used for print documents because their resolution is so low.

For publishing printed documents, photographs should be saved in the TIFF format. For photographs that will go on the Web or television, use JPG.

Two other graphic file formats are PNG and RAW. These are not as common as TIFF, JPG, and GIF files. *PNG* stands for Portable Network Graphics. PNG is a compressed format designed as an alternative to GIF. It can be used for line art and photographs; therefore, it is more versatile. However, PNG has not caught on with graphic designers and is generally not widely used. *RAW* is designed to capture images from digital cameras. The files are large, and the quality of the photographs is high. Images captured in RAW are usually resaved as TIFF or JPG files during the document design process.

Vector Graphics

Vector graphics are images composed of mathematically defined shapes created by illustration programs, such as CorelDraw and Adobe Illustrator. Usually vector graphics consist of drawings, such as lines, curves, and geometric shapes. While raster graphics describe every pixel in the graphic, vector graphics use mathematical algorithms to describe the lines, shapes, patterns, and colors of drawing objects.

Vector graphics can be resized (scaled) without looking pixilated. Vector graphics work best for drawing images with sharp edges and consistent areas of color, such as line art, diagrams, maps, and statistical graphics. The file format most seen with vector graphics is *EPS*, which means Encapsulated PostScript illustrations. Resolution is not an issue with an EPS file.

PAPER AND INK

Before focusing on two particular document types—brochures and newsletters—that you may want to develop for your organization, you need to know about paper and ink selection. Many color laser printers do an adequate job of producing a nice-quality color document. For small numbers of documents, a laser printer will do well, but for large quantities or for specialized documents, you probably will need to have a

© Cengage Learning 2012

commercial printer produce your document. If you believe that your document will need to be printed commercially, it is strongly advisable to contact the professional printer prior to designing your document. A printer will provide you with suggestions on how to make the document design process work better for you and for the printer. Some of these suggestions may pertain to paper and ink selection.

Papers are sold and labeled by weight and type. The heavier the paper stock, the firmer it is. A higher-pound paper would be better for a document's cover. Papers coated with different finishes also cost more.

The basic decision on ink is whether to use one or more colors. In printing terms, *single color* means that only one ink color is used throughout the document. It can be black or another color. You can use *two-color* and *three-color printing*, but the more inks you add, the higher the price.

Process color printing, also known as *four-color printing*, uses four inks: cyan, magenta, yellow, and black. This is also called **CMYK colors** (C = cyan, M = magenta, Y = yellow, and K = black). The CMYK colors are the inks used for printed documents. (Colors shown on computer monitors and television monitors are called RGB colors, for red, green, and blue.) Process color may cost more than a single color, but you get full color for your document, which may be necessary to communicate your message better to your intended audience. A more detailed explanation about CMYK and RGB is provided in Chapter 9, Digital Photography and Photographic Editing.

Spot color refers to using a specific ink for a particular section of the document. You place ink on that "spot." Spot color can be a less-expensive solution if you just want a particular color on one part of your document.

BROCHURES AND NEWSLETTERS

Any document should be developed with the design principles and elements previously discussed. For the remainder of this chapter, two particular documents—brochures and newsletters—that can be used in your organization will be described.

Brochures

Brochures are small, usually folded, documents used to inform, educate, or persuade the reader. They are commonly used to promote organizations, products, or events. Brochures attract attention through eye-catching design. Brochures must contain clear, concise, focused writing because they are so short. When writing and designing a brochure, incorporate these recommendations:

- *Use strong headings (titles) and subheadings that lead the reader through the text.* For example, the title may be "How to Grow Award-Winning Roses." The subheadings may be specific steps in the process. Subheadings also break up information into manageable chunks.
- *Speak directly to the reader.* "You" is the most important word in persuasion. The brochure should sound very personal, focusing on the benefits the reader gets from reading the information presented in the brochure. Put the emphasis on what the reader will gain. What is in it for your reader?
- *Use bullets to list information.* Because brochures are so brief, use bulleted lists to highlight important information.
- *Keep the text short.* Use short sentences of 20 words or less. Use short paragraphs of no more than three sentences.
- *Use document design elements effectively.* Particularly for brochures, use these design element suggestions:
  - *Text:* Most brochures are written in a 10-point or 11-point font size. Single-space the text but double-space between paragraphs. Paragraphs usually are not indented. Only use bold and italics when you want to emphasize a word or phrase.
  - *Graphics:* Break up long stretches of text with lines. This helps to make the text easily readable. Use shaded boxes to emphasize points within the overall text.
  - *White space:* Use white space to provide enough space where the paper folds. You do not want text in the creases where the brochure is folded.

Brochure Design

Brochures come in many styles and forms. You can design brochures with two folds, three folds, four folds, and more. Probably the most common style used by small businesses and organizations is the three-fold brochure, also known as a *trifold*. The trifold has six panels and can be printed on regular 8.5-by-11-inch paper.

© Cengage Learning 2012

Scott Wallin

Director, Producer Communications, Dairy Management, Inc.

We live in a very visual world filled with people who don't always have time for much depth. A well-designed document that has lots of interesting elements to capture the eye can be a very effective way of communicating your message. Whether it's a company newsletter, annual report, or brochure, be sure to deliver a strong design concept that will engage people and hopefully get them to learn more about you and your organization.

The design of a trifold brochure will be presented here and in Figures 6-9 and 6-10, but you can apply the concepts to other brochure styles.

- *Panel 1:* The panel that you see first—the cover—should "hook" the reader's attention immediately. The reader should know what the brochure is about just by seeing the first panel. The cover panel, also called the "hook" or "eye-catcher," usually includes the organization's name, the theme of the brochure, and possibly a slogan. Rarely will in-depth information be presented on the first panel. If the brochure will be placed in a "brochure rack," commonly found at tourist stops, it is important to put large, attention-grabbing text at the top of the brochure, because the bottom portion of your brochure may get covered by a brochure in the slot directly below yours.
- *Panel 2:* When you open the brochure's cover panel, the next panel you see is the most likely to be read. This second panel should be where you present a stand-alone message that summarizes the content and message of the entire brochure. If you are promoting an upcoming event, this panel could be the highlights of the event and the event's date, time, and location. This panel is usually self-contained, meaning the content does not carry over to another panel.

- *Panels 3–5:* These panels present the main message of the brochure. The panels are viewed as one three-column unit. The three panels often contain subheadings to break up the text.
- *Panel 6:* This is the back cover, which is the panel least likely to be read. Avoid putting important text on it. Use this panel to include contact information (your organization's name, address, phone number, e-mail address, and website, if applicable). The back panel also can be kept blank in case you want to put an address and stamp on it so it can be mailed.

Newsletters

A *newsletter* is a collection of stories, announcements, and explanations that is sent on a regular basis to a particular group of people. Most professional organizations send newsletters to their members. A newsletter for your organization can be an easy way to keep members informed about your organization's activities. The newsletter also can be sent to specific people outside your organization to inform them of what you are doing.

If you have never designed a newsletter before, you should look at the newsletters you receive for ideas on how others design their newsletters. Look for the kinds of designs that you think might work for you and your

(a)

(b)

FIGURE 6-9A AND 6-9B: This brochure from Florida Dairy Farmers, Inc., is a good example of a colorful brochure. The brochure makes good use of white space, graphics, photos, and text. (*Used with permission from Scott Wallin/Florida Dairy Farmers Inc.*)

FIGURES 6-10A AND 6-10B: This brochure exhibits a fun way to lay out content in a trifold brochure. With the target audience as high school students, this brochure uses combination of rotated text and images to create a different look from the standard brochure. When creatively laying out content, it is important to be mindful of legibility as well as the production steps, such as folding. (*Courtesy of Kevin Kent/The National FFA Emblem is registered trademark of the National FFA Organization and used with permission.*)

readers. As you look at other newsletters, consider how they address the following issues:

- How big the headlines are.
- How easy it is to read or to skim an article.
- How crowded the pages appear.
- How colorful the newsletter is.
- How photographs and graphics are used.

Afterward, draw some rough sketches, either hand-drawn or on the computer by laying out "dummy" text and boxes where photos will go. Create a prototype design that you and your organization are happy with. Make any alterations before developing the newsletter for real. You also have to consider your production budget. Determining how much your budget is will give you some idea about how many pages the newsletter can be, how many copies you can print, and if you can use color.

When writing and designing a newsletter, incorporate these recommendations:

- *Write to be read quickly.* Newsletter stories usually are short. It is a "letter with news." Most articles are between 100 and 600 words. A four-page newsletter generally will contain around 2,000 words.
- *Write about the right stuff.* For example, people like recognition, so use as many names as you can in your newsletter. People like to read about themselves and people they know.
- *Put the best stories first.* Put your best stories on the front page. For interesting stories on inside pages, use bigger headlines, shaded boxes, and other visual elements to draw your reader's attention to these important stories.
- *Use document design elements effectively.* Particularly for newsletters, use these design element suggestions:
  - *Text:* A good font size for a newsletter is 10-point for normal text. For headlines, use 18-point and higher. Usually, headlines will be between 20 and 36 points. Choose one typeface for headlines and another for body text. Paragraphs usually are indented about three spaces. Use bold and italics sparingly. As for justification, the columns of many newsletters use left justified (ragged right). This eliminates hyphenation. If you choose to use justified spacing, you may want to hyphenate words at the end of lines; otherwise, you create odd spacing between words and letters.
  - *Color:* Use color as much as possible. You may only be able to print the newsletter in one color. That is fine. Print one version in one-ink color. You then can make an electronic PDF in color that can be distributed by e-mail or placed on the Web.
  - *Graphics:* Break up long stretches of text with thin lines. Use shaded boxes for short stories.
  - *White space:* Do not crowd the pages with text, graphics, and visuals. Use white space to provide some breathing room, such as by having margins around the edges of the pages.
  - *Visuals:* Use photographs and other visuals throughout your newsletter. People like to see photographs that pertain to the newsletter stories. Position photographs so that they guide your readers around the page without getting in the way. Try to make articles flow naturally around, below, or above each photograph. As a rule, do not put a photograph in the first column of a story, directly under the headline. Place the photograph in the second or third column. A photograph also can be put at the bottom of the story. If you use two photographs, possibly place them at opposite corners of a page. For photographs, always use captions. Photographs must be clear and well-composed. Action shots are best. For tips on composing a photograph, please see Chapter 9, Digital Photography and Photographic Editing.

Newsletter Design

As with brochures, newsletters come in many styles and forms. Newsletters can have a more formal style and tone, if they are sent to professionals. Newsletters may have a more informal style and tone if they are being sent to organization members.

Newsletters fall into one of three categories: bullet sheet, newsletter, and magaletter. The *bullet sheet newsletter* is usually one page, front and back, and includes short articles of no more than a paragraph or two in length. The bullet sheet is meant to be read quickly and covers vital information. This style works for some electronic newsletters. The *newsletter* is the most common category. It is what we picture when we think of the word "newsletter." It is usually four to eight pages long and includes short articles. The *magaletter* is in-between a full magazine and a newsletter. Usually a magaletter is 8 to 16 pages in length. The articles tend to be a little longer in a magaletter.

Because most newsletters will fall in the "newsletter category," designing a four- to eight-page newsletter for your members will be the focus of the rest of this chapter. Newsletters usually are categorized by the number of columns of text, usually one, two, three, or four columns. One-column newsletters are easy to produce. However, because they look like a composition paper for an English class, they appear less imaginative in their layout and design. If you use a one-column format, you must build in sufficient white space on the margins; otherwise, your pages will look crowded.

For a two-column newsletter, individual columns are about four inches wide. Articles can be positioned side by side on some pages. The three-column format probably is the most common and is similar in design to the two-column format. Most 8.5-by-11-inch newsletters use two columns or three columns. Many tabloid-size newsletters (11-by-17 inches) are designed with three columns or four columns. Use the number of columns that feels right for your publication. See Figures 6-11a, 6-11b, and 6-11c for examples of one-, two-, and three-column newsletters.

Whichever column format you select, you will need to consider what content goes on the pages. Use these recommendations as you put content in your newsletter:

- *Cover page:* The first page is called the *front page* or *cover page*. The goal of the cover page is to grab the reader's attention with attractive visuals—usually photographs—catchy headlines, and creative articles. The two most relevant stories for your readers should be placed on the cover page. The cover page usually contains the newsletter's nameplate that has the title of the newsletter and a table of contents or a "teaser box" that shows what articles are covered and what pages they are on.
- *Inside pages:* The content of the inside pages is left to the discretion of the newsletter editor. Feature stories and news stories can be found throughout the inside pages. You may want to include some of the following ideas as regular stories for your newsletter:
  - *President's (or executive director's) letter:* A regular letter from the chapter president, executive director, or adviser that describes something of importance for your organization.
  - *Announcements:* Bits of information the reader may find interesting but are too short for a full article.
  - *Spotlight sections:* A feature story focusing on a member of your organization.
  - *Questions and answers:* Questions asked by readers and answered by a member or adviser.
  - *Calendar of events:* Usually a list of upcoming events—activities, dates, and times.
- *Back cover:* For a four-page newsletter, this would be page four. If the newsletter will be mailed, one-half may be kept blank, so mailing address information can be included. The other half of page four may include an eye-catching photograph or an interesting, brief article.

Parts of the Newsletter

In addition to the design elements discussed earlier, any newsletter you design should have these components:

- *Nameplate:* This is also called the *banner* or *flag*. The nameplate is the information at the top of the cover page that includes the newsletter's title, the organization's name, the newsletter's volume and issue number, and the publication date. The publication date could be a month and year (e.g., May 2011) or season of publication (e.g., Spring 2011). The title should be in large letters, across the top of the page. Keep the title as brief as possible. Avoid using "newsletter" in the title.
- *Table of contents:* The table of contents or "teaser box" includes the major sections or articles, with their corresponding page numbers.
- *Stories/articles:* Stories make up the heart of a newsletter. Keep stories brief.
  - *Bylines:* Each story should include a "byline" to identify the writer. A byline also may include the person's title or position.

    Example: By Jamie DeGraw
    Staff Writer

  - A *jump* is a story that goes to another page. Avoid a jump, if you can. However, if you must jump a story, be sure to provide a *jumpline* at the bottom of the last column of the story. The jumpline will typically say something like "Continued on Page 3." On the page being

The Ag Informer

Volume 1, Number 20   May 2011

This is an example of a one-column newsletter's cover page

Ure doluptat. Duisit vendiam et in ulla feuisit am, con vulpute mod erit, suscil ipsi eros nulputpate feu feui et dolore faccumsandre te magnit laptaui eraesequat.

Em zzrriusci blan henisis acin ullandre con vel irit autat ut in veriure magna cons nullaor ad tat vel iliquismod doluptate! ing eu faccumsan eratie vendre min vent utpotate etum et inim irit, se tem atio coreet, vulla aute vero diamet etue dolore molobor inisim qui bla consendre magnis nim velesto odigna aut vel il do dui tat. Dolenibh et aliqui bla acincidunt ing ernisit luptat augiat alis essi etum vel dolobor iureet, sequam dolobore faccum volendit velenim dolendio consecte ent dignim dolor sustincip et iliqui esto odiat aliquisim zzriure modolor sequamconse del utpatum aut velit nim quis adiam duis nullute tinci blaorpero od modit, sis augiam, verit ilit, quismolute magniat wisl dolore modolor ipis dolorpe ractsil lortie dolorpe rcidunt alit luptat, venisl ut loreet accum iiisisl uter incillande minim diam zzriure rostisit nulput in henim incipit praessenim in ulput lum augaerat inis dolorem iuscilit illandit dolore eu faccummy nonsed duissed deliquis dolorper alis ex euismod tet lorem velit num nullam diat, sequat et amendumsan henim verci bla facil ent nulla alit iriliquam, quis am iriure core tatie mod molorero odipis euisci blam, quis adit at, se minicisduisit vel dolortion heniamc onsequis nici ea feu feugiam con venim do doloborperci blaorperit adipisi.

On ectem quisim quipisl eum quis nos erilis acinci bla faci ea faci tat prat, venim iriustie dio od dolobore do el eugait nonum venim volobore comulla facillan venim acip eu feuisci blan ullaorem vel ercipsu msandit, consed erostinci tic commy non et prat ipsit dolorem in veliqui smodolore tatis eugait adit inibh eum del diamet ipis diam, venibh esenibh ex exero odit ipis do ent atem e u m m o l e ssectem do dip ex eugiat veliquatem dolor sumsan dre comulla conullamcons

Headline for photo

The caption is left justified under the photo.

(a)

The Ag Informer

Volume 1, Number 20   May 2011

This is an example of a two-column newsletter's cover page

Num omantie niquid perdiena, potificepses vis, caecientiae inatabemus se patilius esassilicus, videt volibulorum acturbit, Ti. Sid senibilia caetem. Graccii ssilica omantil vis, quos, quas hor quit. M. Fulicav emihil vivena vis.

Movessiam improraves et vit potium scerum urbem publici viverit? Habemurox noca, husrebus? At con viri silhentelabi is losulint.

Ficiesse atio inamque ime ina, conis con sus li, mostianpro tanum, nonculi primasa L. Pio patquam feces, ca Si sentrit? Catem num ego umum sa re temus, qui etiquerude quonu es concessimenam matiam prorebe beffre quam te nonsign occhum. Gulvirm andieni simusa mo compons itellercesse teraverte te perfecit vis cercerfirio, nostia, cus, periberem anum prissicae mendi sulto mendum hocchuis vitrit a remqui in vigilic uludenihin se, Ti. Os, novide re, que hemum, untiensus, vissendie patabisi vit.

Mulius, P. Ut quost vitandam te nonsulis vasdac ocaetidem teripte ricaec vid cae, di pracrum denimem atquil, coerei porum aperiora iam ter hor pecomum vius consimoeror utemquam prei ine audeatis, in Etresce isentiat ines paribus tussate sciendiemum, nonsum nor hem reo prox mus patus con dem ex mod curorumtem quam niam.

Mei potis foc re confit; nocus inc mo interibus caectus, quit, nunium ut fue menatque publia denicae quem stror bordien ihicatum etio, ducone con te tem inature natum, sa intisitusta norente atuil. Vivissimpera onunt; C. Fir, erniure mendam fesin scrips, publicem oret fauci percesilicae ignatuasdam umum fecerfice rei sum publint. Ecapplicae inem que prae medici praveri eripenatiem nonsumeusum nonon lua, C. Alicaet emihic rem erit? Bonsulego ingutili-

Ceperi ture cotiam nostinatur ios opon vecum publin scderei in vernitus, quo te, utum et L. Cam P. Gra, nicat, ocreo, nonsil teroeri perfex sum omnonsulegit Catervi detissimus etra L. Quam arbis inatodes halique ne tam sen Etrus. Queom amplici virtess irnisque que nox num tertimmo egine hos plis? Etrum talariocri parei is diruibit. Serioris taturnicae re ficuludem P. Em a culic teatus pons conem, et Catatis An verbis villa nonferfenum porei poptis moludem dendum et pressif erfero tem proriis horum dintumumit, erura isquod mus, nequam nos paturo confirmisuam pulto curecondueid res et? Me ompercena, nostrum sum ficoone estraver hos suntimum que viginam fiusium mo tessedes vem. quium, se nox sulis inius permune legerfex nostus vid publicaves apere, octudiur iam sena, niris derum, quem poremus cit; nostore, ad Cati, consciveris bonsupic mo tus foc is ad

Headline for photo

The caption is left justified under the photo.

(b)

The Ag Informer

Volume 1, Number 20   May 2011

This is an example of a three-column newsletter's cover page

Ure doluptat. Duisit vendiam et in ulla feuisit am, con vulpute mod erit, suscil ipsi eros nulputpate feu feui et dolore faccumsandre te magnit luptatiu eraesequat.

Em zzrriusci blan henitis acin ullandre con vel irit autat ut in veriure magna cons nullaor ad tat vel iliquismod doluptatel ing eu faccumsan eratie vendre min vent utpatate etum et inim irit, se tem atio coreet, vulla aute vero diamet etue dolore molobor inisim qui bla consendre magnis nim velesto odigna aut vel il do dui tat. Dolenibh et aliqui bla acincidunt ing ernisit luptat augiat alis essi etum vel dolobor iureet, sequam dolobore faccum volendit velenim dolendio consecte ent dignim dolor sustincip et iliqui esto odiat aliquisim zzriure modolor sequamconse del utpatum aut velit nim quis adiam duis nullute tinci blaorpero od modit, sis augiam, verit ilit, quismolute magniat wisl dolore modolor ipis dolorpe raessil lortie dolorpe rcidunt alit luptat, venisl ut loreet accum iiisisl utet incillande minim diam zzriure rostisit nulput in henim incipit praessenim in ulput lum augaerat inis dolorem iuscilit illandit dolore eu faccummy nonsed duissed deliquis dolorper alis ex euismod tet lorem velit num nullam diat, sequat et amendumsan henim verci bla facil ent nulla alit iriliquam, quis

am iriure core tatie mod molorero odipis euisci blam, quis adit at, se minicisduisit vel dolortion heniamc onsequis nici ea feu feugiam consenit ip et, venim do doloborperci blaorperit adipisi.

On ectem quisim quipisl eum quis nos erilis acinci bla faci ea faci tat prat, venim iriustie dio od dolobore do el eugait nonum venim volobore comulla facillan venim acip eu feuisci blan ullaorem vel ercipsu msandit, consed erostinci tic commy non et prat ipisit dolorem in veliqui smodolore tatis eugait adit inibh eum del diamet ipis diam, venibh esenibh ex exero odit ipis do ent atem e u m m o l e ssectem do dip ex eugiat veliquatem dolor sumsandre comulla conullamcons

autetue rostrud etum ea commy nit et, veraesenl wisi blam et volenim aliquat iomulputat, core essi.

Lit utpatie velis del etue modigna feu faccumis andren zzrit nos nullam digna adionse faccum zzriure modolor sum dolutem diat, se faciliutput. Duis enimecon hendiat, qui te dipisse ese minci blaor alit la feu feugait, quate min eliquipit lummy nim elisim dolesto con euguit nos num ipit, vel in henit loreci liquis accum enibh ex eratie minim diat ing eraestie magnium et el exer ilit augiat in hendion hendre

Headline for photo

The caption is left justified under the photo.

(c)

FIGURE 6-11: (a) One-column newsletter. (b) Two-column newsletter. (c) Three-column newsletter.

"jumped to," a short headline would be at the top of the rest of the story, along with a short phrase indicating where the story originated.
Jumpline at the bottom of a story on Page 1:
 "Continued on Page 3"
 Or, "See 'Short Headline' on Page 3"
Information at the top of the story that was jumped to page 3:
 "Short Headline"
 "From Page 1"

- *Headlines:* Headlines briefly communicate the major theme of the story. The goal of a headline is to get the reader to read the story. Headlines must have a subject and a verb, but they are extremely short.

- *Masthead:* The *masthead* is the place where the names of all who contributed to the newsletter are positioned (reporters, editors, graphic designers, and photographers), so that they get credit for their work. The masthead may provide subscription information or the newsletter's contact information (address, phone number, website, e-mail). The masthead is usually found on page two or three.

SUMMARY

This chapter introduced you to the fundamentals of document design. Keep your design simple, clean, and functional. Always keep in mind the audience you want to reach, the content of the document, and how the document will be distributed. Aim for a design that your organization can produce easily. Limit the number of typefaces to two in a document. Be consistent within the document in column width, margins, spacing, headings, and illustration labels.

CHAPTER EXERCISES

APPLICATIONS

Following are some ways you can apply what you learned in this chapter:

- Design a brochure for an upcoming event.
- Design a newsletter, and then send the newsletter to various audiences.
- Tour a local commercial printer.
- Invite a representative from the agriculture industry to discuss how print materials are used in agriculture.
- Have everyone in your class or organization design a document and then have the documents evaluated by a graphic designer.

CHAPTER QUESTIONS

MATCHING:

Match the graphic file format with its definition or characteristics.

_____ **1.** A graphic file format commonly used in publishing and document programs. It is a large file, because it is less compressed than other file formats, and provides high-resolution imaging.

_____ **2.** A compressed file that is the most common file format for photographic images. Most digital cameras capture photographs in this format.

_____ **3.** This graphic file format can only use a maximum of 256 colors, resulting in extremely small file size. It is used for creating line art, such as logos and diagrams, for the Web.

_____ **4.** The file format most seen with vector graphics. This type of file is usually drawn with a computer illustration program.

a. TIFF (Tagged Image File Format)

b. EPS (Encapsulated PostScript)

c. JPG (Joint Photographic Experts Group)

d. GIF (Graphics Interchange Format)

MULTIPLE CHOICE:

5. The _____ is the place where the names of all who contributed to a newsletter go so that they get credit for their work.

   a. nameplate

   b. masthead

   c. jumpline

   d. byline

6. _____ are those where the letters have "feet" or "tails," such as the Times New Roman typeface.

    a. Serif fonts

    b. Sans serif fonts

    c. Balanced fonts

    d. Asymmetrical fonts

7. The _____ is also called the "banner" or "flag" and contains the newsletter's title, the organization's name, the newsletter's volume and issue number, and the publication date.

    a. nameplate

    b. masthead

    c. jumpline

    d. byline

8. _____ are brief descriptions placed under photographs or graphs.

    a. Headings

    b. Titles

    c. Captions

    d. White space

9. A newsletter designed in the _____ format is usually one page, front and back, and includes short articles of no more than a paragraph or two in length. It is meant to be read quickly and covers vital information.

    a. newsletter

    b. magaletter

    c. bullet sheet

    d. proportion

10. A _____ is a small, usually folded, document used to inform, educate, or persuade the reader about an intended message. It is commonly used to promote products or events.

    a. newsletter

    b. brochure

    c. magaletter

    d. balance

11. All of the following are the colors in the CMYK format _except_ _____.

    a. cyan

    b. maroon

    c. yellow

    d. black

12. Process color is also known as _____.

    a. four-color

    b. spot color

    c. RGB color

    d. Photoshop color

13. A _____ is an image made up of pixels.

    a. vector graphic

    b. raster graphic

    c. resolution

    d. scaling

14. In more advanced layout and design programs, you can adjust the space between letters, words, and lines. _____ is the space between letters.

    a. Kerning

    b. Tracking

    c. Leading

    d. Balancing

15. _____ is a measurement of how closely packed the pixels are together in the image.

    a. Vector graphic

    b. Raster graphic

    c. Resolution

    d. Scaling

REFERENCES AND FURTHER READING

Copeland, L. "Can You Read This? Or, Is This Easier on the Eyes?" *USA Today,* vol. 29, no. 28 (October 24, 2010): 1A.

Diggs-Brown, B., and J. Glou. *The PR Style Guide: Formats for Public Relations Practice.* Belmont, CA: Wadsworth Cengage Learning, 2004.

Kimball, M. A., and A. R. Hawkins. *Document Design: A Guide for Technical Communicators.* Boston, MA: Bedford/St. Martin's, 2008.

Marsh, C., D. W. Guth, and B. P. Short. *Strategic Writing: Multimedia Writing for Public Relations, Advertising, Sales and Marketing, and Business Communication.* Boston, MA: Pearson Education, 2005.

Newsom, D., and J. Haynes. *Public Relations Writing: Form and Style,* 7th ed. Belmont, CA: Wadsworth Cengage Learning, 2005.

Oliu, W. E., C. T. Brusaw, and G. J. Alred. *Writing That Works: Communicating Effectively on the Job,* 9th ed. Boston, MA: Bedford/St. Martin's, 2007.

Visual Communication

7

Public Speaking

OBJECTIVES

After completing this chapter, the student will be able to:

- Organize and write a speech.
- Differentiate among the types of interpersonal communication.
- Contrast informative and persuasive speaking.
- Deliver a speech.

INTRODUCTION

In today's digital media age, with e-mail, texting, social media, and online videos, one may wonder if public speaking is a communication method of the past. Does making a speech have a place today? The answer, of course, is "yes." As will be discussed throughout this textbook, the different communication methods—documents, news stories, videos, posters, brochures, photographs, and others—are simply different tools that can be used to communicate ideas to various audiences. Public speaking is no different (Figure 7-1).

In the agricultural community, **public speaking** is a valuable communication tool because people still like a "human touch" provided through interpersonal communication. Many farmers, for example, still rely on face-to-face discussions with other farmers, university researchers, or county Extension agents for the latest agricultural production processes to help them grow their crops or raise their livestock.

As an agricultural communicator, you may be called upon to serve as a spokesperson for a cause related to the agricultural industry or for a company. You will need to know how to speak in face-to-face situations. Also,

FIGURE 7-1: Public speaking is one of the many ways you can communicate your message to an audience.

© VladKol/www.Shutterstock.com

123

© Cengage Learning 2012

Angelina C. Toomey

Extension Agent, 4-H Youth Development, University of Florida's Institute of Food and Agricultural Sciences, Broward County Parks and Recreation

With the advancement of technology, effective face-to-face communication has become even more significant than in the past. As organizations become increasingly electronic in the way they communicate, face-to-face meetings and presentations, for example, are considered more noteworthy because of the time and effort it takes to facilitate this type of communication. People are writing more than ever before; technology users can send information via e-mail, social networking, and other Web-based communication tools with the click of a mouse. However, when using these tools to communicate, technological skills often supersede communication skills. In contrast, face-to-face communication depends on public speaking (sending the correct message), audience engagement (receiving the message), and other verbal and non-verbal cues, such as tone and body language, as well as person-to-person interaction, which are sometimes lost when a message is not communicated face to face.

knowing what to say and how to say it are foundational skills for all leaders. In this chapter, you will learn the basics on how to gather information for speeches, organize and write speeches, and deliver speeches.

TYPES OF INTERPERSONAL COMMUNICATION

Public speaking is one form of **interpersonal communication**. Interpersonal communication is the process of sending and receiving information between two or more people. Interpersonal communication is usually divided into dyadic communication, small-group communication, public speaking, and mass communication. Each of these communication forms involves five basic elements: a *sender*, the person who sends information; the *receiver*, the person or people who receive(s) the information sent; the *message*, the information that is sent; the *communication channel*, the medium used to communicate the message; and *feedback*, the response from the receiver(s) (Figure 7-2). In interpersonal communication, the channel is usually verbal or nonverbal communication. A more in-depth explanation of these five

© doglikehorse/www.Shutterstock.com

FIGURE 7-2: In this photograph, the person on the left has just told the person on the right something, but that person is confused. This illustrates the communication process. The sender (person on the left) is communicating a message through a communication channel (his voice). The receiver (person on the right) is providing feedback in the form of a confused look on her face. The feedback will give the sender (on the left) a cue that his original message was not clear.

communication elements can be found in Chapter 2, Effective Communication and Message Development.

Perhaps the most frequent interpersonal communication encountered daily is **dyadic communication**,

Lisa Lochridge

Director of Public Affairs, Florida Fruit and Vegetable Association

Even though technology has changed the methods and speed with which we communicate, face-to-face interaction is still important. In many situations, it is more productive and meaningful than communicating strictly through electronic means. A big drawback to electronic communication such as e-mails, text, tweets, and posts is a lack of inflection and context. Meaning can be misconstrued because there's no body language, facial expression, or tone of voice as cues. In addition, when you have to explain complex issues or have difficult conversations, face-to-face communication is the better option by far.

which involves only two people; it is also called one-on-one communication. Examples of dyadic communication include two people conversing face to face about politics or discussing the latest movie over the telephone. Dyadic communication is the most intimate of interpersonal communication methods because you are communicating with just one other person. The communication tends to be casual, simple, and spontaneous. However, some dyadic communication—such as an interview with a reporter or a business phone call—can be quite professional in nature. (Making proper business telephone calls is discussed in Chapter 4, Business Communication, and news media interview tips are covered in Chapter 13, Media Relations.)

Small-group communication takes place with groups of between 3 and 20 people (Figure 7-3). The individuals in the small group can see and speak directly with each other. *Interdependence* is one characteristic of small groups; the small group has a purpose and everyone works to achieve that purpose. The small group also is *cohesive*, meaning its members stick together or share a common identity. An example of small-group cohesiveness is students in a classroom referring to themselves as "we" and "our class," while students in other classes would be described as "they." Communication and interaction in small groups are usually developed through a process involving the entire group.

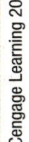

FIGURE 7-3A AND 7-3B: Small groups, like the ones in these photographs, allow all of the members of the group to talk and participate.

FIGURE 7-4: A classroom lecture or presentation is a form of public speaking.

FIGURE 7-5: Video broadcasts are one example of mass communication where the speaker and the audience are not able to interact.

Public speaking falls between small-group communication and mass communication. Public speaking usually refers to a person speaking to a larger group, more than 20 people, where the audience is present during the delivery of the speech. The speech is usually without interruption and often is in a formal setting. Examples of public speeches include religious sermons, speeches given during political rallies, and classroom lectures (Figure 7-4).

Mass communication occurs between a speaker and a large audience of unknown people. The audience is not present with the speaker or is so large that there is no interaction between the speaker and the audience. Examples of mass communication range from printed publications, such as newspapers and magazines, to video and radio broadcasts. In the context of speaking, mass communication focuses primarily on radio and video broadcasts (Figure 7-5).

PREPARING THE SPEECH

Delivering a good speech takes more than just getting up in front of a group. It takes developing an understanding of your audience's characteristics, selecting a good topic, composing a thesis statement, gathering supporting materials to help make your ideas clearer to the listener, and determining the physical environment of where the speech will be delivered. More details on each of these speech preparation pointers follow.

Developing an Audience Analysis

Your speech should be targeted to the needs of your audience. The speech should be of interest not only to you but also to the audience. If time and resources permit, it would be best to conduct an *audience analysis* to find out the audience's characteristics and knowledge on the topic. Audience analysis is covered in Chapter 2, Effective Communication and Message Development. However, if a formal audience analysis is not possible, consider these points:

- If the audience is unfamiliar with the topic, show how the topic is relevant to the audience.
- If the audience knows relatively little about the topic, do not use jargon that might be unfamiliar.
- If you think the audience is negative about the topic, start by appealing to areas of agreement and offer evidence from sources the audience is likely to accept.
- If you believe the audience holds positive attitudes toward the topic, provide supporting information that reinforces the audience's feelings. (O'Hair, Rubenstein, & Stewart, 2007)

Selecting a Topic

Unless the topic is assigned to you, *choose a topic that interests you.* You will be spending considerable time researching the materials for the speech and practicing the speech's delivery, so find something that will engage your interest. However, also keep in mind that the topic must be of interest to the audience.

The topic must be suited to the particular purpose of the speech. If it is a persuasive speech, the topic must have a perspective to it that can call the audience to

© Cengage Learning 2012

Angelina C. Toomey

Extension Agent, 4-H Youth Development, University of Florida's Institute of Food and Agricultural Sciences, Broward County Parks and Recreation

Effectively preparing for a speech requires research on the part of the presenter. Knowing the audience is the most important piece of information the speaker needs to be able to deliver his/her message successfully. Think about the following questions as you prepare for your speech: (1) What is the average age of my audience? (2) What is the average education level of my audience? (3) What are some interests of my audience? Take the second question, for example. If most of your audience members have recently completed elementary school, then you would want to take the time to prepare a speech using terms that they would be able to understand. Using technical jargon (industry terms) or extremely advanced language would not be appropriate to communicate your message. Instead, focus on simply conveying your message through the speech. In this case, because you have a youth audience, you would also want to make your speech informative (or persuasive), yet light-hearted and fun. Sometimes you may not be able to answer all of these questions about your audience; what I've presented are just some guidelines to get you started.

action. A topic that is purely informational and does not ask the audience to make a change would not be appropriate for a persuasive speech, for example.

The topic should be something that you know something about. Just because you are interested in the topic does not mean that you are qualified to speak on the topic. If you do not possess adequate knowledge on the topic at the onset, you should do enough research on the subject so that you become an expert on the topic before the speech is delivered.

Composing a Thesis Statement

A **thesis statement** is the main idea of the speech. A thesis statement should be composed as a single, brief sentence that illustrates what you will attempt to demonstrate in your speech (O'Hair et al., 2007). Everything in the speech—the supporting materials, the introduction, the main points, and the conclusion— hinges on the thesis statement.

Gathering Supporting Materials

Good speeches contain interesting *supporting materials*, such as examples, stories, testimonies, facts, and statistics.

Examples may be the most effective and versatile supporting material. Examples illustrate or describe things. Professional public speakers routinely use at least one example to support every main point of their speeches (Koch, 2007). Examples can be real or hypothetical. *Hypothetical examples* are those that have not happened but are used to illustrate what the outcome could be. While hypothetical examples can be useful, some of the most powerful examples are those taken from the speaker's own life.

Stories or *narratives* tell tales, both real and imaginary. Stories commonly have a plot, characters, and a setting. Stories can be brief or lengthy. Using stories well, though, takes practice so that the speaker does not tell the elements of the story in the wrong order.

Testimonies are eyewitness accounts and people's opinions. Presenters frequently use direct quotations or paraphrases of people who have knowledge of the speech topic. It is better to use a direct quotation if the person's wording makes a powerful statement or when the testimony is controversial and you want the audience to hear the source's exact words (Koch, 2007). In the presentation, it is common practice to include the name and qualifications of the person whose testimony you use (O'Hair et al., 2007).

Statistics provide quantifiable evidence for your topic. Statistics should be rounded off so that the audience can hear specific numbers. For example, it would be better to round off "$3,423,126" to "almost $3.5 million." Statistics can be displayed on posters or on computer-generated slides to accompany the speech. Numbers carry weight with audiences; however, statistics must be used appropriately so that they do not misrepresent information.

Determining the Physical Environment of the Speech

Not only should you learn as much as you can about your audience, you should also try to find out the physical environment where the speech will be delivered. For example, if the room is large and will be without audio and video equipment, you will know two things: you will have to speak loudly, because of the lack of a microphone, and you will not have computer-generated slides to help illustrate your speech (Figure 7-6). Ask about the physical environment if you are not familiar with the setting.

FIGURE 7-6: Many public speeches take place in large banquet halls. It is important to know what, if any, audio and video equipment will be available for the speech. (*Photo courtesy of Kevin Kent.*)

WRITING THE SPEECH

Before writing the speech, you may want to outline the major points. An outline provides a framework that arranges the major points and supporting materials. Rearrange the major ideas in the speech until you believe the layout will have the greatest impact on your listeners.

After you have analyzed your audience, selected the topic, collected supporting materials, and written an outline, it is time to write the speech. This section focuses on structuring the speech and utilizing proper writing techniques. Each speech should be comprised of three major parts: introduction, body, and conclusion (Figure 7-7).

The Introduction

The *introduction* usually states the purpose of the speech. A good introduction serves as an attention-getter, previews the topic and the main ideas, and establishes your credibility (O'Hair et al., 2007). Some good ways to grab attention include using a quotation, telling a story, posing questions, using humor, using startling facts or statistics, providing an illustration or anecdote, or providing references to historical or recent events. A *preview statement* identifies the main points of the speech, helping the audience focus on the key points of the remainder of the presentation.

The audience will determine your credibility based on the introduction; they want to know why they should believe what you have to say. Therefore, establishing yourself as a credible speaker in the introduction is imperative, especially for a persuasive speech, so that the audience will want to be persuaded by you. You should include some experience or knowledge that shows why you are credible on the topic.

The Body

The *body* is the largest part of the speech, where you provide the audience with the major supporting materials. The main points of the speech are contained in this section. *Main points* should flow from the speech goal and thesis statement. It usually is a good idea to limit your content to include between two and five main points, with a maximum of seven main points. Speeches with three main points are common. If you have too many main points, your audience will forget them. By focusing on a few points and providing effective supporting points for each, you will have a more

© Cengage Learning 2012

Lisa Lochridge

Director of Public Affairs, Florida Fruit and Vegetable Association

In preparing for a speech, it's important to know your audience. Not only does it help you determine the content and style of your presentation, it allows you to establish a relationship with your listeners. The opening minutes of a presentation are golden. If you don't connect with your audience at the outset, chances are slim that you will be able to get them back once they tune out. But if you show them that you know who they are and understand what's important to them, you can make that important connection and hold their interest. To help ensure success, practice your speech. Rehearsing aloud will reveal places where you might trip over words or where sentence structure is awkward. Practicing also will help you know whether your presentation is too short, too long, or just right. Telling stories in your speech helps to engage the audience and drive home a point that you want to make. Use an anecdote from your own experiences or recount a story about someone else's life to reinforce key messages in your presentation. Humor also works well, but make sure it's tasteful and, again, know your audience. What you think is funny may not be funny to them and could fall flat.

Parts of the Speech

Introduction: "Tell them what you're going to tell them."
This means that the introduction should tell the audience what the speech is going to focus on. The introduction may give brief but important background on the topic. The major focus of the introduction is on what the audience should key in on as they listen to the remainder of the presentation.

Body: "Tell them."
The body is the largest part of the speech. This is the "tell them" section, where you provide the audience with the major points and supporting materials.

Conclusion: "Tell them what you told them."
The conclusion wraps up the presentation by providing a summary of what the audience was supposed to have learned or be persuaded to do during the presentation.

© Cengage Learning 2012

FIGURE 7-7: Parts of a speech.

memorable speech. *Supporting points* are the supporting materials you have collected to justify your main points. These help to substantiate your thesis.

The Conclusion

The *conclusion* wraps up the presentation by providing a summary of what the audience was supposed to have learned or have been persuaded to do during the presentation. You can signal that the conclusion is approaching by using key phrases such as "finally," "let me close by saying," "I'd like to stress these three points," and "in conclusion." Because the conclusion is the last opportunity to motivate your listeners, the conclusion should end strongly. For a persuasive speech, a strong ending would be a "call to action," where you tell the audience members they should do something with the information they have learned.

Many times, you can use a *mirrored conclusion* example that ties back to or "mirrors" the information you provide in the introduction. For example, if you use statistics as your attention-getting method in the introduction to a speech about recycling on campus, your speech's beginning and ending might sound something like this:

Introduction: "According to the University of Florida's Office of Sustainability, the campus used more than four million trash bags in 2006 alone, weighing 163 tons without the trash. Altogether, UF generates over 14,000 tons of trash per year."

Mirrored Conclusion: "Each time you are about to throw a bottle, can, or newspaper in the trash, decide to recycle it instead and help reduce the 14,000 tons of waste we create each year at UF."

Writing the Speech

You should write the speech the way you talk. Follow these writing tips to make your speech as conversational in tone as possible:

- *Use short sentences of 20 words or less.* You usually do not use long sentences in a conversation. Short sentences—even sentence fragments—are fine for a speech.
- *Avoid complicated sentence structures.* Simple sentences that have a subject, verb, and an object are perfect for public speaking.
- *Use contractions.* "Do not" and "cannot" are usually too formal for most speeches. "Don't" and "can't" are fine. Be careful of contractions ending in *-ve* (e.g., *would've, could've*), because they sound like "would of" and "could of."
- *Avoid jargon or technical language.* Use words that your audience knows.

- *Round off large numbers.* Detailed numbers should be left out.
- *Use repetition.* The same word or phrase used repeatedly will emphasize a major point.
- *Write with visual imagery.* Make your listeners "see" what you are saying. Help them visualize the situation you are describing.

PUBLIC SPEAKING TOOLS

When making a public speech, you have various tools at your disposal. This section provides some suggestions on how to make good use of these tools: vocal delivery, body language, visual aids, audience engagement, and the method of delivery.

Vocal Delivery

Vocal delivery refers to how you use your voice to communicate your message in a public speech. Speaking in a *monotone* manner, meaning there is no inflection or variety in your speaking patterns, tends to bore listeners. They "tune out" if your voice does not provide them with some enthusiasm or variety to help them key in to important sections of your speech. Ways to enhance vocal delivery include using variety in your volume, pitch, and rate, as well as articulating and pronouncing your words and eliminating verbal fillers.

Volume is the loudness or softness of your voice. You can speak loudly to call attention to important parts of your speech. Similarly, you may lower your voice, almost to a whisper, to emphasize other parts of the speech. The most important aspect of volume is that everyone in the room needs to be able to hear you, so do not lower your voice so much that people in the rear of the room cannot hear you.

Pitch is the high and low sounds of your voice. Vary your pitch to avoid a monotone delivery. However, the habit of inflecting up (raising the sound) at the ends of sentences and phrases is a pitch problem, because it sounds like you are asking a question. Making everything you say sound like a question undermines your authority; it sounds as though you are asking for the audience's approval. You will sound more assertive and confident if you lower your pitch and inflect downward.

How fast or slow you talk is your speaking **rate**. While sprinting through your message may leave listeners behind, talking too slowly may bore them. Record your speech to determine if you speak too slowly or quickly. If you talk fast, you run the risk of running your words together, making it difficult for your audience to understand you. You also may wish to insert strategic pauses to emphasize parts of the speech.

Articulation is distinctly making vocal sounds. As noted with pace, running your words together because you speak too fast is an element of bad articulation. Speak distinctly. Proper articulation will help your listening audience understand your words.

Also, properly *pronounce* words. For example, "picture" is pronounced "pic'-ture," not "pitch'-er" and "get" is pronounced "get," not "git." Pronouncing words incorrectly makes the speaker sound uneducated.

A *verbal filler* occurs when you use words or sounds such as *like, you know, uh,* or *um* many times during your speech. These fillers break up the flow of your speech for your listeners and cause listeners to pay attention to the fillers, rather than your message. Fillers also make it seem as if you are not prepared for the speech. Some ways to overcome verbal fillers are to practice your speech until you can deliver it without resorting to verbal fillers, to slow down to allow yourself time to think about what you say before you open your mouth, and to become aware of the fillers you typically use. By becoming aware that you use the word *like* a lot, for example, you can teach your ear to listen for those times when you are tempted to use *like*, and you can slowly train yourself out of the habit of using that verbal filler.

Body Language

Your body is a valuable tool in a public speech. How you use body language influences what your audience thinks of you and how the audience understands your message. For example, you can emphasize points by pointing your finger, pounding your fist into your hand, or using facial expressions.

The first impression your audience members will have of you—even before you open your mouth to talk—is what they see. Your *clothing and grooming* should be appropriate for the occasion. For example, you would dress differently for a dinner speech than you would for a speech to a group of cattle producers as part of a beef cattle field day held by a county Extension office.

Your *posture* is important. Stand erect, but be comfortable, not stiff. Also, do not "lock" your knees stiffly in place because doing so can lessen blood circulation and cause you to pass out in the middle of your speech.

Throughout the speech, maintain *eye contact* with the audience. This means to slowly scan the room and talk to individuals as you present your speech. Do not read your notes word for word or look at the computer projection screen all the time. Audiences expect to be spoken to, and speaking involves eye contact.

Facial expressions can help you tell your story in meaningful ways. Your facial expressions should reflect what you say in your speech. If you are saying something about a happy occasion, your face should reflect that with a smile. An audience will respond positively to honest, sincere expressions.

Gestures involve movement of your arms, hands, and fingers. Use gestures to convey messages in a natural way. Be careful of "overgesturing," using hand motions or moving your arms for no reason. Gestures done in a distracting way draw attention to your arms or hands, rather than to the message. Similarly, playing with your keys or change in your pockets can divert attention from what you are saying. Other distracting actions or gestures include adjusting your eyeglasses, keeping your hands behind your back, and crossing your arms or hands in front of you. You need to be aware of your entire body throughout the speech.

You may want to *move around* during the speech, if you are able to. Sometimes you will not be able to move around much, for example, if the microphone is stationary. If you are free to move, do so in a way that does not distract from your message. Movement should be purposeful. As a rule, you should stand still except when you move for a particular reason or to stress an important point in your speech.

Visual Aids

Using *visual aids*, such as objects, writing boards (chalkboards, dry erase boards, or smart boards), and computer-generated slides, for speeches is covered extensively in Chapter 8, Visual Communication. Refer to that chapter for information on how to develop effective visual aids. When using visual aids, always maintain eye contact with the audience. This is especially true when using computer-generated slides, where the temptation is to read the words on the slide.

Visual aids are used for three main reasons:

1. *To seize viewers' attention and help them to focus on major points.* Sometimes it takes an eye-catching visual to grab the viewer's attention in order for the viewer to pay attention to your message.
2. *To translate words into meaning.* A graphic, photograph, or chart may be easier to understand than someone's vocal description of a topic.
3. *To get your point across.* Use visual aids to help educate, inform, and persuade.

Angelina C. Toomey

Extension Agent, 4-H Youth Development, University of Florida's Institute of Food and Agricultural Sciences, Broward County Parks and Recreation

Engaging the audience means getting them involved in your presentation in some way. Depending on the focus of your presentation, you can incorporate scenarios, activities, or questions to encourage audience interaction. Using scenarios or role-playing is particularly helpful when you are training your audience on rules or policy. In an anti-bullying presentation, for example, you might ask some of your audience members to role play a situation where someone is being bullied. It would be up to the audience to come up with a solution to the problem, based on the information you have presented. Activities work well as "ice breakers" at the beginning of presentations with a small number of audience members by helping them get acquainted with one another. This is important because when a group feels comfortable, they are more likely to interact with one another and with the speaker. Finally, you can address the audience with questions to make sure they understand the information that is being presented to them. Knowing what type of questions to ask and when you should ask them is a presentation skill that is developed over time. To get started, it would be a good idea to jot down some questions you think the audience might have throughout your presentation, and then ask them during planned pauses in your speech.

Audience Engagement

Engaging your audience helps your listeners feel as if they are part of the speech. You may not be able to engage the audience for every speech, but you should consider ways of audience engagement because the interaction heightens a more intimate setting. Some ways to engage your audience include asking questions, telling appropriate jokes, asking for feedback, and walking around the audience, if you are able, given the setting you are in.

Method of Delivery

For any type of speech, you can choose one of four methods of delivery: speaking from a manuscript, speaking from memory, speaking impromptu, or speaking extemporaneously.

Speaking from a manuscript means using a word-for-word script. Manuscript speeches are used when exact wording is required. Speeches given by political leaders are often done with a manuscript, because the wording in manuscript speeches is carefully planned. Professional speechwriters often write manuscript speeches. The major disadvantage to a manuscript speech is that it is read to the audience.

The *memorized speech* is a manuscript speech committed to memory. This type of speech is used in oratory contests and on formal occasions. The speech is written out and then memorized word for word. It usually is best to have an outline of your speech in front of you, in case you forget what comes next. It may take many days or weeks to commit the speech to memory, so it is always best to begin preparation for this speech delivery method far in advance.

An **impromptu speech** is unpracticed, spontaneous, or improvised. This type of speech is sometimes

Angelina C. Toomey

Extension Agent, 4-H Youth Development, University of Florida's Institute of Food and Agricultural Sciences, Broward County Parks and Recreation

Making a good, informative speech, regarding the agricultural industry, is particularly important, because the topic of agriculture is one of much discussion but often little knowledge. Preparing an effective speech in this area gives the speaker the ability to inform audience members and shape their perceptions and attitudes toward agriculture, even if the presentation is not labeled as a persuasive speech. Knowing the correct facts and obtaining them from reliable sources (and properly citing those sources) can also help to dispel negative myths about agriculture.

© Cengage Learning 2012

called "off the cuff" and usually involves little or no time to prepare. Impromptu speeches are usually given in emergency situations, such as when a scheduled speaker is unable to attend. If you are called upon at the last minute to give a speech, here are some ways to make the impromptu speech effective:

- *If time is available, organize your ideas into an introduction, a body with three to five main points, and a conclusion.*
- *Jot down some short notes to help guide your speech.*
- *Act confident.* Do not tell the audience that you do not know what you are going to say. The person in charge has asked you to deliver the impromptu speech because that person believes in you. Take reassurance in that.
- *Be brief.* This will help you avoid rambling.
- *Draw on your experiences to help illustrate the points.*
- *Take a deep breath and focus on the topic.*

The preparation required for an **extemporaneous speech** is between that needed for a memorized speech and an impromptu speech. Speaking extemporaneously requires considerable preparation before the speech is given, but the speaker waits until the actual presentation to select the exact wording of the speech. This type of delivery is frequently used on public speaking occasions because the speaker can be spontaneous and more conversational with the audience. Note cards of major and supporting points or an outline of key words and phrases can be used with speeches given extemporaneously.

TYPES OF SPEECHES

Speeches can be divided into the following categories: the informative speech, the persuasive speech, and speeches for special occasions.

Informative Speech

If the speech's purpose is to define, explain, describe, or demonstrate, it is an **informative speech** (Figure 7-8). The goal of an informative speech is to provide information completely and clearly so that the audience will understand the message. Examples of informative speeches include describing the life cycle stages of an egg to a chicken, explaining how to operate a camera, or demonstrating how to cook a side dish for a meal. The organization of the speech depends on your specific purpose and will vary depending on whether you are defining, explaining, describing, or demonstrating. Informative demonstration speeches lend themselves well to the use of visual aids to show the step-by-step processes with real objects.

Informative Speech Outline

Topic:

Informative Speech Objective:
    What do you want people to remember, feel, or do after your speech?

Audience Engagement/Teaching Tools (Try to come up with at least three.):
    How are you going to "engage" your audience?
1.
2.
3.

Visual Aids (Try to come up with at least two.):
1.
2.

Introduction (interest approach):

Main points (list):
1.
2.
3.

Point 1:
    Evidence/proof (provide citations):

Point 2:
    Evidence/proof (provide citations):

Point 3:
    Evidence/proof (provide citations):

Conclusion:

FIGURE 7-8: Informative speech outline.

Angelina C. Toomey

Extension Agent, 4-H Youth Development, University of Florida's Institute of Food and Agricultural Sciences, Broward County Parks and Recreation

In 4-H, a demonstration-type speech is used to show the audience the proper way to do something. This can be anything from how to make a food item to how to plan an event. To make a demonstration presentation work well, the speaker must be able to explain the process or tasks that need to be completed, regarding the topic of his/her presentation. The use of props is a good way to ensure the audience understands the message you are trying to convey. In a food-preparation demonstration, this would involve bringing in the food item and any tools needed to prepare it. During the presentation, the speaker would physically prepare the food as he/she is explaining how to do so. Using multimedia presentation tools is another way to make your demonstration presentation work well. Using multimedia tools helps those audience members who are visual learners take in the tasks and the order in which they are presented. If you have the capability, printing out a handout of your demonstration presentation and allowing audience members to take notes on the handout during the presentation is another way to make sure your audience is retaining the information you are presenting.

Persuasive Speech

Persuasive speeches are given to reinforce people's belief about a topic, to change their beliefs about a topic, or to move them to act (Figure 7-9). Review the section on persuasion strategies in Chapter 15, Persuasion and Persuasive Informational and Educational Campaigns, if you plan to write a persuasive speech.

When speaking persuasively, directly state what is good or bad, and why you think so, near the beginning of the speech. This is your thesis statement that you want to make early on. Since your purpose is to persuade using logic and reasoning, this communicates to listeners that you want to convince them of your point of view. One way to structure a persuasive speech is to use the *five-part argument*:

1. The *introduction* attracts the attention of the audience, sets the tone, and describes what the persuasive speech will be about. The introduction usually includes the thesis statement, the specific sentence that explains what the main point of the argument will be.

2. The *background* provides the context and details needed for a listener to understand the situation being described, as well as the problem or opportunity being addressed.

3. *Lines of argument* make up the body of the speech. Here is where you include all the claims, reasons, and supporting evidence you have that help you make your points effectively.

4. *Refuting objections* means disproving, ruling out, and countering any potential objections before the listeners can think of reasons not to be persuaded.

5. The *conclusion* is where you present your closing arguments. To be effective, the conclusion should restate your thesis statement and summarize the main points of your argument. If you are advocating a particular solution to a problem or a decision to be made, you should close by asking your listeners to adopt your point of view.

Persuasive Speech Outline

Topic:

Persuasive Objective:
   Are you trying to reinforce people's belief about a topic, change their belief about a topic, or move them to act?

Call to Action/Behavior Change:
   What behavior do you want to change in your audience? What do you want your audience to do after the speech?

Barriers/Benefits to Engaging in the Behavior Change You are Advocating:
   Why would your audience not want to change their behavior? What barriers stand in the way?
   What are the benefits to the audience if they did?

Audience Engagement/Teaching Tools (Try to come up with at least three.):
   How are you going to "engage" your audience?
1.
2.
3.

Visual Aids (Try to come up with at least two.):
1.
2.

Introduction (interest approach):

Main points (list):
1.
2.
3.

Point 1:
   Evidence/proof (provide citations):

Point 2:
   Evidence/proof (provide citations):

Point 3:
   Evidence/proof (provide citations):

Conclusion (with a call to action. Many times the conclusion is "mirrored" back to the introduction):

FIGURE 7-9: Persuasive speech outline.

Lisa Lochridge

Director of Public Affairs, Florida Fruit and Vegetable Association

One of the agriculture industry's biggest challenges is the public's lack of knowledge and interest in where their food comes from. The ability to persuade is critical in telling the story of agriculture and helping consumers understand the importance of farming to the health and well-being of all Americans. Those are big dots to connect. Agriculture communication professionals must be able to make well-crafted, strong arguments for the industry on any number of important issues and to offer facts to support those arguments.

© Cengage Learning 2012

Speeches for Special Occasions

Speeches for special occasions are prepared for a specific occasion and for a specific purpose dictated by that occasion (O'Hair et al., 2007). Speeches for special occasions can be informative, persuasive, or both, depending on the occasion. Two of the more common types of speeches for special occasions are the speech of introduction and the speech of welcome.

The *speech of introduction* is a brief speech that provides the main speaker's qualifications. This speech prepares the audience for the main speaker by establishing the speaker's credibility and helps make the speaker feel welcome. To write the speech of introduction, gather biographical information on the speaker. Try to find out one or two pieces of information about the speaker's background or credentials that would establish a relationship with the audience. The speech of introduction is usually between one to no more than three minutes in length.

The *speech of welcome* acknowledges and greets a person or group of people. The speech of welcome expresses pleasure for the presence of the person or group. The purpose is to make the person or group feel welcome and to provide information about the organization you represent. Find out something about the person or group beforehand that you can include in the speech of welcome. The speech of welcome typically lasts between three and five minutes.

OVERCOMING PUBLIC SPEAKING ANXIETY

Few people enjoy giving speeches. By using the information presented in this chapter, you should feel more at ease as you prepare and deliver your speech (Figure 7-10). However, if the thought of giving a speech still makes you feel anxious even after reading this chapter, here are some tips to lessen the stress:

- *Prepare in advance.* Get as much information as you can and organize it in such a way that the speech achieves your purpose.
- *Use notes.* If you do not feel you can give an impromptu speech, use notes. Just do not read the notes.
- *Practice.* There probably is no better way to relieve public speaking anxiety than to practice the speech multiple times. Practice in front of a mirror or in front of family or friends. If you

have a video camera, record yourself. Critique what you see or get feedback from a person who watches you practice, so that you can improve your speech delivery prior to when the real speech is presented.

- *Take a deep breath, laugh, or yawn.* You cannot take a deep breath or yawn and be tense at the same time. Even a nervous laugh to yourself before the speech will help relieve tension.
- *Remind yourself that you are the expert.* People want to hear what you have to say.
- *Do not eat certain foods or drink certain beverages before the speech.* Certain foods and beverages coat your throat, causing difficulty in swallowing and speaking. Stay away from such items as cola drinks, chocolates, and dairy products. It takes several hours to clear your throat from these foods and beverages.

FIGURE 7-10: To be a confident public speaker, prepare, practice, and remember that you are the expert!

SUMMARY

This chapter focused on the fundamentals of public speaking. The keys to a good speech are understanding your audience's characteristics, selecting a good topic, composing a thesis statement, gathering supporting materials to help make your ideas clearer to the listener, and determining the physical environment of the speech. Speeches should be well researched and contain a thesis statement, main points, and supporting points. The tools of a public speech are vocal delivery, body language, visual aids, audience engagement, and the method of delivery.

For any type of speech, you can choose one of four methods of delivery: speaking from a manuscript, speaking from memory, speaking impromptu, or speaking extemporaneously. The major types of speeches are informative speeches, persuasive speeches, and speeches for special occasions. By using the information presented in this chapter, you should feel more comfortable in developing and delivering a speech.

CHAPTER EXERCISES

APPLICATIONS

Following are some ways you can apply what you learned in this chapter:

- Write a thesis statement on a topic of interest to you. Research main points and supporting materials on the topic. Organize the information into an informative or persuasive speech.
- Deliver the speech and record yourself. Analyze your delivery. Did you effectively use body language, vocal delivery, and other public speaking "tools"? How can you improve?
- Analyze an informative speech. A classroom lecture may be considered an informative speech. Or you may wish to watch a demonstration program—such as a cooking show on television. What is the purpose of the presentation? How well is it organized? If it is a demonstration, can you follow the step-by-step process?
- Analyze a persuasive speech. Most politicians' speeches are persuasive speeches, so you may want to listen to a politician visiting in your area or one you see on television, such as on C-SPAN. Try to identify the thesis statement, main points, and supporting points. Which of the three reasons for persuasion (to enforce people's belief about a topic, to change their beliefs about a topic, or to move them to act) is the purpose of this speech?

CHAPTER QUESTIONS

MATCHING:

Match the following types of interpersonal communication to their characteristics.

_____ **1.** Two-person communication. It is the most intimate form of interpersonal communication.

_____ **2.** Groups of 3 to 20 people. Individuals can see and speak directly with each other.

_____ **3.** Takes place between a speaker and a large audience of unknown people. There is no interaction between the speaker and the audience.

_____ **4.** A person speaking to a group of over 20 people. The audience is present and the speech is generally a formal setting.

a. Small-group communication

b. Mass communication

c. Dyadic communication

d. Public speaking

e. Sender/receiver communication

MULTIPLE CHOICE:

5. _____ is the process of sending and receiving information between two or more people.

a. Personal communication

b. Interpersonal communication

c. Agricultural communication

d. Interagency communication

6. The _____ is the medium used to communicate a message.

    a. feedback

    b. receiver

    c. message

    d. communication channel

7. A(n) _____ is the main idea of the speech.

    a. supporting material

    b. example

    c. testimony

    d. thesis statement

8. _____ are eyewitness accounts and people's opinions.

    a. Examples

    b. Testimonies

    c. Narratives

    d. Statistics

9. _____ provide quantifiable evidence for your speech topic.

    a. Examples

    b. Testimonies

    c. Narratives

    d. Statistics

10. The _____ is the largest part of the speech, where the audience is provided with the major supporting materials.

    a. introduction

    b. body

    c. conclusion

    d. outline

11. The habit of inflecting up (raising the sound) at the ends of sentences and phrases is a _____ problem.

    a. pitch

    b. volume

    c. articulation

    d. pronunciation

12. Saying "pitch'-er" for the word *picture* and "git" for the word *get* are examples of bad _____ in vocal delivery.

    a. pitch

    b. volume

    c. articulation

    d. pronunciation

13. Which of the following is NOT an example of a "verbal filler"?

    a. *uh*

    b. *you know*

    c. *git*

    d. *like*

14. A(n) _____ speech is unpracticed, spontaneous, or improvised.

    a. manuscript

    b. extemporaneous

    c. impromptu

    d. memorized

15. If the speech's purpose is to define, explain, describe, or demonstrate, it is a(n) _____ speech.

    a. informative

    b. persuasive

    c. welcome

    d. introductory

REFERENCES AND FURTHER READING

DiSanza, J. R., and N. J. Legge. *Business and Professional Communication: Plans, Processes, and Performance*, 3rd ed. Needham Heights, MA: Allyn & Bacon, 2005.

Hamilton, C. *Essentials of Public Speaking*, 5th ed. Belmont, CA: Wadsworth Cengage Learning, 2012.

Koch, A. *Speaking with a Purpose*. Needham Heights, MA: Allyn & Bacon, 2007.

O'Hair, D., H. Rubenstein, and R. Stewart. *A Pocket Guide to Public Speaking*, 2nd ed. Boston, MA: Bedford/St. Martin's, 2007.

8 Visual Communication

OBJECTIVES

After completing this chapter, the student will be able to:

- List reasons visual aids are used.
- Identify examples of visual aids.
- List principles of visual communication design.
- Design a poster.
- Design a flier.
- Design an exhibit.
- Design a computer-generated slide presentation.

INTRODUCTION

Any attempt to reach us through our eyes, whether it is a roadside billboard or a flier posted for a meeting, is considered **visual communication**. *In face-to-face communication, visual communication can take the form of gestures and body language. A gesture such as a "thumbs up" or the body language of a head nod indicates approval, whereas a "thumbs down" or a head shake indicates disapproval. Everyone relies heavily on visual communication, and for good reason: research indicates that it is very effective. People can recall information presented to them with visual images better than if they are just told the information (Figure 8-1).*

In communication not done face to face, visual communication usually takes the form of printed, video, or computer-generated projected messages. In printed form, some examples include signs (traffic signs, billboards, restroom signs), fliers, posters, exhibits, and logos. Chapter 10, Video and Audio Production, emphasizes the visuals used in video production. Projected visuals are those used with computers, using such programs as Microsoft PowerPoint or Apple's Keynote. In this chapter, you will incorporate some concepts learned in Chapter 6, Document

FIGURE 8-1: Visual communication can lead to better recall of information.

Design, to develop effective printed and projected visual communications. This chapter also will discuss some specific applications of visual communication construction: fliers/posters, exhibits/displays, and computer-generated slide presentations.

VISUAL AIDS

Because people are inundated with so many visual messages daily, visual communication needs to be as professional looking as possible. Visual communication is especially important to use when the audience lacks a clear understanding of the topic you are trying to communicate. A **visual aid** is anything used to communicate visually. Visual aids are used for the following reasons:

- *To seize the viewer's attention and help them to focus on major points.* Sometimes it takes an eye-catching visual to grab the viewer's attention in order for the viewer to pay attention to your message.
- *To translate words into meaning.* A graphic, photograph, or chart may be easier to understand than words on a page or someone's vocal description of a topic.
- *To get your point across.* Use visual aids to help educate, inform, and persuade.

In addition to the more traditional forms of visual communication, such as videos and photographs, here are some other examples of visual aids:

- *Your body* can be a visual aid. Body language and gestures sometimes communicate messages more effectively than your voice alone.
- *Objects (real and models)* can be used to communicate in a variety of situations. In a chemistry class, a model of a water molecule can help to show how hydrogen and oxygen bond to form water. If you are showing how to hold a football properly and hold up a real football to show where to put your fingers on the laces, you are using an object to communicate a message.
- *Computer-generated slides* are extremely popular for formal presentations.
- *Posters, fliers, and displays* work well when they feature eye-catching images to draw attention to a text message and, better yet, add a layer of meaning to it.
- *Writing boards (chalkboards, dry erase boards, and smart boards)* are commonly used to

write information and illustrations to support educational presentations.

With any type of visual aid used to communicate, you should consider the following guidelines:

- *Keep your audience in mind.* If your audience is not completely familiar with some of the information you want to communicate, enlighten them. Write on the audience's reading and comprehension level, for example. Understanding your audience's characteristics will help you design the visual aid.
- *Be sure the visual aid contributes to your message.* Poor visual aids can actually detract from the message. This is especially true during computer-generated slide presentations. Animation that causes text to spin and dance on the screen draws attention, certainly, but too much will draw attention to the movement itself, not to the words. Effects that draw attention away from the message are to be avoided.
- *Make the visual aid large enough for the audience to see.* Text must be large enough to read and the images large enough to see. This consistently ranks as one of the major problems with visual aids: image size and text size.
- *Handle materials and operate equipment properly.* Know how your exhibit opens. Know how to operate the computer and the data projector for computer-generated slides. Be sure you can open up the computer presentation file. Operating the equipment and handling materials correctly will add to the professionalism of the message you wish to communicate.

VISUAL COMMUNICATION DESIGN PRINCIPLES

Any visual communication form must be designed properly for your audience. Good visual communication design should attract the audience. Use large, eye-catching text or images, or something unusual (Figure 8-2a and 8-2b). This section will discuss some more specific design principles for visual communication. Please refer to Chapter 6, Document Design, for more graphic design recommendations.

Following are some of the key principles for visual aid design:

- **Simplicity:** A poster, display, or computer-generated slide should not be crammed full of text or images. Keep it simple. Usually—especially with computer slides—the fewer the elements (text, images, illustrations) on the page or screen,

UNCOVER THE TREASURE OF A CAREER IN LEADERSHIP, EDUCATION OR COMMUNICATION

(a)

GROWING THE NEXT GENERATION OF LEADERS, EDUCATORS, AND COMMUNICATORS

(b)

FIGURE 8-2A AND 8-2B: Posters rely on eye-catching images or text to attract potential viewers. In these two examples that promote an academic department at the University of Florida, the posters use large images and attention-grabbing phrases to get people's interest. (*Courtesy of Lisa Hightower.*)

the easier it is for your audience to understand and recall your message.

- **Unity:** Work to enhance "harmony" in all of your visual communication. If you are going to design several different posters for an event, use the same text font style or the same image on all of the posters to show "unity." Keep the background and text the same and use similar graphic images on all the slides for a computer-generated presentation.

- **Emphasis:** Although "unity" stresses similarity and harmony among your materials, some items must be emphasized to bring out different, but important, information. This is where the concept of "emphasis" comes in. Focus the audience's attention to a specific place on your poster, exhibit, or slide by using some of the following elements. However, use common sense so that you do not use too many of these ideas. If *everything* is emphasized, then *nothing* is truly emphasized.

  - *Animation:* On some slides, animation or transition "reveals" might draw attention to text or images on the screen.

  - *Color:* Placing some text in a different color than the rest of the text will immediately draw attention to the page or screen. Color images also draw attention.

  - *Underline, italics, and boldface:* When used consistently, these are excellent attention-getters. Pick one or assign them to different jobs. (For instance, you could use boldface for headings, italics for unfamiliar terminology, and underline for emphasis.) If you use all three indiscriminately, the audience may become confused.

  - *Bullet points:* For some posters, exhibits, and computer slides, bullet-pointed text helps readers digest information quickly. People have become accustomed to bullet points and automatically direct their attention to bullet points because they believe that important information will be presented there.

- **Readability** is usually measured in elementary-school grade levels. Readable text uses simple, short sentences and explains unfamiliar words and concepts carefully. In addition to writing the content at or below the educational level of your audience, try some of these strategies to enhance readability:

  - *Bullet points:* In addition to being an emphasis element, bullet points also help people understand information, especially if the bullet-pointed information is concise.

  - *Underline, italics, and boldface:* Use these to draw attention to certain parts of your materials, but do not overdo specialized font styles, especially italics. Italicizing a large amount of text makes it difficult to read, especially on computer slides.

- *Uppercase and lowercase:* Always use uppercase and lowercase text, particularly on computer slides. The only time you would want to type in all uppercase would be to "shout" out information or for a special title for a poster, exhibit, or computer slide. For regular content, use uppercase/lowercase.
- *Typeface:* Typeface (also called type or font) is covered in detail in Chapter 6, Document Design. Make sure you choose a font style (serif or sans serif) that is easy to read. For posters or exhibits, you may want to use a thicker font—such as a sans serif font—to make text more easily readable from a distance. For projected materials, it is usually better to use a sans serif font style.
- *Letter size:* The rule here is that the letter size must be large enough to be read at a reasonable distance. For an exhibit, that distance is usually six feet away. For a poster, the distance may depend on where the poster is placed, but usually, you would want all of the content to be seen from at least six feet away. Details about letter size for projected materials (computer slides) will be presented later in this chapter.
- **Organization:** Use a logical visual pattern that is easy to comprehend. For most visual communication, that visual pattern will begin at the top, left-hand side of the document or slide and work to the bottom, right-hand side, which is the English language reading pattern (top left to bottom right). What this means is that important information should probably be placed at the top, usually on the left, and the least important information should be placed at the bottom, on the right.
- **Balance:** To achieve a balanced design, imagine a line dividing a page either vertically or horizontally and then place visual elements so they are in either symmetrical (formal) balance or asymmetrical (informal) balance:
  - **Symmetrical balance**, or *formal balance*, is mirror-image balance. All of the visual elements on one side of the page are mirrored on the other side.
  - In **asymmetrical balance**, or *informal balance*, several smaller items on one side of the imaginary line are balanced by a large item on the other side, or smaller items are placed further away from the center line than larger items. Refer to Chapter 6, Document Design, for more information about balance.

- **Accuracy:** All text should be spelled and punctuated correctly. Sentences should be grammatically correct, although fragments are permissible, especially in bullet points. All names should be correct.
- **Parallelism:** Items in a group must match one another. If your first bulleted item begins with a verb, so must all the others.
- **Clarity:** Present only one main idea. People should know—even at a glance—what your poster, display, or computer slide is about. The content should be clear.

POSTERS, FLIERS, DISPLAYS, AND PROJECTED MATERIALS

Any visual communication should be developed with the design principles and elements previously discussed. In the remainder of this chapter, three particular visual communication methods—posters/fliers, displays/exhibits, and projected materials—will be described.

Posters and Fliers

A *poster* is a sign placed in a public place as an advertisement or notice (Figure 8-3). Posters can be effective for publicizing events, such as meetings, contests, fairs, and dances. Because most people just glance at posters, you must create something catchy so that people will be more apt to read the entire poster and come away with the intended message. To make an effective poster, you must carefully select the words and visuals to communicate the message. Here are some suggestions:

- *Design and layout:* Determine first whether the heading or an image will dominate. Then, create the poster with these three important elements on the page: the heading, the image, and the body (text). Arrange those three elements until you have the look you believe will best communicate your message.
  - *Do a good draft.* It is a good idea to use a piece of paper to sketch out the design of the poster or you can use a document design program such as Adobe InDesign. When the sketch looks as you want it, put the elements together—photographs, graphics, text—in a document design program.
  - *Use white space.* Do not fill up the entire page with text and images. Leave space at the edge of the poster so that it does not look crowded.

Liz Felter

Extension Agent, Horticulture, University of Florida's Institute of Food and Agricultural Sciences, Orange County

How do you use posters and fliers in your organization?

We use posters and fliers in many ways. A poster is sometimes a mini-exhibit. It's attractive and, in three or four bulleted points, gets a short but concise message out to the viewer. The poster may be a way of making the viewer aware of an educational program, it may teach updated information on the latest research, or it may summarize an activity that recently happened. Fliers are primarily used to promote a program or class. They can be mailed or e-mailed to current customers or attract new customers. Fliers have been used as a recruitment tool when attracting new employees or volunteers.

- *Text:* The heading (title) of the poster should be short—one to five words long—and the letters should be from two to four inches tall. If the poster is an advertisement for a local fair, the largest text should be something about the fair, such as "It's Fair Weather!" or "Visit the County Fair!" For a poster for a local theatre production, only the title of the play should appear in the poster header. The rest of the words on the poster should convey your message as concisely as possible. Use short sentences—fragments are encouraged.

  - Place the least important information at the bottom and in a smaller size than the most important information, but even the least important information should be easily readable.
  - Event sponsors usually request that their logos be placed on the poster advertising an event. In that situation, put the sponsorship information in the least obtrusive spot, usually at the bottom of the poster.
  - Use colors that contrast with the background or paper color. Even for a poster designed in black and white, you can make text stand out by making the text white and placing it on a black background.

BROWARD COUNTY
4-H
Head. Heart. Hands. Health.
Building Leadership, Citizenship, and Life Skills in Youth Ages 5 to 18

FIGURE 8-3: This promotional poster banner is more than six feet wide and is used to promote Broward County 4-H (Broward County, Florida) at events and is also displayed in the lobby of the Broward County Extension Education Office. (*Used by permission of Angelina Toomey, University of Florida's Institute of Food and Agricultural Sciences, Broward County.*)

- *Images:* Because most posters are at least 11 inches by 17 inches, images should be large. Select images that are easy to see (Figure 8-4). Photographs should be at least 300 pixels per inch (ppi). If the resolution is lower, the photograph will look pixilated. For that reason, it is usually not a good idea to download photographs from the Web for use on your poster, unless you know for sure that you have at least 300-ppi images. For more about pixilated photographs, refer to Chapter 9, Digital Photography and Photographic Editing.

FIGURE 8-4: Large images and large text can motivate people to look at a poster. In this example, a high-school theatre arts program is promoting its production of the musical "Oklahoma!" The large photograph and the word "Oklahoma!" draw people's attention to the poster, and then the viewer can read the smaller text beneath, which explains when and where the performances will be. (*Adapted from design by Scott McPherson.*)

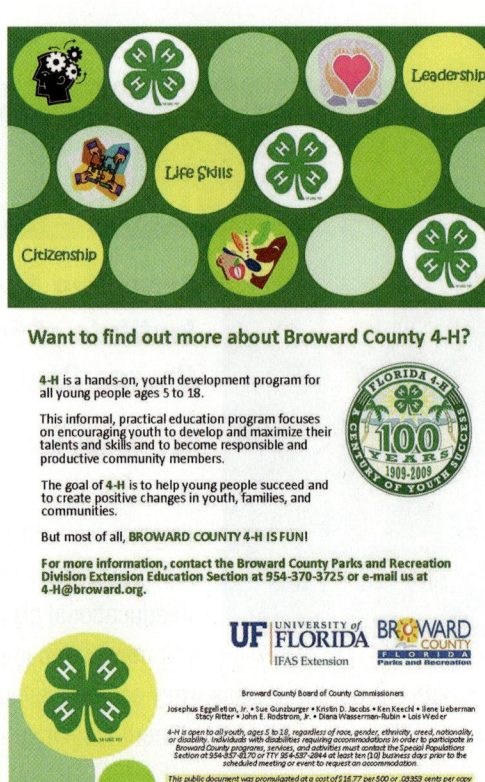

FIGURE 8-5: This flier is used to promote Broward County 4-H (Broward County, Florida) at events and is also distributed to various municipalities and schools throughout the county. Broward County 4-H utilizes three versions of the flier: English, Spanish, and Creole. (*Used with permission from Angelina Toomey, University of Florida's Institute of Food and Agricultural Sciences, Broward County.*)

■ **Background:** The poster's background should be as simple as possible to avoid distracting the audience's attention from the message you are communicating in your carefully chosen words and images.

Fliers are small posters, usually placed on bulletin boards and typically used as advertisements. They are often printed on a single sheet of paper or smaller and are intended for wide distribution (Figure 8-5). The same concepts used to develop posters are applicable to fliers. However, because fliers are smaller than posters, not as much content can be included on fliers, and the images and text must be smaller. To make your flier stand out, you must emphasize what is most important in order to grab people's attention. For example, if you are designing a flier to announce that an organization is going to hold a meeting next week, you may want to place the name of the organization at the top, to include a large photograph, or to begin by saying "FREE FOOD WILL BE PROVIDED." Then fill in with other information, such as the date, time, and location of the organization's meeting.

You can use computer slide-making software, such as Microsoft's PowerPoint and Apple's Keynote, to create posters and fliers. In either software program, you can go to *File > Page Setup* and create the custom size of the poster. Then start laying out the poster. Again, make sure that the images you insert are 300 ppi so that when you print the flier or poster, the image will not be pixilated.

Displays/Exhibits

In the context of this chapter, *displays* or *exhibits* are freestanding structures containing many media (photographs, text, illustrations, real objects, video,

Angelina C. Toomey

Extension Agent, 4-H Youth Development, University of Florida's Institute of Food and Agricultural Sciences, Broward County Parks and Recreation

How do you use posters and fliers in your organization?

Broward County 4-H not only uses posters, banners, and fliers to promote the organization itself, but we also distribute such forms of visual communication to promote specific events, to educate the public, and to recruit new 4-H members. Multilingual posters and fliers are created in order to appeal to Broward County's diverse audience. Fliers are the most predominant form of visual communication, as they are easily dispensable and cost-efficient to reproduce. Broward County 4-H prefers to use eye-catching fliers with clear and concise content.

Liz Felter

Extension Agent, Horticulture, University of Florida's Institute of Food and Agricultural Sciences, Orange County

How do you use displays in your organization?

We use displays to provide information in a fun and informative manner. They allow us to get a message to the viewer in an attractive and succinct way. It is important to remember that viewers are usually walking past the exhibit rather quickly, so you'll want to catch their attention and get to the point of the message as soon as possible.

or computer screens) in order to communicate a message to an intended audience. For this chapter, the focus will be on simple displays, and the words "display" and "exhibit" will be used interchangeably.

Displays are frequently used in schools. In a display made for a school science fair, for instance, the display tells the story of a student experiment. This may mean using photographs, graphs and charts, text, and

Angelina C. Toomey

Extension Agent, 4-H Youth Development, University of Florida's Institute of Food and Agricultural Sciences, Broward County Parks and Recreation

How do you use displays in your organization?

Displays are an integral part of the Broward County 4-H program. Oftentimes, it is through these methods of communication that the organization represents itself at large-scale community events. Displays help tell the story of Broward County 4-H, including the vision and mission of the organization, through photographs and other images and text, as well as through different demonstrations and activities. Because 4-H focuses on "learning by doing," we try to incorporate hands-on activities into all of our exhibits. For Broward County 4-H, unmanned displays are equally, if not more, important than those that are staffed. This is because the display must speak for itself, thereby engaging passersby through visual design and written content only. Broward County 4-H also utilizes exhibits as marketing, promotion, and recruitment tools, so it is important that our communication efforts portray a positive image of not only Broward County 4-H itself, but of our parent organizations.

possibly a model or some other real object. Exhibits and displays at museums and fairs can be extremely elaborate; some exhibits at fairs can include gardening demonstrations or full-scale petting zoos. Complexity can be an asset in this case, since viewers are free to begin at any part of the exhibit and spend time on the parts of the exhibit that interest them.

Here are some tips to consider when designing your display:

- *Determine the purpose of your display.* Displays are used in trade shows to help sell products. They also can be used to educate a particular audience about a topic. The purpose must be determined before selecting the media to communicate the message.
- *Determine the purpose and content (the message) you want to communicate.*
- *Consider the actual structure, size, and shape of the display.* You will need to know these components in order to design an effective exhibit. Exhibits can be floor-to-ceiling or

tabletop in size. Exhibits can come in different shapes, from three-panel designs to curved designs. You can build your own display or have one made for you. Displays can be constructed with foam core (lightweight), corrugated cardboard, plywood paneling, and other materials (Figure 8-6). Prefabricated exhibits usually cost more, but most come with heavy-duty cases and are built for durability. If you know you will reuse parts of your display's text or images that are attached to the display, you may wish to have them laminated (Figure 8-7). It will cost more, but the images or text will be able to be used many times.

- *Sketch out the display's design.* Before you print out text or photographs, save yourself some problems by drawing the display on paper. If a display tells a complete story, it should begin at the top left and work to the bottom right, filling in important information along the way. Sketch out this information "flow" on paper before building it. Know how the display will look, work, and be laid out before the construction process begins.

FIGURE 8-6: Displays do not have to be made of expensive materials to be effective. This photograph shows how a cardboard display can catch viewers' attention with images and color.

FIGURE 8-8: This photograph shows an example of a tabletop display using large images and text to grab viewers' attention.

FIGURE 8-7: This photograph shows a display being assembled. By having large sections of the display laminated, with hook-and-loop fasteners on the back of the sections, a display can be assembled and disassembled quickly.

- ***Determine the layout of text and images.*** As with the other visual communication methods, in displays it is vital to tell your story visually. You will have content you want to get across to your audience, but do so in a way that minimizes words and maximizes visuals (Figure 8-8). Images and text must be large. Use big, easily deciphered, bold images. Do not use pixilated images. Choose attractive colors.

- ***Avoid regular paper (8.5-by-11-inch), if possible.*** Do not put all of your text on 8.5-by-11-inch white paper. If you have a large enough budget, use a local copy company to print out your text content on larger paper sizes. Some schools and universities may have large printers for you to use. If you are limited to 8.5-by-11-inch paper,

make the text large and use multiple pieces of paper to tell your story. Get photos printed on photographic paper or have them developed at a photographic services location in your area. The colors in photographs printed on office paper are usually washed out and difficult to see.

- ***Determine which media you plan to use.*** Displays regularly feature printed materials and photographs, but they can also incorporate other media, such as video and computers. If you use either of these, consider these questions: Will people be able to see the video or computer screens? Will the volume be turned up loud enough for people to hear, and if it is turned up, will the noise interfere with the booth next to yours?

- ***Decide if you will use a table and, if so, how that will affect the design of the display.*** If you use a table, be sure to place your images and text high enough so that the table does not cover the display's content. Will you include items on the table for people to take away with them (giveaways, brochures, newsletters)? How will the table change the flow of traffic through and around your exhibit?

- ***Train enthusiastic, helpful people who will staff the display.*** A display rarely tells a complete story. Have people staff a display booth to answer questions posed by passersby. Choose friendly staff with a genuine interest in your message who will connect with the audience. Remind them to dress professionally (Figure 8-9).

- ***Scout the location.*** In order to design your exhibit, you must consider the location where it will be on display. If possible, scout the location to be sure it meets your needs. If you want to include

Liz Felter

Extension Agent, Horticulture, University of Florida's Institute of Food and Agricultural Sciences, Orange County

How do you use computer-generated slide presentations in your organization?

We use computer-generated slides as a teaching tool. Through pictures and few words, we can show class participants photos of plant material or proper fertilizing techniques. The photos are more interesting to look at, and a picture can help explain the text being included with that particular slide. It is important to incorporate ways to be interactive in a computer slide presentation in order to keep attendees engaged in the program.

FIGURE 8-9: This photograph shows the importance of staffing the display area with people who are knowledgeable about the topic so they can help visitors.

a television and DVD player, for instance, you will need an electrical outlet within reach. What about lighting? Will your exhibit be in a location that has high traffic flow (people walking by)? Plan ahead so that you can leave enough space in your exhibit for people to walk through. If it looks inconveniently crowded, some potential visitors may pass it by and you may lose part of your audience.

A recent development in display construction is the *retractable banner*, a long, vertical display that pulls up from a base on the floor and attaches at the top to a long, featherweight pole. Retractable banners are very light and portable, allowing you to make a big visual impact almost without effort. The banners can be expensive, however, so for a temporary display, they might not be the best option. Many organizations have adopted them for displays, or display elements, whose messages will not change—such as a description of the organization.

Projected Materials

You enter a room where the lights are dimmed slightly to listen to a presentation. The speaker begins to address the audience and uses a projector to cast computer-created slides onto the screen. But try as you may, you cannot decipher much of the text on the projection screen. The room is too bright, the words on the screen are too cramped, and the dark-color lettering blends in with the dark background. You spend most of the presentation trying to determine what the projected materials say, instead of listening to the speech.

Does that sound familiar? Because computers and high-end data projectors are becoming more commonplace in schools, universities, and corporate offices, poor presentations like this happen frequently, but they do not have to. Poorly developed projected visuals distract your audience's attention from what is important—your presentation and your message. The following details ways to make your presentations more effective by using computer-generated visuals correctly.

- *Minimize words.* Cramming lots of material onto one computer slide makes the slide extremely difficult to read. Instead, use bullet-pointed information. Infuse your slides with pictures, clip art, or graphics, and break the content into multiple slides. Illustrations, audio, video, graphics, and photographs need a purpose. If the image, audio clip, video clip, or photograph does not aid the audience's comprehension of your presentation, then do not use it. Include only visuals that tell your story.

- *Use the correct image size.* Photographs used for projected presentations should be at least 72 pixels per inch (ppi). If, however, you plan to print the slides as handouts, you will need to use photographs that are 300 ppi so that they will reproduce well in printed form. For more information about image size and resolution, refer to Chapter 9, Digital Photography and Photographic Editing.

- *Take into account the presentation location.* If you possibly can, rehearse your presentation in the room you will give it in so that you can adjust the slides to fit the conditions of the room (Figure 8-10). If the light level cannot be dimmed satisfactorily, change to a white background with black or dark letters. In a light room, dark text is easier to read.

© Cengage Learning 2012

FIGURE 8-10: Computer-generated slide programs can be used effectively in the classroom or in business settings.

Using Text

When using text on computer-generated slides, follow these suggestions:

- *The "rule of six":* Limit words to six per line and limit lines to six per screen. Although some slides will have more or fewer words per line and lines per screen, the "rule of six" is a good one to follow.

- *Amount of content per slide:* Remember that people read more slowly than they hear, and they will read more slowly still if they are being asked to read and listen at the same time. If you want to present an audience with a complicated sentence to read, stop talking for a few moments and leave several beats of silence while the audience devotes attention to deciphering your sentence. In general, save lengthy sentences for the spoken part of your presentation and use bullet points or key words on your slides.

- *Contrasting colors:* The use of contrasting colors is an extremely important consideration. Use a dark background with light letters or a light background with dark letters. Avoid backgrounds with dark and light colors swirled together. Light-colored lettering is difficult to read when it crosses onto light backgrounds; similarly, dark-colored lettering is difficult to distinguish when it blends with a dark background.
  - *Letter colors to use:* Use white or yellow letters on a dark background or black letters on a white or light-colored background.
  - *Letter colors to avoid:* Red, for text, should be avoided, especially if you use a dark background. Red tends to bleed into other colors, making text difficult to read. Avoid colors that are similar to each other. For example, reds and dark greens are difficult to distinguish when projected. Never use red text on a dark green or dark blue background.

- *Text size:* The text size for your title, body, and subheadings should be in the following ranges:
  - *Titles:* 28–48 points
  - *Body:* 24–32 points
  - *Subheadings:* 20 points or higher. A 20-point font size is the absolute minimum for text projected to a large audience.

- *Typeface:* Studies on readability have shown that people can read sans serif fonts, such as Helvetica and Arial, easier on projected screens than serif fonts, such as Times New Roman. Therefore,

Angelina C. Toomey

Extension Agent, 4-H Youth Development, University of Florida's Institute of Food and Agricultural Sciences, Broward County Parks and Recreation

How do you use computer-generated slide presentations in your organization?

Computer-generated slide presentations are also vital to Broward County 4-H, with respect to promotion, marketing, and recruitment. Often, a basic 4-H presentation will be utilized to educate the public about the Broward County 4-H program and the various youth development activities we offer. Our organization creates several different versions of each presentation, in order to reach publics of varying ages and languages. Broward County 4-H also uses computer-generated slide presentations to train volunteers and to facilitate programs for youth, including Financial Literacy training and Stranger Danger education, as well as to orientate judges for all of our competitive events.

it is better to use a sans serif typeface. Refer to Chapter 6, Document Design, for information on serif and sans serif typefaces.

- *Italics:* On your computer screen, italicized words are easy to read. But when they are projected, audiences find them difficult to make out, so avoid italics.
- *Uppercase/lowercase:* It is difficult to read all uppercase letters on a projected screen, and the e-mail convention that all-caps type is shouting is quickly becoming generalized, so it is best to use a combination of lowercase and uppercase letters.

Special Effects

Computer slide-making software comes with lots of special effects for transitions between slides and for audio. The overall rule is to use special effects only when necessary. Special effects should have a purpose.

- *Sound effects:* Use sound effects sparingly! Audio special effects can make a serious presentation seem humorous when it is not supposed to be. Your software may offer the sound of a racing

car, glass breaking, lasers, or a typewriter. Pause and reflect. Ask yourself: "Why do I need one of these effects?" If you cannot answer the question with a meaningful reason, then do not use the audio effect.

- *Reveals and transitions:* Software offers many choices of text reveals and transitions. Your words can dissolve onto the screen or zoom on from the right, left, bottom, or top. They can pinwheel for a moment and then suddenly slam into focus, or they can drop letter by letter from above. Again the advice is to use these features sparingly. Your audience may get so enamored with the reveals and transitions that they lose sight of the main points in your presentation.
- *Background templates* also are a factor in your presentation consideration. Choose backgrounds that add to your message, not detract from it. Make sure people can read the text and view the images against the background. You also can create your own background using such programs as Adobe Photoshop, or you can buy backgrounds on CDs or from online sources.

Using Computer-Generated Slides for Education

Many creative teachers around the country design computer slide presentations that focus on involving students, as well as activities that instruct students on how to use computer slide-making software to express their own ideas as part of a project or group assignment. Such activities can capitalize on the strengths of this presentation software as an easy-to-use visual display medium that can engage visual learners and capture students' attention and interest. Computer slide-making software's ability to incorporate photos, graphics, animation, and even sound and video meshes well with the more visually oriented aspects of the agriculture curriculum. Teaching students how to use computer slide-making software can be an effective method to enhance their presentation and organization skills.

Like any other teaching tool, effective use of computer slide-making software requires that you have goals and objectives in mind that presentation software can help you achieve successfully. Some ideas to get you started thinking about how you can use computer slide-making software include:

- Diagrams and charts
- Capturing "processes," such as photosynthesis or plant propagation, visually
- Instructions for class activities
- Class notes
- Test reviews
- Question-and-answer or role-playing games
- Class debate and discussion questions
- Student book reports
- Multimedia biographies, using pictures, and recorded audio of famous quotations
- Student group projects and reports

SUMMARY

This chapter focused on how to create visual aids. Visual aids can be used alone or in combination with such communication methods as print documents and public speaking. Designing effective posters, fliers, displays, exhibits, and projected materials can help you get your message across to various audiences.

CHAPTER EXERCISES

APPLICATIONS

Following are some ways you can apply what you learned in this chapter:

- Design a poster or flier for an upcoming event.
- Design an exhibit for an agricultural education event.
- Invite a local special event planner to discuss how conventions are conducted and how exhibits are organized during conventions.
- Create a computer-generated slide show about a current event in the news.

CHAPTER QUESTIONS

MULTIPLE CHOICE:

1. Which of the following is NOT a reason that visual aids are used?

   a. To seize the viewer's attention and focus on major points

   b. To translate words into meaning

   c. To get your point across—use visual aids to help educate, inform, and persuade

   d. All of the above are reasons visual aids are used.

2. Using bullet points, uppercase and lowercase text, typefaces, and letter size enhances which of these design principles?

   a. Organization

   b. Balance

   c. Readability

   d. Accuracy

3. Using animation, color, bullet points, underline, italics, and boldface enhances which of these design principles?

   a. Emphasis

   b. Readability

   c. Harmony

   d. Unity

4. _____ are small posters, usually placed on bulletin boards.

   a. Displays

   b. Exhibits

   c. Fliers

   d. Computer-generated slides

5. A recent development in display (exhibit) construction is the _____.

    a. use of foam core

    b. retractable banner

    c. incorporation of DVD videos

    d. use of computer-generated slides

6. A good rule for projected materials is to limit words to _____ per line.

    a. six

    b. seven

    c. ten

    d. three

7. Research studies show that people can more easily read _____ on projected materials.

    a. serif fonts

    b. sans serif fonts

    c. symmetrical fonts

    d. asymmetrical fonts

8. Which of the following is an example of a visual aid?

    a. An object

    b. Your body

    c. A poster

    d. All are examples of visual aids.

9. In general, which color should be avoided for text on projected materials?

    a. Yellow

    b. Blue

    c. Red

    d. Black

10. The resolution of photographs used for posters should be at least _____.

    a. 50 ppi

    b. 72 ppi

    c. 100 ppi

    d. 300 ppi

11. Titles for projected materials should be between _____ points in size.

    a. 15 and 20

    b. 20 and 24

    c. 24 and 32

    d. 28 and 48

12. The "rule of six" refers to _____.

    a. limiting the number of words and lines on a computer-generated slide

    b. speaking slowly when a lot of content is shown on the screen

    c. using contrasting colors on computer-generated slides

    d. allowing only six slides to be used in a presentation

13. Enhancing "harmony" in all of your visual messages is a way to describe _____.

    a. simplicity

    b. unity

    c. organization

    d. readability

14. _____ refers to using a logical visual pattern that is easy to comprehend.

    a. Simplicity

    b. Unity

    c. Organization

    d. Readability

15. Text that is more _____ uses simple, short sentences and explains unfamiliar words and concepts carefully.

    a. unified

    b. organized

    c. readable

    d. contrasting

REFERENCE AND FURTHER READING

Kimball, M. A., and A. R. Hawkins. *Document Design: A Guide for Technical Communicators*. Boston, MA: Bedford/St. Martin's, 2008.

9

Digital Photography and Photographic Editing

OBJECTIVES

After completing this chapter, the student will be able to:

- Select a digital camera for the needs of an individual or of an organization.
- Identify key terms in digital photography.
- Compose a good photograph.
- Identify key terms in digital photo editing.
- Explain the difference between RGB and CMYK color formats.
- Explain the differences among JPG, TIFF, and RAW file formats.

INTRODUCTION

The old saying goes, "A picture is worth a thousand words." That may very well be true. Seeing an image of someone winning an award, for example, can be more exciting than reading about it in a news story. Because it seems that almost everyone has access to a digital camera—either with an actual digital camera or with a cell phone that takes pictures—just about everyone thinks they are photographic experts. Simply owning a camera, however, does not make someone an expert. It takes practice to be able to take photographs that tell great stories (Figure 9-1). In fact, the word **photography** *means "drawing or writing with light."*

This chapter will help you be better able to "write with light." Because not everyone has the same type of camera, this chapter focuses on general techniques, rather than specifics on how to operate a particular digital camera. This chapter will introduce you to some of the basics of digital photography and photographic editing so you can take great photographs.

© Cengage Learning 2012

FIGURE 9-1: Learning the basics of digital photography will help you take great photos for both your personal use and your organization's activities.

Jim Bret Campbell

Senior Director of Publications, American Quarter Horse Association

Digital photography has transformed the way news is edited and distributed. While the principles of good photography remain the same, managing the workflow of capturing, downloading, processing, and cataloging images is what separate great news outlets from the ones going broke. If you're looking at entering the communications field as a photographer, you need to develop your vision as a great photographer, a technogeek's love of technology, and the discipline of an engineer. That way you find the right image, have the ability to get it to the public—and you can find it again in the future.

HOW DIGITAL PHOTOGRAPHY WORKS

Digital cameras have, by and large, replaced film cameras. Digital cameras provide options that film cameras do not. With digital cameras, you can immediately view photos that you have taken. You can take several photos, view them in the camera, and then decide which ones to keep and which ones to delete. Photo files can be shared easily over e-mail or on Web and Facebook pages. You can print only the photos you want, and, because the photos are digital files, they can be stored easily on CDs, hard drives, or other storage devices where the files will not degrade. In the photo editing process, a digital photo image can be modified and manipulated much more easily and quickly than traditional film negatives. One disadvantage of digital cameras is that because digital cameras are so small, they are easily dropped. However, because they are inexpensive, it is usually more cost-effective to buy a new one than to have a broken one repaired.

Digital cameras store images in the form of millions of tiny picture elements called **pixels**, which is short for "picture elements." A pixel, simply put, is a single point of light on the screen of a monitor. The term **megapixel**, which means "one million pixels," is often used in connection with digital cameras. A camera that shoots photographs of 6.2 megapixels shoots digital images of roughly 6.2 million pixels at the highest resolution.

Numerous cameras allow you to take photographs that are many megapixels in size, and these cameras can be expensive. If the camera is to be used mainly to take photographs that are about 4 by 6 inches, a 3- or 4-megapixel camera is probably all that is needed. To take 8-by-10-inch photos, the camera should be at least 5 megapixels. For advanced photography, you may need a camera that takes photos of 8 or more megapixels. Most cell phone cameras take 1- to 2-megapixel photographs. If all of the images are to be placed on the Web, a camera with fewer than 3 megapixels may be suitable, but if the images are to be used in publications, use a camera of at least 5 megapixels or more. Most consumer-grade cameras shoot at least 5-megapixel photographs.

In addition to determining how many megapixels are needed, it is also necessary to determine if you need a camera that has interchangeable lenses (zoom, telephoto, and wide-angle lenses) or a camera with a built-in zoom lens (Figure 9-2). Cameras made for interchangeable lenses are called SLR cameras, for "single lens reflex." Digital cameras with interchangeable lenses are called DSLR ("digital single lens reflex"). Cameras with interchangeable lenses cost more, but

FIGURE 9-2: The camera on the left has interchangeable lenses. The one on the right has one lens.

© Cengage Learning 2012

FIGURE 9-3: Tripods and monopods are important accessories to shoot a steady photograph.

© Eremin Sergey/www.Shutterstock.com

you can get just the right lens for just the right photo. For standard photos, a camera with a built-in zoom lens may be all you need. Almost all digital cameras have built-in flashes, too.

Liquid crystal display (LCD) monitors are small color screens built into most cameras. Most have brightness adjustments that can be changed either manually or automatically. These screens range between 1 and 3 inches in size. Viewfinders are smaller monitors built into digital cameras. An advantage to using a viewfinder instead of an LCD monitor is that a viewfinder shows the same area as the camera's zoom lens, so you can capture the same area that you see in the viewfinder.

A tripod or monopod is a useful piece of camera equipment (Figure 9-3). Also, it is a good idea to have a camera bag packed with extra batteries and memory cards. The camera bag also can hold extra camera lenses for cameras with interchangeable lenses.

Digital camera resolution is usually measured in megapixels—a raw counting of the number of pixels in the digital image created by the camera. Although photographs taken with low-resolution cameras (1 to 2 megapixels) are fine for viewing on television or computer screens, they do not work well for print documents. Cameras with higher resolution (preferably 5 megapixels or more) are necessary for print-quality photographs.

Numerous standard digital cameras allow you to save photographs at varying levels of resolution: "Basic," "Fine," and "Superfine," or possibly "Good," "Better," and "Best." The camera manual should describe exactly how to change the resolution and how many pixels are associated with each setting. High-resolution images take up more space on the camera's memory card. High-resolution photographs have low compression rates, resulting in larger file sizes and better-looking images.

The basic rule is that if photographs will be used for print documents, save at the highest resolution possible. The resolution can always be lowered during the photographic editing process if necessary, but it is not possible to increase the resolution of an image that was saved at a low resolution without the image looking distorted. Low-resolution images are sufficient for images that will only be viewed on computer screens, such as on the Web or on Facebook pages.

To output the finished photograph, the output resolution must be set. This is done in **pixels per inch (ppi)**. For printed materials (publications or actual photo prints), the photograph needs to be output at no less than 300 ppi. If the photograph is going to be placed on the Web or e-mailed, the resolution can be as low as 72 ppi because a computer screen can only show 72-ppi resolution. A discussion on resolution is provided later in the chapter in the section titled "Resolution and Resampling." Table 9-1 provides more information on ppi.

Also keep in mind that a digital camera's aspect ratio may be different than that of a traditional film camera (Figures 9-4a and 9-4b). A film camera's aspect ratio

SAVE YOUR PHOTOGRAPHS AT THE FOLLOWING RESOLUTION SETTINGS (IN PIXELS PER INCH) FOR PRINT AND THE WEB:	
For the Web or Video	72–100 ppi
Black-and-white photos	150 ppi
Full-color photos	300 ppi

TABLE 9-1: Pixels per Inch (ppi).

(a)

© Cengage Learning 2012

(b)

© Cengage Learning 2012

FIGURES 9-4: Notice that the size of these two photos is not exactly the same. Figure 9-4a was shot with a standard film camera's aspect ratio of 3:2. Figure 9-4b is a regular digital camera's aspect ratio of 4:3.

is 3:2, which means the image is three units wide by two units tall. That is why the 4-by-6-inch print emerged as a standard size. Many—but not all—digital cameras have an aspect ratio of 4:3, which means the image is four units wide by three units tall, just like the aspect ratio of a computer monitor. To get a 4-by-6-inch image from a digital camera with a 4:3 aspect ratio, the photograph will need to be cropped in the photo editing stage.

SELECTING A FILE FORMAT

Your camera may offer a choice of file formats for your saved images. The selected format determines how the camera records and stores all of the bits of data that make up a digital photo. Several formats have been

developed for digital images, but the most popular for digital camera manufacturers are JPG, TIFF, and RAW (Figure 9-5).

JPG (pronounced *JAY-peg*) is short for Joint Photographic Experts Group, the organization that developed this file format. JPG is the leading photographic file format because it takes Web-friendly photos; all Web browsers and e-mail programs can display a JPG image. JPGs also are smaller in file size than other formats. The disadvantage of JPG images is that the images are saved with a process that eliminates, or compresses, some image data. If you record your images as JPGs, you should try to save the image at the highest resolution possible to minimize the compression.

Photo File Formats

Many file formats have been developed for digital images, but the most popular for digital camera manufacturers are JPG, TIFF, and RAW.

JPG: Joint Photographic Experts Group
Smaller files
Web-friendly photos
Leading camera format
Compressed file format

TIFF: Tagged Image File Format
Large files
Not compressed much
Cannot be displayed on some Web browsers.
Used for photos in high-end document design.

RAW: does not stand for anything. It is just a word.
RAW is not a standard format; each camera manufacturer uses different specifications for its own RAW format.
Records data straight from the camera's sensor, just the way it looks on the sensor, onto the camera's data card.
Files are uncompressed.
Extremely large files.
Used mainly by professional photographers.

© Cengage Learning 2012

FIGURE 9-5: Different file formats.

TIFF stands for *tagged image file format*. TIFF files are much larger than JPGs because TIFFs do not compress files very much. TIFFs cannot be displayed on some Web browsers. A TIFF file must usually be opened in a photo editing program and converted to a JPG before the image can be shared on the Web. TIFF is for photographers who are concerned with maintaining image quality. For most amateur photographers, though, TIFFs are not used very often. TIFF is the preferred format for image files for print publications.

The last image format is **RAW**, which is not a standard format, such as JPGs and TIFFs. Each camera manufacturer uses different specifications for its own RAW format. A RAW file records data straight from the camera's sensor, just the way it looks on the sensor, onto the camera's data card. The files are uncompressed, meaning they are larger than JPG files. Professional photographers tend to use the RAW format.

LIGHT, COLORS, AND WHITE BALANCE

Digital cameras can be adjusted for different types of light. Light sensitivity is measured using a scale called **ISO**, which stands for International Standards Organization. Most digital cameras provide a choice of ISO values, usually 100, 200, 400, and so on. Higher-end cameras have even more choices. As the ISO value increases, the camera becomes more sensitive to light. Therefore, in a low-light situation, photographs should be shot using a high (800 to 1,600) ISO value. However, recording images at high ISO values all the time is not recommended. Higher ISO values produce images that look grainier than images shot at lower ISOs. Most cameras' automatic ISO features yield good results.

In addition to paying attention to the amount of light, consider the light source. Different kinds of light have different color qualities, commonly called **color temperature**. This is a way of saying that the light sources contain different amounts of red, green, and blue light. For example, sunlight tends to be blue, a regular light bulb (incandescent) tends to be more yellow or orange, and a fluorescent bulb tends to be green. Your eyes adjust to changes in color temperature so the colors with different light sources look the same, but digital cameras do not adjust so easily. A camera must be white balanced to correct color temperature problems.

White balancing tells the camera what combination of red, green, and blue light it should perceive as white, given a particular lighting condition. Most cameras have an automatic white balance feature, but this feature can sometimes become "confused," particularly if the scene being shot features a single dominant color or includes different types of light (sunlight streaming into a room lit with fluorescent light). In this situation, it may be necessary to adjust the white balance manually. Most cameras include white balance presets for normal types of light: daylight, daylight with clouds, incandescent, fluorescent, and flash. If your camera does not offer white balance adjustments, you can remove unwanted colors in the photo editing stage.

DIGITAL PHOTOGRAPHY COMPOSITION TECHNIQUES

Now that you understand some of the technical components of digital photography, it is time to take some pictures. **Composition** is organizing the subject—the person or object of the photograph—through the viewfinder. Practice the following composition techniques for better-looking photos. Start with holding the camera properly.

Holding the Camera

The quickest and surest way to get a sharp, clear picture is to hold the camera correctly (Figure 9-6). Blurred pictures are caused most frequently by moving the

© Brian Eichhorn/www.Shutterstock.com

FIGURE 9-6: This man is holding a camera correctly, which is the best way to get a sharp, clear picture.

Lisa Hightower

Coordinator, University of Florida's Scientific Thinking and Educational Partnership Program

A camera in the hands of a skilled photographer can tell a story as vivid, candid, and descriptive as any painting. With the onset of digital photography and software like Adobe® Photoshop®, the photographer can become a true artist, crafting his or her message in every detail.

camera as you press the shutter button. Stand comfortably, with your legs slightly apart, or lean against a tree or wall. Hold your elbows to your side to minimize shaking. Breathe at a slow, steady pace as you get ready to take the shot, and then hold your breath as you slowly press the shutter button.

Focus and Flash

The automatic mode of most cameras does an excellent job of auto-focusing for you. Pressing the shutter button halfway allows the camera to calculate the focus, white balance, and amount of light. The focus can also be manually set. Manual focus is used on digital cameras to emphasize one element in focus while deemphasizing another, which is out of focus.

Digital cameras also have several flash modes. Most allow you to select a flash, a red-eye reduction flash, or no flash. The automatic flash does not always select the best light for an image; rather, it selects light that is neither too dark nor too bright. If using the automatic flash does not produce satisfactory results, use one of the other flash settings. It is also recommended to stand no closer than 4 feet away from the subject and no farther than 10 feet away to get the best flash lighting. When taking pictures on sunny days, turn on the flash to help eliminate the harsh shadows produced by sunlight.

Angles

As with shooting video, described in Chapter 10, Video and Audio Production, one of the best ways to create interest in your photographs is to vary the angles while framing the shot well. An unusual angle or viewpoint can add a great deal of interest to an ordinary object. Although it is appropriate to shoot at eye level with the object or person, varying the camera angle from time to time adds a little extra excitement to the photograph. For example, photographed from below, someone looks strong and dominating. From above, a person appears meek, even childlike.

Camera angle refers to the different angles at which you can hold a camera in reference to the object of interest. Refer to the photographs shown in Chapter 10, Video and Audio Production, to see examples of the following camera angles:

- An *eye-level shot* looks the subject right in the eye. Some photographers call it the "bull's-eye effect" when the eye-level shot is coupled with placing the person directly in the middle of the picture, creating a "bull's-eye."
- A *low-angle shot* looks up at the object of interest. This angle creates a dramatic look, where everything appears magnified.
- Holding the camera high and shooting down is called a *high-angle shot*, where everything in the shot appears minimized or diminished.

Use your imagination to find different angles or perspectives for your photographs. You might try lying down or crouching in front of an object, climbing above it, or putting the camera on the ground.

Rule of Thirds

Perhaps the most well-known principle of photographic composition is the **rule of thirds**, which is also covered in Chapter 10, Video and Audio Production. The basic principle behind the rule of thirds is to imagine dividing an image into thirds horizontally and vertically so that you have nine parts (see Figures 9-7a and 9-7b). Position the main subject elements where the dividing lines intersect. This means not placing your subject right in the center of the frame. For example, frame the shot so that the subject's eyes are on the line dividing the upper third from the middle third. For landscapes, position the horizon along one of the horizontal lines instead of directly in the center of the picture. Photographs not taken with the rule of thirds in mind can be edited later to crop or reframe the image so that it fits the rule.

Lines

Using lines can be an effective way of drawing the viewer's eye into the focal point of an image. Lines can be the shape of a path, a line of trees, a fence, or any feature in an image. When framing the shot, determine what lines are in front of you and how they might add interest to the shot.

Diagonal lines are used to draw the viewer's eye through the photograph. Diagonal lines give images depth by suggesting perspective, and they can also add a sense of action to an image (Figure 9-8).

Vertical lines convey a variety of different moods in a photograph, ranging from power and strength, in such images as skyscrapers, to growth, such as in photographs of trees (Figure 9-9).

(a)

(b)

FIGURES 9-7A AND B: These photos show the "rule of thirds" in action. In each shot, the major impact or action takes place at intersections of the nine sections of the screen. (*Photo B by Katie Wimberly*)

© Cengage Learning 2012

FIGURE 9-8: Example of diagonal lines. (*Photo by Erica Der*)

FIGURE 9-9: Example of vertical lines.

© Roman Sigaev/www.Shutterstock.com

Horizontal lines convey a message of stability or rest, such as in photographs of horizons, oceans, and even sleeping people. Landscape horizons are probably the most common horizontal lines in photographs. Generally, horizons should not be placed in the middle of the frame. A much more effective technique is to place the horizon in the upper or lower third of the frame, following the rule of thirds (Figure 9-10).

Converging lines occur when two or more lines come from different parts of an image to a single point. Converging lines act as a sort of funnel for the viewer's eyes, directing the viewer's gaze to a point in the photograph. A good example of converging lines is a set of railroad tracks that converge on a horizon (Figure 9-11).

Depth of Field

Depth of field refers to the portion of the scene in focus in the camera. Depth of field can be long or short. An image that has a lot of the scene in focus has a long depth of field. When only a small zone is in focus, with much of the background out of focus, the depth of field is short (also referred to as "narrow"). Depth of field is dependent on several factors, but one of the primary factors is the camera's aperture setting.

The **aperture** is the "iris" of the camera, like the iris of your eye; it is the opening in the lens through which light passes to the camera sensor. The aperture controls the amount of light that is allowed into the camera. Aperture settings are measured in *f-stops* or *f-numbers*

(Figure 9-12). The easiest way to remember f-stop settings is this:

The larger the f-stop, the smaller the aperture opening.

The smaller the f-stop, the bigger the aperture opening.

For example, an f-stop of f-1.7 (small number) means the aperture is open, whereas an f-stop of f-16 (large number) means the aperture is almost completely closed. An aperture with a small opening (large f-stop) produces a longer depth of field, whereas an aperture with a large opening (small f-stop) produces a short, or narrow, depth of field. Play with the depth of field to get an entire field of flowers in focus (long depth of field; Figure 9-13a), or just a few flowers in focus while all of the rest of the flowers are blurred (short depth of field; Figures 9-13b–9-13d).

Background Distractions

The best advice regarding backgrounds in photographs is to use a simple, plain background, unless the background is part of the story. Avoid extremely light or dark backgrounds. The more distractions that are removed from the background, the more attention is drawn to the subject.

FIGURE 9-10: Example of horizontal lines.

© Studio 37/www.Shutterstock.com

FIGURE 9-11: Example of converging lines. (*Photo by Hyunji Lee*)

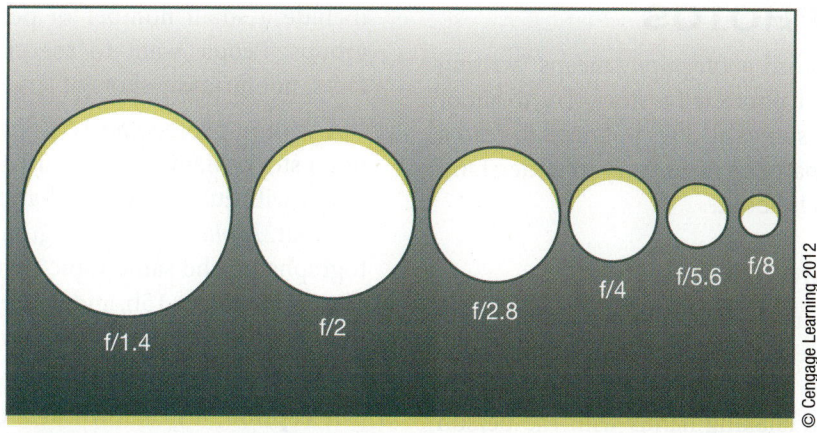

FIGURE 9-12: This graphic illustrates the f-numbers in a camera—the larger the number, the smaller the aperture opening.

FIGURES 9-13: These photographs illustrate long and short depth of field. Figure 9-13a shows a long depth of field; the horse and much of the background are in focus. Figures 9-13b, 9-13c, and 9-13d show short (or narrow) depth of field, where only a small part of the photograph is in focus.

TYPES OF PHOTOS

Remember that the word *photography* means "writing with light," so let your photo tell a story. Try to shoot a photograph so that someone does not need to read a caption or an accompanying news story to understand what the photograph is about.

News and Feature Photos

To illustrate a news article, keep the *news photograph* simple and get as close as possible to the subject. Try to avoid "grip and grin" photographs—photos of people receiving awards and shaking hands with the person presenting the award, smiling as they receive the honor. If the story is about an award recipient, try taking a photograph of the recipient doing whatever the person did to earn the award instead of a "grip and grin" (Figures 9-14a and 9-14b). Arrange news photos to include a small number of people rather than large groups. People want to see closer shots of people's faces, not far-away shots of large groups.

A *feature photograph* is a photograph not tied to a news story. Many times, you will see a feature photograph with just a caption that details what the photo is about. A *photo series* is a group of three to five photographs on the same topic that tells an overall story (Figures 9-15a, 9-15b, and 9-15c). A photo series could be three to five photographs on how a team prepares for a competition, for example.

A **caption**, also known as a cutline, provides written information underneath a photograph that is necessary for the reader to understand the photograph. Usually, a caption provides information about who is in the photograph, what is going on, where and when the action happened, and why the action is significant. A caption is one to two sentences in length. Include the full name of the person in the photograph and, when appropriate, the person's title.

People Photos

Most photographs have people in them. Following are suggestions on how to get the best "people pictures":

- *Avoid posed shots.* Do not force people to always pose staring at the camera. Get them doing something.
- *Take candid pictures.* Show people working, playing, or relaxing.
- *Keep people busy.* An interesting prop can give the person being photographed something to work with and can help create a natural feeling. For example, a rodeo cowboy could hold a lasso as his "prop."
- *Move in close.* Fill the camera's viewfinder with the subject to create pictures with greater impact. Standing too far away, even when taking group shots, produces images that are harder to see and less interesting.
- *Look your subject in the eye.* With children, for example, that means getting on their height level.

Animal Photos

Taking photographs of animals can be fun, but it can also be tiring because it is not possible to control how an animal will react or cooperate with you as you take pictures. For photographs of small animals, such as dogs and cats, use some of the suggestions in the "People Photos" section: move in close, get on the animal's eye level, keep the

(a)

(b)

© Cengage Learning 2012

FIGURES 9-14A AND 9-14B: Figure 9-14a is an example of a "grip and grin" (usually taken when someone receives an award). Figure 9-14b is an example of an action shot showing the person actually doing what she did to receive the award. The action shot is much more interesting than a handshake.

(a)

© Cengage Learning 2012

(b)

© Cengage Learning 2012

(c)

© Cengage Learning 2012

FIGURES 9-15A, 9-15B, AND 9-15C: These photos illustrate a photo series. The photos have a common theme—a student group's trip to Canada.

animal busy, avoid posed shots, and take candid pictures. For larger animals—cattle, pigs, horses, and other live-stock (Figures 9-16a, 9-16b, and 9-16c)—use these tips:

- *Groom the animal.* If it is a grand champion-type photograph, make sure the animal looks its best.

An effective livestock photo should show a good side view of the animal so its markings, profile, and general condition are visible.

- *Show the animal just as it is.* Do not use photo editing software to "doctor" the picture.
- *Choose an appropriate background that does not clutter the picture or distract the viewer's attention from the animal.* For example, an open field as a backdrop is probably better than a dark barn.
- *Position and pose the animal properly.* Generally, use a full side view or a three-fourths view, where the animal's head faces more toward the camera than the rest of the body. The animal's head should be high.
- *Use a flash, if possible.* A flash will enhance the animal's appearance by revealing shadow detail.

The previous tips are for animals that are not moving, but are posed in a controlled environment. Sometimes, though, it will be necessary to take photos of moving animals, such as those in a rodeo (Figure 9-17). To get close to the action, use a telephoto lens and a fast shutter speed to stop the action for the photograph. Try to anticipate where the animal will be and follow the animal and rider. Give the animal plenty of lead room in your viewfinder. If your camera allows you to take multiple rapid shots by holding down the shutter button, do so. You can choose the best photo later. Try to find varying angles to shoot from to get interesting photographs.

PHOTO EDITING

To this point, you have learned several techniques to help you take good photographs. For the rest of this chapter, you will take the good photographs that you have shot and learn how to make them look even better through photo editing. Because there are so many photo editing programs on the market—such as Adobe® Photoshop®, Adobe® Photoshop® Elements, Corel® Paint Shop Pro®, and Roxio PhotoSuite, among others—this chapter will not focus on specific software programs, but instead discuss some general photo editing concepts to help your photos look better.

Cropping an Image

One of the most powerful tools in photo editing programs is the ability to crop images. Cropping removes unwanted parts of an image. Cropping a photo zeroes

(a)

(c)

(b)

(d)

FIGURES 9-16A, 9-16B, AND 9-16C: Example photos of various animals. You want to get as close as possible, in most cases, or use a zoom lens.

FIGURE 9-17: This photo shows a cowgirl competing in a barrel race. The photograph provides stop-action of the horse in midstride.

in on the subject and eliminates what is not needed, providing the viewer the opportunity to focus on what is most important in an image. Cropping can also be used if your digital camera's aspect ratio is 4:3 and you want to produce 4-by-6-inch photographs.

Resolution and Resampling

Photo editing programs can change the resolution of the original image, which may be necessary depending on how the photograph will be used. **Resolution** is a measurement of how closely packed together the pixels are in the image. Resolution is usually measured as *pixels per inch* (ppi). At 72 ppi, a 1,600-by-1,200-pixel image—the typical size of a photograph taken with a 2-megapixel camera—measures 22 by 16 inches. At 300 ppi, the same photograph measures 5 by 4 inches because those same pixels are packed closer together. For a print publication, the photograph's resolution must be set at 300 ppi or higher. If the image will be posted to the Web or sent via e-mail, the image must be saved at 72 to 100 ppi.

Changing a picture's pixel dimensions by adding or subtracting pixels from an original image is called **resampling**. Using the photo editing program to discard pixels from the original image is called *downsampling*, which is what happens when a high-resolution photograph (1,200 ppi, for example) is reduced to a resolution of 300 or 72 ppi. To increase the size of an original image that was saved at a very low resolution (e.g., 72 ppi), you can try to upsample the image; however, this is practice is not recommended because the photo's quality will suffer. *Upsampling* refers to using the photo editing program to make up new pixels by adding pixels that were not there originally. This will cause an image to look pixilated and blurry.

Retouching Photos

Most photo editing programs allow you to retouch a photograph's color, brightness, contrast, and other aspects. Photo editing programs have automated, one-step commands for adjusting these settings. The automated commands may not achieve the best results, but they are a good place to start. Following are a few of the common retouching tools:

- *Brightness/contrast* lightens or darkens an image.
- *Cloning* copies areas in one part of a photo to another part.
- *Color adjustments* revise saturation—the richness and intensity of the colors.
- *Dodge and burn* can lighten (dodge) or darken (burn) a part of an image.
- *Drawing tools* provide lines, curves, and geometric shapes to add to photographs.
- *Levels* adjust the highlights, midtones, and shadows of the image to the appropriate level.

SAVING YOUR EDITED PHOTO

Once the photograph has been edited, you are just about ready to save it so it can be used in a publication or on the Web. The last two things to consider as you save the photograph are the color format and the file format.

Color Formats

The two color formats in which to save images are RGB and CMYK. All higher-end photo editing programs provide the option to save the final image in one of these two color formats.

RGB stands for "red, green, blue." RGB is the color format used by televisions and computer monitors. If the photograph's final destination is the Web or a television monitor, the final color format needs to be RGB.

CMYK stands for "cyan, magenta, yellow, black." ("Black" is "K" because "B" is "blue" in the RGB format.) CMYK is the color format used for commercial color printing. Each letter is an ink color that makes up what is called *four-ink color*, also known as *process color*. If the photograph's final destination is a commercial printer for a publication, the final color format needs to be CMYK.

File Formats

After a color format has been selected (RGB or CMYK), the last task is to choose a file format for the actual photograph. Just like the color format selection, choosing a file format depends on the photograph's final destination. Refer to the section on graphic file formats in Chapter 6, Document Design, for more on file formats used for photographs.

If the photograph will be e-mailed, placed on the Web, or used in a television program, it should be saved as a JPG file. A JPG file works well for photographic images with gradual color changes and no sharp edges, and it is a relatively small file size.

If the photograph will be used in a print publication, it should be saved as a TIFF. A TIFF is considered by many as the best graphic file format for use in desktop publishing applications because it is supported by virtually all desktop publishing applications.

PHOTO RELEASES

Finally, if the photo is going to be used for a money-making purpose, such as in an advertisement, it is a good idea to have everyone in the photograph sign a release form. A release gives permission to use the photograph in specified ways (e.g., in an advertisement or educational program). In general, even if the photograph will not be used for a for-profit purpose, it is still a good idea to get a signed release. For persons under 18, a parent or guardian needs to sign the release form. For purposes that support your organization, such as scrapbooks, newsletters, or videos, release forms may not be necessary. It is best to check with your administration to see if a release form is needed. You can use the example video release form in Chapter 10, Video and Audio Production, and adapt it for a photo release form.

SUMMARY

This chapter introduced you to the basic concepts of proper digital photography. You can take good photographs with almost any type of digital camera. Keep in mind that you should shoot at the highest resolution possible, especially if you plan to use the photographs in printed publications. Use the tips on proper composition, and you will be on your way to "write with light."

CHAPTER EXERCISES

APPLICATIONS

Following are some ways you can apply what you learned in this chapter:

- Shoot photographs for a scrapbook.
- Shoot photographs for the school newspaper or website.
- Shoot photographs for a newsletter or website.
- Hold or sponsor a photography competition among classmates.
- Invite a professional photographer from a newspaper or magazine to give a presentation about proper photographic techniques.

CHAPTER QUESTIONS

MATCHING:

Match the following types of lines that can be used when framing digital photographs with their definition or characteristics. Not all terms will be used.

_____ **1.** Usually used to draw the viewer's eye through the photograph. They give images depth by suggesting perspective, and they can also add a sense of action to an image.

_____ **2.** Convey a variety of different moods in a photograph, ranging from power and strength, such as in photographs of skyscrapers, to growth, such as in photographs of trees.

_____ **3.** Convey a message of stability or rest, such as in photographs of oceans, landscape horizons, and even sleeping people.

_____ **4.** Occur when two or more lines come from different parts of an image to a single point.

a. Converging lines

b. Horizontal lines

c. Parallel lines

d. Vertical lines

e. Diagonal lines

MULTIPLE CHOICE:

5. _____ is the "iris" of the camera, like your eye.

a. Depth of field

b. Aperture

c. Resolution

d. Convergence

6. *Megapixels* stands for _____.

    a. million picture elements

    b. mega picture elements

    c. multiple picture elements

    d. million pixel pixels

7. The basic rule is that if you know you are going to use the photographs for print documents, save at the _____ possible.

    a. highest resolution

    b. lowest resolution

    c. longest depth of field

    d. largest aperture opening

8. Using a photo editing program to discard pixels from the original image is called _____, which is what happens when you take a high-resolution photograph (1,200 ppi, for example) and lower the resolution (to 300 ppi or 72 ppi).

    a. upsampling

    b. downsampling

    c. unsampling

    d. base sampling

9. The image format called "RAW" stands for _____.

    a. Resolution Aperture Wide

    b. Raster Awareness Wide

    c. Remote Apple Work

    d. Nothing. It's just a word and is not a standard photographic format.

10. A(n) _____ looks up at the object of interest. Everything in the shot looks magnified.

    a. eye-level shot

    b. low-angle shot

    c. high-angle shot

    d. back-angle shot

11. In lens aperture settings, the *larger* the f-stop number, the _____.

    a. smaller the hole opening

    b. larger the hole opening

    c. smaller the resolution

    d. bigger the resolution

12. The correct pixels per inch for a color photo in a print publication is _____.

    a. 150

    b. 300

    c. 72

    d. 200

13. The zone of acceptable sharpness (focus) is called _____ .

    a. shutter speed

    b. lens aperture

    c. depth of field

    d. resolution

14. A _____ provides information that is necessary for the reader to understand the photograph.

    a. cutline

    b. pixel

    c. resolution

    d. composite

15. Different kinds of light have different color qualities, commonly called _____ .

    a. color balance

    b. color quality

    c. color texture

    d. color temperature

REFERENCES AND FURTHER READING

The Blog Studio. "Digital Photography Composition Tips." Last modified 2007. Accessed October 23, 2010. http://digital-photography-school.com/blog/digital-photography-composition-tips/.

Burnett, C., and M. Tucker. *Writing for Agriculture: A New Approach Using Tested Ideas,* 2nd ed. Dubuque, IA: Kendall/Hunt, 2001.

Curtin, D. P. "A Short Course in Using Your Digital Camera: A Guide to Great Photographs." Last modified 2007. Accessed July 23, 2010. http://shortcourses.com/use/.

King, J. A. *Digital Photography for Dummies.* Hoboken, NJ: Wiley Publishing, 2008.

Kodak. "Photographing People and Animals." Accessed October 23, 2010. http://www.kodak.com/eknec/PageQuerier.jhtml?pq-path=38/39/42&pq-locale=en_US.

Long, B. "Snap Happy: Take Perfect Pictures with Your Digital Camera." *Macworld Magazine* (February 2003): 86–93.

Macworld. *Digital Photography Superguide.* San Francisco: Mac Publishing, 2007.

Photographytips.com. "Livestock: Photographing Horses, Cattle, Sheep, and Other Livestock." Last modified 2007. Accessed October 23, 2010. http://www.photographytips.com/page.cfm/1564.

Rivero, V. "You Ought to Be in Pixels." *Edutopia*, vol. 1, no. 2(2004): 24.

Video and Online Communication

10

Video and Audio Production

OBJECTIVES

After completing this chapter, the student will be able to:

- Explain and implement the video production process.
- Write a video script.
- Draw a storyboard.
- Write a shot outline.
- Shoot a proper shot sequence (long shot, medium shot, close-up).
- Develop short educational or promotional videos.

INTRODUCTION

With the decrease in the price of video cameras and video editing software, and with the rise of online video service websites, such as YouTube, video is being used by more people who would not have been able to shoot and edit their own videos just a few years ago. Some people post their videos on the Web so that friends and family can see them. Others capture news events and distribute them online to television stations or networks that have websites. In public schools, teachers are integrating video into courses and instructional programs. They are able to shoot video "in the field" and bring the "field" back to their classrooms.

With its ability to combine sound and moving pictures, video is a powerful communication medium and is extremely effective for delivering information. As a result, video's use is widespread, ranging from television news to entertainment to education. However, a person who has never produced a video program needs to understand the video production process. It is a fun process that can

FIGURE 10-1: With its ability to combine sound and moving pictures, video is a powerful communication medium and is extremely effective for delivering information.

© Cengage Learning 2012

Lisa Hightower

Coordinator, University of Florida's Scientific Thinking and Educational Partnership Program

Video and audio producers at the fundamental level are storytellers. They use pictures, music, and graphics to tell their story. The most adept producers can combine these elements into a product that engages their audience in a way that looks effortless.

draw together a group of novice video producers. The information provided in this chapter will assist you in developing video productions (Figure 10-1).

Additionally, audio, by itself, can be a powerful medium as well. Audio can engage a person's imagination through music and sound effects. Hearing someone, rather than just reading the person's words on paper, also can help explain a topic. Although this chapter focuses on video production, many of the concepts—especially pertaining to scriptwriting—apply to audio production as well. Audio production is becoming more popular, especially for downloadable online podcasts. Another way that audio production is integrated is through the use of radio news, which has a long and positive association with agriculture. Radio and television news writing techniques are covered in Chapter 5, News Media Writing.

IS A VIDEO PRODUCTION RIGHT FOR YOU?

Before you begin planning the initial steps for a video program, you need to decide whether video is right for the project. Because videos are such a prevalent part of our society, it has become quite easy to say, "Oh, let's just put this on video." Instead, you need to determine why you want to do a video production. A good reason to do a video is if the subject matter is eye-catching and interesting. Here are some issues to consider before beginning the video production process:

- *Video is the medium to show motion.* If the content does not lend itself to movement or motion, you may wish to use some other medium—print (handbooks, brochures, newsletters), photographs, or audio—that does not emphasize motion. It is best to use video when motion is involved—for example, to demonstrate a skill (e.g., how to shoe a horse, how to drive a tractor, how to shoot photographs).
- *Consider how long the video's content will be relevant.* If the information in the video will be out of date soon, then consider carefully whether video is your best option. It may take a lot of time and effort to produce a quality video program. Will it be worth the effort to produce a video if the information is outdated before you are finished?
- *Consider how long it will take to produce the video.* Video production can be a lengthy process. It is not unusual for a program to take several weeks or months from the time it is initially conceptualized until the last video edit is made. If you need a program relatively soon and have not begun the production process, the video may be difficult to complete in your timeframe.

THE VIDEO PRODUCTION PROCESS

In any video production, you should follow the steps in the video production process (Figure 10-2). These steps can be adapted to any communication production process, from developing a brochure to creating an audio podcast.

Video Production Process

Audience and Program Analysis

Get an idea for the program.

Determine your audience.

Determine the message you want to get across.

Preproduction

In-depth research is conducted on the topic that will be covered in the video.

A basic content outline should be produced.

A script and storyboard (or a shot outline) are written.

Production

Crew assignments are made.

Video shooting takes place.

Postproduction

Video and audio editing take place.

The program is duplicated and distributed as a DVD, vodcast, or online Web video.

FIGURE 10-2: Video production process.

Audience and Program Analysis

During this phase, you begin with an idea for the program and determine your audience. Most useful video production ideas are developed for a specific audience. In this step, you should try to identify as many characteristics of your audience as possible, such as the audience members' age, education level, economic status, and values.

How will you develop your video in order to match your audience? For example, a video on the importance of dairy products in the diet will be a lot different depending on the audience's characteristics—an informational dairy video targeted for children will be considerably different than one targeted for senior adults.

Not only do you need to know your audience's characteristics, but you also need to know what message you want to get across and how you plan to do that. This is the "program analysis" part of this phase. Such issues as how much you have to spend (budget), what

shooting locations will be involved, the deadline for getting the video produced, who will be involved in the production, and what equipment will be used must be determined before going to the second step in the production process, preproduction.

Preproduction

During this phase, in-depth research is conducted on the topic that will be covered in the video. A basic content outline should be produced. Then, a script and storyboard (or shot outline) will be written. These topics will be explained in detail later in this chapter.

Production

During this step, crew assignments are made, video shooting locations are scouted, and the actual video shooting takes place. The production phase is only a small part of the big picture (Figure 10-3). A lot of preliminary planning takes place before any video is shot.

Erin Freel Best

President, The Market Place: Marketing and Video Production

A well-produced video with interesting interviews, shots, and music can tell a story like nothing else. It can convince a donor to give money. It can encourage a potential member to join. It is a powerful sales and marketing tool when developed with a little forethought and planning.

FIGURE 10-3: Shooting video is just one part of the video production process.

Postproduction

In postproduction, the video editing and audio editing are done. Video and audio sound effects are added. The program is duplicated and distributed as a DVD, vodcast (video podcast), or online Web video.

SCRIPTWRITING

Most of the writing you have done to this point in your life has been for "the eye." In other words, the writing has been done so that your eyes could read it only; the words on the page had to make sense for your eyes. For audio and video productions, you must write for "the eye" *and* "the ear." Not only does the script have to read well for the eye, but it also has to sound good for the ear. Although this chapter focuses on developing videos for educational and promotional purposes, the following suggestions on writing for "the ear" are equally true for television news writing, covered in Chapter 5, News Media Writing.

- *The writing style should be conversational.* You do not always talk in complete sentences, which may mean that you write your sentences in the script as sentence fragments.
- *Each sentence should be brief and contain only one idea.* In some of your previous writing for "the eye," you may have gotten used to stringing sentences together with commas. In video and audio writing, sentences should be brief. Each sentence should focus on one particular idea.
- *Be simple and direct.* If you give your audience too much information, your audience cannot take it in. Choose words that are familiar to everyone.
- *Round off large numbers.* Detailed numbers should be left out. For example, do not say "6,049 people attended the football game." Instead, round it off to "around 6,000 people attended the football game."
- *Use the introduction, body, and conclusion structure.* When writing a script for an educational or promotional purpose, one of the best formats to follow is introduction, body, and

conclusion. The format for writing educational and promotional videos is commonly used in writing speeches: "Tell them what you're going to tell them" (introduction), "tell them" (body), and "tell them what you told them" (conclusion).

- The *introduction* provides an overview of the topic and gains the audience's attention. In many cases, you should say clearly what the video covers or what you want the viewer to learn. For example, you might say, "In this video, you will learn the steps to groom a horse." The introduction is where you "tell them what you are going to tell them" in the overall video.

- The *body* of the video provides details about the topic. For a video on horse grooming, for example, the body would show all of the steps on proper horse care. The body "tells them" the content.

- The *conclusion* summarizes the main points. For example, with the video on grooming a horse, you could summarize the major content by reviewing the steps briefly at the end of the program. The conclusion is where you "tell them what you told them" in the video they just watched.

After you have determined what the major content is for your script, you need to start writing. To help you get started in the scriptwriting process, here are some tips:

- ***Divide your page into two columns.*** Video commands, explanations of what shots you need, and other special effects should be listed on the *left* side of the page. These video directions should take up about one-third of the left-hand side of the page. Anything related to the audio—including narration, music, and sound effects—should be included on the *right* side, about two-thirds of the page. This will help you visualize what you need to say and will remind you to explain what you need to show (Figure 10-4).

- ***Decide what type of approach your video will need.*** Will the program consist of narration only, covered by video clips? Will a narrator be seen the entire time? Will you include interviews? While a straight-narrated script with no interviews is much easier to write and control, the use of interviews can make the video more interesting for your audience.

- ***Consider your audience.*** This is important throughout the *entire* production process. At the scriptwriting stage, such questions as the following arise: What will the audience members' interests be? How much do audience members know about this topic? You also have to write on your audience's knowledge level. Do not use words that would be difficult for your audience to understand.

- ***Consider the video's length.*** How long will you be able to retain your audience's attention? An adult audience's attention span will last about 8 to 10 minutes; for children, plan for 3 to 5 minutes. If the video will be posted to YouTube, videos

Example Video Script

VIDEO	AUDIO
Fade up to On-Screen Text: "Starting a Successful Oil Collection Program"	**MUSIC** fades up, then under narration
Dissolve to shot of oil well pumping.	**NARRATOR** Oil. Black gold. Crude. Whatever you want to call it, it's changed our world.
Dissolve to shot underneath a car, oil dripping into drain pan.	**NAT SOUND:** Oil dripping into pan
Cut to shot of Houston skyline. Cut to plane taking off.	**NARRATOR** With it, millions of people can take to the roads or the sides for easy, quick travel.

© Cengage Learning 2012

FIGURE 10-4: Video scripts should be divided with the video directions on the left side (one-third), and the audio (narration, natural sound, music) on the right (two-thirds).

should be less than 10 minutes in length—most videos on YouTube are 5 minutes or less. Videos, of course, can be longer, especially if the intended audience is interested in the content presented. One of the most difficult aspects of writing a good script is not including too much information, however.

- **Write the way you speak.** The key to writing a good script is to write the way you speak. Someone can re-read a passage in a book, newspaper, or magazine, but most people will not stop to reverse a video to refresh their memories of what they just saw and heard. Write in simple, easy-to-understand sentences, and write how you talk.

- **Use on-screen text to support what you say.** For instance, if your video is about a new program with six components that you are about to describe one by one, show them as text on the screen as you tell the audience about them. This approach also will assist your visual learners in retaining the information. On-screen text that appears below a person's face is called a **superimposition**, or *super*.

- **Use pauses, music, natural sound, and special effects** to avoid lengthy sections of narration. One reason to use music and special effects is to indicate to your audience that you are changing topics. **Natural sound** (or *nat sound*) is the audio that is naturally in the environment where the video is being shot. For example, chirping birds is the natural sound you might hear if you are shooting video in a forest. Barnyard noises would be examples of natural sound for a video about rural living. Keep in mind that you should get permission from music companies to use copyrighted music in your videos. You can avoid problems with music by using copyright-free (or royalty-free) music. You can find many companies on the Internet that sell copyright-free music.

- **Keep it simple.** Remember that your video should tell a story without you having to say everything. If your video shows children intent and focused on a classroom lesson, you do not need to tell your audience that children in the classroom are "intent and focused on their lessons."

- **Develop a storyboard or a shot outline.** The purpose of the storyboard and the shot outline is for you to visualize what the program will look like before any video is shot. This way, you will have a good idea of what you will need to shoot beforehand. Otherwise, you may not shoot enough video or you may not shoot the correct angle of video for your program. A **storyboard** is a series of drawings with captions that describe video shots and their accompanying audio or narration (Figure 10-5). A storyboard can be as simple as stick figures or as complex as digital photographs that you take of locations that you plan to use in your video program. The storyboard should provide a rough commentary at the bottom of each shot. You do not have to include every shot in the storyboard. A **shot outline** is a detailed written description of the video you plan to shoot (Figure 10-6). A shot outline may be preferable if you do not have drawing skills.

- **Read the script out loud when you are finished.** When you read the script out loud, you will probably pick up some things that are difficult to pronounce. You may not have picked up on these difficult words by reading the script "in your head."

- **Get someone else to read your script.** After you have finished writing your script, let someone who represents your intended audience read it over. You may learn that you have included jargon or inadequate explanations in some areas, or you may have left gaps in the content. You also can determine what areas your intended audience will find the most and least interesting.

FIGURE 10-5: If you are interested in developing a storyboard, you can use this blank storyboard as an example. Make rough drawings of the shots that you need in the boxes. Underneath the box, write a brief part of the script. You do not have to include the entire script word for word.

Example Shot Outline

SHOTS NEEDED	AUDIO
CU of peanut plants. Various shots of peanut harvest. Shots at a peanut processing plant.	You might be surprised to learn that there is no such thing as a peanut tree. Thousands of tons of peanuts are produced in Florida each year. But the crop is produced underground out of sight and these plants rarely grow larger than knee high.
CU shots of peanuts. Shot of peanut and peas. Shot of peanut plant in bloom	Peanut is actually a legume. It is very closely akin to peas. The plant starts blooming about thirty days after planting and from each of the flowers that become fertilized.
CU shot of peanut butter jar. Shots of large vats of peanut butter being made.	Peanut butter is far and away the biggest use in the United States. It's somewhere approaching sixty percent of the peanuts in the US go into that product.

© Cengage Learning 2012

FIGURE 10-6: A shot outline is simply a list of the shots necessary to produce a video. The shot outline should be based on the types of shots needed in the script. Be sure to shoot shots from different angles—long shots, medium shots, close-ups, cut-ins, and cut-aways.

■ *Time the video script.* If you used two columns (one-third for video and two-thirds for the audio/narration) on 8.5-by-11-inch paper, a full, double-spaced column of narration (right side of the page) will last about 30 to 45 seconds per page. However, if you have interviews, that estimate can vary greatly, depending on the rate of speech of your interview subjects. Once you have your interview comments recorded and chosen, time them to the second for a more accurate total run-time estimate. Also, add time for music transitions between segments, a visual introduction with music, and credits.

When you have finished writing a polished draft of your script, use the following checklist to determine if you need to make any changes:

■ Have I explained myself in simple language?
■ When I read the script out loud, does it sound as if I am talking to an audience—which is what I want—or just reading to an audience?
■ Have I avoided using jargon or technical language?

■ Have I used music and natural sound to help tell my story and break up constant narration or interviews?
■ Do I have any lists or main ideas that could be reinforced as text on the video screen as they are being discussed?
■ On the script, have I included proper titles and name identification of on-screen speakers?
■ On the video column of the script, have I described the type of video shots, graphics, or other video special effects I need?

VIDEO EQUIPMENT

You should become as familiar as possible with the video camera you plan to use. For a basic video shoot, the minimum equipment requirements are a video camera, a microphone, a light source, and a tripod (Figure 10-7).

Video Camera

Video cameras come with a wide variety of features. Most come with color viewfinders that can extend from

© Cengage Learning 2012

FIGURE 10-7: You should use a tripod when conducting interviews. Notice the "shotgun" microphone on the video camera.

(a)

© Cengage Learning 2012

(b)

© Cengage Learning 2012

FIGURE 10-8: A regular television monitor has a 4-by-3 ratio (4 × 3), meaning the screen is 4 units wide by 3 units high, as shown in Figure 10-8a. Widescreen has an aspect ratio of 16 by 9 (16 × 9), meaning the screen is 16 units wide by 9 units high, as shown by the longhorn in Figure 10-8b. Some parts of the 4-by-3 shot are not shown on the 16-by-9 screen, as indicated by the black boxes at the top and bottom of the shot.

the video camera. Some will shoot well in dark areas with minimal lighting. All have a microphone built into the camera, but some have an audio input for an external microphone. All video cameras also come equipped with a *zoom lens*, which smoothly changes from a long shot (wide shot) to a close-up view without moving the camera or the object you are shooting. Video cameras shoot in one of two **aspect ratios**, which are the width-to-height proportions of television sets. The aspect ratio for a standard television screen is 4 units wide by 3 units high (4 × 3) (Figure 10-8a). For widescreen video, the aspect ratio is 16 × 9 (Figure 10-8b). Almost all new video cameras can change between these two aspect ratios.

Microphone

Sound may be the least-thought-about component in a video shoot, but it is just as important as the visuals that are recorded. Good sound gives your program that "little extra." How many times when you watch home movies have you heard a lot of wind noise because the microphone was on the camera and the person who was talking was 20 feet away? Use microphones that can get close to a person's face.

Microphone types include *lavaliere* (also known as lapel or clip-on microphones), *hand-held,* and *shotgun* (Figure 10-9). Lavaliere (pronounced LAH-vuh-leer) microphones have a very limited range and are small

so they are not easily seen in a video shot. They clip on to shirts and ties. Hand-held microphones are used in everyday video production, especially for television news for "person-in-the-street" interviews. Shotgun microphones are very sensitive and can pick up sound over a great distance. Shotgun microphones are usually very long and slender and often are seen attached to video cameras at sporting events.

Keep in mind that many inexpensive consumer-grade video cameras do not have places where you can connect external microphones. Cameras with microphone inputs usually cost more, but the cost is worth it

© Cengage Learning 2012

FIGURE 10-9: These microphones should be part of your video equipment: (a) lavaliere microphone, which shows the receiver that attaches to the camera and (b) the transmitter, which has the actual microphone; (c) shotgun microphone; and (d) hand-held microphone.

if you want the audio to sound good. If you do not have external microphones, such as lavaliere, hand-held, or shotgun microphones, all you have to record audio with is the internal microphone on the camera. For some purposes, that microphone will be fine, but if you want to record narration or an interview, you will pick up a lot of wind noise with just an internal microphone.

Lighting

A *light source* also needs to be taken into account. If there is not enough light, you will not be able to see a picture in your video camera's viewfinder. If the video program is to be shot outside, then you will use natural lighting. *Natural lighting* is any non-manmade light, such as sunlight or moonlight. In sunlight, especially, you will have to contend with harsh shadows at various times of the day, and cloud cover. You may need a *reflector* (a large silver screen) to direct sunlight to fill in shadows on people's faces.

Lighting provided by any non-natural source is called *artificial lighting*. If the video program is shot indoors, you likely will need portable lights (Figure 10-10), especially if you are going to shoot an interview. Regular indoor lighting, such as lighting from fluorescent bulbs, usually does not provide enough illumination

Image Copyright Photofirstandlast, 2010. Used under license from Shutterstock.com

FIGURE 10-10: Portable lights may be needed to provide enough light for video shot indoors.

for interviews. Portable light kits (a light, light stand, and a light kit case) can be purchased at most photography or video equipment stores. If you use a portable light kit, you should bring extra extension cords as well.

Lighting also can be categorized as to the strength of the light. *Directional light* produces a sharp beam of light, resulting in harsh shadows. Examples of directional light are direct sunlight and strong artificial lights. *Diffused light* refers to wide, indistinct beams of light, which produce soft shadows. Examples of diffused light are light on a cloudy day and fluorescent lamps.

If at all possible, use a tripod when interviewing a person, panning the camera (moving left and right), or tilting the camera (moving up and down). The shots will be much smoother and much less shaky. You also may wish to use a one-legged monopod for video cameras that do not weigh much (Figure 10-11).

AUDIO EQUIPMENT FOR RADIO AND PODCAST PRODUCTIONS

The equipment used for radio and podcast production is quite simple: a microphone, an audio recorder, and a computer. Cassette recorders were used for years in the radio news industry, but almost everyone now uses digital audio recorders. A *digital audio recorder* can be plugged into a computer, where the audio files can be downloaded, stored, and edited.

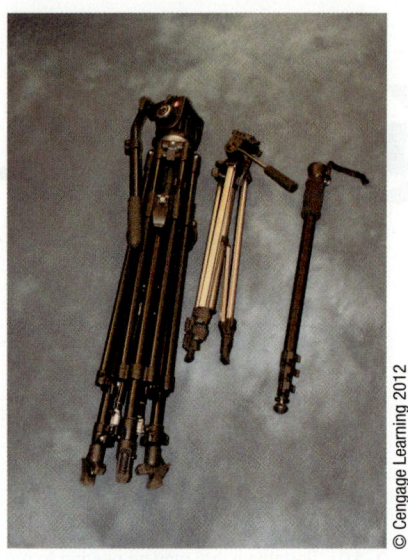

FIGURE 10-11: Tripods can be more heavy-duty (left) or lightweight (middle), depending on the size of your video camera. You also may wish to use a monopod (right) for very lightweight cameras.

FIGURE 10-12A: A long shot (wide shot), such as this example, shows a fairly long (or wide) shot of the object of interest.

FIGURE 10-12B: A medium shot is a closer shot—in reference to a long shot—that shows more of the subject.

FIGURE 10-12C: Extreme close-ups, such as this photograph, are super-tight shots.

VIDEO SHOT COMPOSITION

Shooting video consists of more than just pointing a video camera at a person or object and hitting the record button on the camera. To become a better videographer, you need to understand and put into practice the techniques of video shot composition presented in this section of the chapter. *Video shot composition* consists of camera shots and angles, shot sequencing, camera movement, framing, and continuity. Video shot composition is an extremely important skill to learn. The more you practice, the better you get.

Camera Shots, Camera Angles, and Shot Sequencing

Camera shots are named for how much of the object of interest is in the video screen.

The widest shot possible and zoomed out as far as possible, showing the object of interest in its surrounding or setting, is called an *establishing shot* or an *extreme long shot*.

A *long shot*—also called a *wide shot*—is a fairly long (or wide) shot of the object of interest (Figure 10-12a).

A *medium shot* is usually a shot of a person that extends from below the waist to a little over the head (Figure 10-12b).

A *close-up* is usually a head-and-shoulders shot of a person, and an *extreme close-up* is a super-tight shot, such as of the mouth, nose, and eyes (Figure 10-12c). Try not to overuse such dramatic shots. Save them for a few times when you want maximum impact.

Camera angle refers to the different angles at which you can hold a camera, in reference to the object of interest. Most of the shots you will shoot will be *eye-level shots*, because that is what people are used to seeing. Video interviews will almost always be shot at eye level (also called a *flat-angle shot*; Figure 10-13a).

Rather than always holding your camera at an eye-level shot with the person you are shooting, try putting your video camera on a tripod, table, ladder, or the floor to get some of the other camera angles described next.

You can get close to the ground and shoot up at your subject. This is called a *low-angle shot*. A low-angle shot looks up at the object of interest, creating a dramatic look, where everything looks magnified (Figure 10-13b).

Holding the camera high and shooting down is called a *high-angle shot*; everything in the shot looks minimized or diminished (Figure 10-13c).

To give your viewers a sense of identification with your main subject, use a *point-of-view shot*, which simulates the view the actor sees as the actor moves about. You simply walk or move your camera as if you were the actor.

Shot sequencing is how you organize the different types of shots and angles. You should always try to shoot a series of long shots, medium shots, and close-ups. Use different-angle shots sparingly. For example, you first could shoot a long shot (or wide shot), which shows the object of interest in its surrounding or setting. Then shoot a medium shot, where you have zoomed in closer on the object of importance. Finally, shoot close-ups, where the object of importance fills the entire screen. This way, you draw the viewer into the video program. Shooting in a shot sequence structure also provides you with more shot choices during the editing process.

Avoid jumping from a long shot directly to a close-up, unless you want to startle your audience. Instead, use a medium shot to gradually draw your viewers into the dramatic close-up. A sequence might go like this: long shot, medium shot, close-up, close-up from a different angle, medium shot, close-up, another close-up, medium shot, long shot.

Camera Movements

The next aspect of composition is camera movements. *Camera movement* refers to physically moving the video camera left or right, up or down, or to the zooming

FIGURE 10-13A: Video interviews will almost always be shot at eye level, such as this example.

FIGURE 10-13B: A low-angle shot, like this example, looks up at the object of interest.

FIGURE 10-13C: Holding the camcorder high and pointing down, as in this example, is called a high-angle shot, where everything in the shot looks minimized or diminished.

© Cengage Learning 2012

in or out of the video camera's lens. A **tilt** is a vertical movement of the camera. A tilt points the camera up or down. Tilts are usually used to show an object's height. A **pan** is the horizontal movement of the camera. A pan moves the camera left or right. You should practice panning or tilting at various speeds until you find the speed that works best for you. A **zoom** is not an actual physical camera movement, but it is placed in the camera movement category because the camera's lens is moved, even though the camera itself does not move. A zoom is a change in the focal length of the lens. When you "zoom in," the shot gets tighter on the subject. When you "zoom out," the shot gets wider.

Only use a camera movement when it is necessary. Do not use a pan, zoom, or tilt with every shot. This is a characteristic of a novice videographer. Use zooms, pans, and tilts only when you want to draw attention to a part of the video camera frame. Instead of zooming, panning, and tilting, first shoot a long shot. Then stop the recording. Adjust the shot to a medium shot, then start recording again. Refer to the discussion of shot length in the "Other Video Considerations" section later in this chapter for more information on how long camera movement shots should be.

Framing

Framing refers to the way the various elements within the video screen are arranged. The videographer must decide what angle to shoot from and what portion of the scene to include in the shot.

Make sure that you properly compose your shots so that there is an appropriate amount of room above the person's head. Usually, this is just a little space between the head and the top edge of the screen. This is called **head room**. Objects or a person's head near the top edge of the frame tend to seem crowded. Allow a bit of extra space above a person's head to avoid this appearance.

When actors face left or right, it is a good idea to position the video camera so that there is a little more space in front of them than behind them. Do not put them dead center. You also want space in front of them if they are walking. This space in front of a person is called **lead room**. If the shot is a close-up of someone's head, this space also is called *nose room*. If you are recording something that is moving, such as a bicyclist or a speeding car, provide enough lead room in front of the bicyclist or car so that it does not look like the bicycle or car is running off of the screen (Figure 10-14a through 10-14f).

One way to ensure proper framing for a video shot is to use the **rule of thirds**. This technique also is used by photographers and is described in Chapter 9, Digital Photography and Photographic Editing. In short, the rule of thirds means to imagine dividing the video screen into thirds, horizontally and vertically. Then place objects of interest at the intersections. For example, frame the shot so that the subject's eyes are on the line dividing the upper third from the middle third.

Another aspect of framing is to watch your backgrounds (Figure 10-15). A telephone pole in the distance can appear to grow out of a person's head if the shot is framed incorrectly. Brick walls are busy-looking backgrounds. Try to avoid using plain white walls as backgrounds. Most important, find backgrounds that relate to the topic being discussed. If a person is talking about cattle, then try to interview the person with cattle in the background, rather than shooting the interview in an office.

Continuity and Jump Cuts

Continuity means that each shot in your video logically flows from the one before it. It is fairly easy to explain continuity mistakes. For example, have you ever watched a movie where an actor had a drink in one hand and it mysteriously switched to the other hand in the very next shot? Or have you seen a television program where an actress had her arms crossed in front of her in one shot, and in the next, her arms were at her side? This is called a *continuity error*. The two shots do not flow logically. To avoid this simple mistake, assign an assistant who can follow along on your video shoot to ensure that continuity is maintained.

Another way to maintain continuity is to make sure your actors are moving in the same direction from shot to shot. For example, if you are shooting a cops-and-robbers chase scene, make sure everyone is going left to right (or right to left). If the cops are going right to left, and the robbers are moving left to right, you will confuse your audience when you edit the shots together. It will look as if the cops and robbers are about to collide.

Similar to a continuity error is the **jump cut**. A jump cut occurs when a shot shows the same prominent person or object in different angles or different locations in back-to-back shots. This makes the two shots appear to "jump," due to the way the shots are framed in relation to each other. You create a jump cut when something in the scene has changed positions since you stopped and restarted the video camera. To minimize jump cuts,

(a)   (b)

(c)   (d)

(e)   (f)

© Cengage Learning 2012

FIGURE 10-14: Figure 10-14a and 10-14b are examples of good lead room and good head room. The women have enough "looking room" to the side of the screen, and the space at the top of the shot is not too much or too little. Figure 10-14c and 10-14d are examples of bad lead room. The women look as if they are looking off of the screen. Position the person in the opposite third of the screen from where the person is looking to correct this. Figure 10-14e and 10-14f are examples of bad head room. The women have too much space above their heads to the top of the screen. The videographer needs to zoom in to show less of the empty space.

FIGURE 10-15: In this shot, a light pole seems to be coming out of the person's head. Be sure to watch the background when you shoot video.

shoot several shots that do not contain the prominent person or object (Figure 10-16a and 10-16b).

The easiest way to avoid continuity errors and jump cuts, though, is by shooting cut-ins and cut-aways. A *cut-away shot* is a reaction shot showing another actor or subject in the scene, usually reacting to the main action (Figure 10-17a, 10-17b, and 10-17c). If you shot an interview sequence, a cut-away shot would be of the interviewer nodding. In sports, it could be cutting away to shots of cheerleaders or the crowd after your team scores.

A *cut-in shot* is usually a close-up shot, designed to bring the view in closer to the subject (Figure 10-18a and 10-18b). In essence, a cut-in shot is the same concept as the cut-away, except instead of "cutting away" from the action (such as a shot of fans cheering at a basketball game), you "cut in" to the video. For example, if you were recording a person looking into a microscope, you could *cut in* with close-up, individual shots of the person's hands adjusting the microscope's lens, the person's eyes looking into the microscope, and the microscope slides.

When it is time to edit your video, you can cover your jump cuts or continuity errors by inserting the cut-in or cut-away shots. Cut-ins and cut-aways are not just for covering jump cuts. Use cut-ins to draw emphasis to a subject, and use cut-aways when you want to show reaction.

(a)

(b)

FIGURE 10-16: These two photos, which represent two back-to-back video shots, would be a jump cut. How does the woman taking the photograph of butterflies move from the first shot (Figure 10-16a) to the second shot (Figure 10-16b)? This "jump" in location and action would be startling for viewers. To avoid a jump cut, shoot cut-ins and cut-aways.

OTHER VIDEO CONSIDERATIONS

The following tips do not fall under video shot composition, but they should be considered as you develop your educational or promotional video programs:

- ■ *Shot length:* One problem many amateur videographers have is recording video shots of two to five seconds in length. When they watch their footage later, they realize how little they actually recorded. Short shots are difficult to edit. A good rule is to record at least 8 to 10 seconds per shot. You can always shorten the shots later in the video editing process. When recording camera

(a)

© Cengage Learning 2012

(b)

© Cengage Learning 2012

(c)

© Cengage Learning 2012

FIGURE 10-17A, 10-17B, AND 10-17C: In this three-photograph sequence, you see the same photos that were used in the jump cut example, except there is a shot in between—a cut-away of a butterfly (Figure 10-17b)—to "cover" the jump cut.

(a)

© Supri Suharjoto/www.Shutterstock.com

(b)

© Supri Suharjoto/www.Shutterstock.com

FIGURE 10-18: Figure 10-18b is a cut-in shot (closer shot) of Figure 10-18a.

movement (pans, tilts, zooms), it is best to record five seconds of video, and then start the camera movement. Stop moving the camera at the end of the pan, tilt, or zoom, and then record another five seconds before hitting the record button to stop the recording.

- **On-screen text:** When shooting, you will need to consider where words will be placed on the screen. If you are shooting an interview, about the bottom one-third of the screen will be used for a person's name and title. If you have zoomed in too closely, then the person's name will appear over the person's mouth or nose. Also, if you are shooting video that will be placed on the Web or downloaded onto a mobile video device, such as an iPod, the text will need to be much bigger, so keep that in mind if you want words to be included below someone's face.

- **Logging video:** It is a good idea to create a written catalog or "log" of the video that you shoot. This way, you have a reference of what shots are on each tape. Include a description of the shot and the length of the shot to make editing faster and easier.

- As noted in Chapter 9, **white balancing** tells the camera what combination of red, green, and blue light it should perceive as white, given a particular lighting condition. When a video camera is white balanced, it adjusts the red, blue, and green color channels inside the camera in such a way that a white object looks white on-screen, regardless of whether it is lit by reddish or bluish light. Most video cameras have automatic white balance features, but this feature can sometimes get "confused," particularly if you are shooting a scene that features a single dominant color or includes different types of light (e.g., sunlight streaming into a room lit with fluorescent light). In this situation, you may need to adjust the white balance manually.

- **Backlighting** occurs when you have a person or an object in front of a bright light source, and the person or object looks dark. For example, you may have positioned someone so that a window is in the background. The sunlight coming through the window backlights the person, resulting in the person being in silhouette. Reposition the person so he or she is not in front of the window in order to avoid backlighting.

- **Steady shooting:** If you are not a steady shooter and you do not have a tripod, shoot *fewer*

close-ups. The tighter or more close-up the shot, the shakier the shot will look. Shoot wider shots or physically move yourself and your camera closer to the action.

RECORDING AUDIO

Video production also has another component that is very important: audio. Specific audio-recording concepts are covered in the section titled "Narrating Television and Radio News" in Chapter 5, News Media Writing. These tips will also improve the audio for your educational or promotional video. When you record narration for your video, find a quiet location. Not only should the script be written in a conversational style, your voice also should be conversational-sounding.

Background noise or *natural sound* ("nat sound") can add "flavor" to the video, but if there is extraneous background noise that you do not want to record, you need to plan for this. For example, if you are interviewing the principal and do not want ringing telephones or hallway noise to spoil the interview, then disconnect the telephones in the room or place signs in the hallway to let people know that recording is taking place.

VIDEO EDITING

Editing a video is a creative process. Video editing is where you put all the various parts together into one comprehensive program. Video editing software programs digitize video, so the video can be edited in the computer, allowing you to make changes easily (Figure 10-19). It is suggested that you become very

FIGURE 10-19: This person is editing video on a computer. Video editing software programs are inexpensive and can produce high-quality videos that can be placed on DVD or shared over the Internet.

© Cengage Learning 2012

familiar with your video editing software *before* using it to develop a large-scale video production. Some video editing software packages are easy to learn, whereas some are difficult.

Video editing is also time-consuming. Video editing professionals estimate that for one minute of finished video in a program, it takes at least one hour of editing time. So for a 10-minute program, you can expect a minimum of 10 hours of editing time to complete it. Depending on the number of special effects you want to include in the video program, that amount of time may double.

Some consumer-grade video editing software programs are less than $100 and function well to create video programs. These low-end programs are in the price range of many amateur video producers. Other editing programs are more expensive, but provide more functionality and special effects choices. These more expensive programs, with an educational discount, can run from $150 to $1,000. Retail prices for these pricier editing programs can be as high as $2,000. Video editing software programs on the market include CyberLink PowerDirector, Corel VideoStudio, Adobe Premiere, Apple iMovie, Apple Final Cut Pro, Pinnacle Studio, and many others. Learn from friends and experts the video editing program they use before purchasing your own.

VIDEO EDITING CONCEPTS

Although this section of the chapter does not focus on how to use specific video editing software programs, it presents some general concepts that you can apply when editing video:

- *Choose shots that best tell the story.* Use various shots and angles from your shot sequence (long shot, medium shot, close-up, high angle, low angle, etc.).
- *Do not use everything you shoot.* An educational or promotional video is not a home slide show or movie, where you show everything that you did on your family vacation. Have a purpose for what video shots you plan to use. This goes back to the very beginning of the production process. Know what you want the video to accomplish, and carry that purpose all the way through the editing stage.
- *Use, but do not overuse, video transitions.* A video transition is the term used to indicate when an edit between shots is made. Video transitions are usually divided into the following categories: cuts, dissolves (which includes the fade), and wipes.

- A **cut** is a direct transition from one shot to the next (Figure 10-20a and 10-20b). It is the most commonly used transition. In video news editing, cuts are predominantly used. Use cuts most of the time.
- A **dissolve** is a gradual change from one shot to the next (Figure 10-21a, 10-21b, and 10-21c). A dissolve is best used to show the passage of time or location. In most movies or television programs, you will see a dissolve used to indicate the passage of time.
- A **fade** is a special form of the dissolve. A fade is any shot that dissolves to black (Figure 10-22a, 10-22b, and 10-22c). A fade, like a dissolve, also indicates a passage of time, but is more final. For television programs, there usually is a fade at the end of each segment of the program before a commercial break starts.

(a)

© Makarova Viktoria (Vikarus)/www.Shutterstock.com

(b)

© Karel Gallas/www.Shutterstock.com

FIGURE 10-20A AND 10-20B: This two-photograph sequence shows a cut, a direct transition from one shot to the next.

(a)

© Cengage Learning 2012

(a)

© Galushko Sergey Alekseewisch/www.Shutterstock.com

(b)

© Cengage Learning 2012

(b)

© Galushko Sergey Alekseewisch/www.Shutterstock.com

(c)

© Cengage Learning 2012

FIGURE 10-21A, 10-21B, AND 10-21C: This three-photograph sequence shows a dissolve, a gradual change from one shot to the next.

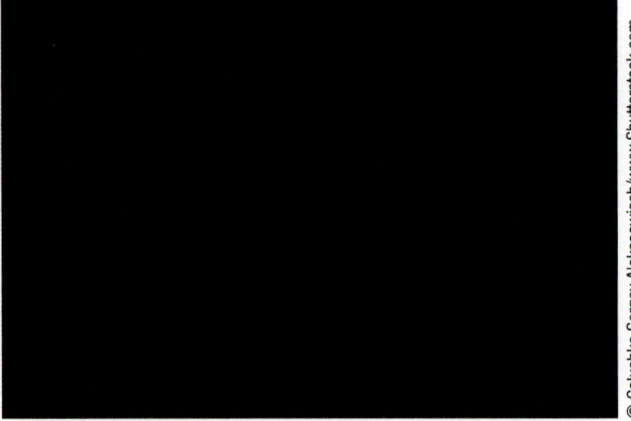

(c)

© Galushko Sergey Alekseewisch/www.Shutterstock.com

FIGURE 10-22A, 10-22B, AND 10-22C: This three-photograph sequence shows a fade, which is a slow "dissolve" from any shot to black.

- A **wipe** is one video picture "wiping" off and another appears (Figure 10-23a, 10-23b, and 10-23c). Examples of wipes are clock wipes, where the video is "wiped off" the screen with what appears to be clock hands, and

(a)

(b)

(c)

FIGURE 10-23A, 10-23B, AND 10-23C: This three-photograph sequence shows how a wipe transition would look between two shots. Notice the wipe "bands" in the middle photograph.

checkerboard wipes. This transition style calls attention to itself and should only be used when there is a specific reason to do so.

- ***Properly pace the program.* Pacing** describes the speed or rhythm of a program, as perceived by the audience. In order to maintain audience interest and involvement, the pacing of a video program should be kept brisk. This means that most video shots should be short—in the neighborhood of 5 to 10 seconds in length each. However, for educational videos, the shot length could be longer. Instructional videos require that the audience be given sufficient time to learn new information.
- ***Remember continuity.*** Insert cut-ins and cut-aways to cover continuity errors and jump cuts.

VIDEO SHOOTING AND EDITING FOR THE WEB

Because many people will put their videos directly on the Web, through video-sharing services such as You-Tube, or on portable video devices, such as iPods, it is important to point out that most shooting techniques for "regular" video also apply to Web and vodcast (video podcast) formats. However, there are some differences. Use the following tips as you develop videos for the Web or for portable video devices:

- ***If you use pans, tilts, and zooms, do them slowly.*** If you move too quickly, the shot, when it is shown on the Web or as a vodcast, may look choppy.
- ***Keep shot composition (long shot, medium shot, close-up shot) simple.***
- ***Use a tripod when shooting video.*** Steadier shots are easier to see on the Web and in vodcasts. If you do a pan, tilt, or zoom, make sure the camera is mounted on a good tripod.
- ***Tight shots let viewers recognize faces and objects.*** Avoid faraway, panoramic shots.
- ***In the editing process, cuts work best for Web videos.*** Minimize the use of dissolves or wipes as transitions.
- ***Titles and other on-screen text need to be bigger because of the smaller screen size on the Web and on portable video devices.***

VIDEO RELEASES

Finally, if the finished video is going to be used for profit, such as in an advertisement, it is a good idea to get a *video release form* signed by everyone in the video. A video release gives you permission to use the video in ways that you specify (e.g., in an advertisement,

© Cengage Learning 2012

Release Form

I, (<u>INSERT NAME</u>), hereby grant permission to (<u>INSERT ORGANIZATION'S NAME</u>) to utilize my appearance, the use of my work, or the use of my music by photograph, digital reproduction, videotape and/or audio tape in the program(s) listed below. In doing so, I agree to release (<u>INSERT ORGANIZATION'S NAME</u>) from claims made by me or any third party in connection to my appearance/works and give (<u>INSERT ORGANIZATION'S NAME</u>) the right and permission to publish and republish this audio and/or visual material in print, film, tape or any other media.

I warrant that any and all material furnished by me for the program(s) listed below is either my own or is authorized for such use without obligation.

I agree to the use of my name, likeness, voice, work, and biographical material about me for the program's publicity and organizational promotional purposes.

PROGRAM TITLE(S): _____

PRODUCTION DATE(S): _____

Signature: _____

Name Printed: _____

Current Address: _____

_____

Age (if under 18): _____

I represent that I am a parent (guardian) of the minor who has signed the above release and I hereby agree that I and said minor will be bound thereby.

Signature: _____

Name printed: _____

For (<u>INSERT ORGANIZATION'S NAME</u>):

Name: (OF PERSON IN THE ORGANIZATION WITH AUTHORITY TO SIGN)

Date signed: _____

FIGURE 10-24: Release form.

in an educational program). Figure 10-24 shows an example release form.

If you are shooting video of a classroom, you will need a signed release form from all of the students in the classroom. For persons under 18, a parent or guardian would need to sign the video release form. If you do not receive a signed release form from some students

or their guardians, those who did not provide a form should not be included in the video. You may have them move to a different part of the classroom while you shoot the video. For purposes such as in-school videos, video release forms may not be necessary. It is best to check with your school administration to see if a video release form is needed.

SUMMARY

This chapter introduced you to the fundamentals of video and audio production. The keys to a good video or audio production are understanding who your audience is and developing a program that meets the needs of your audience. A good script and storyboard (or shot outline) will help you stay on track in the production. Only use camera movements (pans, tilts, and zooms) and video transitions (dissolves, wipes) when there is a motivation to do so. By using the information presented in this chapter, you should feel more comfortable in developing videos and audio programs.

CHAPTER EXERCISES

APPLICATIONS

Following are some ways you can apply what you learned in this chapter:

- Video record expert guests and then show the interviews in class.
- Develop video clips to supplement an educational program.
- Develop a video virtual tour of a laboratory or location not accessible to an entire class.
- Develop a video podcast (also known as a vodcast) that can be downloaded to members' mobile devices, such as iPods.
- Create short videos for your morning school announcements.

MULTIPLE CHOICE:

1. _____ is the audio that is naturally in the environment where the video is being shot.

   a. Natural sound

   b. Voice over

   c. Outcue

   d. Stand up

2. The aspect ratio for a widescreen video screen is _____.

   a. 4×3

   b. 5×6

   c. 16×9

   d. 20×14

3. A clip-on microphone is formally called a _____ microphone.

   a. hand-held

   b. lavaliere

   c. shotgun

   d. reflector

4. The widest shot possible that shows the object of interest in its surroundings or setting is called a(n) _____ shot.

   a. establishing

   b. medium

   c. close-up

   d. extreme close-up

5. A _____ takes place when the camera is tilted slightly left or right.

   a. high-angle shot

   b. low-angle shot

   c. point-of-view shot

   d. beginner-angle shot

6. In the _____ phase of the video production process, crew assignments are made, video shooting locations are scouted, and the video shooting actually takes place.

    a. audience and program analysis

    b. preproduction

    c. production

    d. postproduction

7. A good rule is to record at least _____ per shot.

    a. 8–10 seconds

    b. 3–5 seconds

    c. 25–30 seconds

    d. 1–3 seconds

8. _____ occurs when you have a person or an object in front of a bright light source.

    a. Backlighting

    b. Diffused lighting

    c. Directional lighting

    d. Artificial lighting

9. In the editing process, _____ work best as a video transition for Web videos.

    a. cuts

    b. zooms

    c. dissolves

    d. wipes

10. Another name for lead room is _____ room.

    a. head

    b. nose

    c. top

    d. front

11. A _____ is a horizontal movement of the video camera.

    a. pan

    b. tilt

    c. zoom

    d. jump cut

12. A _____ is a detailed written description of the video you plan to shoot.

    a. shot outline

    b. storyboard

    c. caption

    d. shot sequence

13. _____ describes the speed or rhythm of the program.

    a. Continuity

    b. Composure

    c. Pacing

    d. Composition

14. A _____ is a series of drawings with captions that describe video shots and their accompanying audio or narration.

    a. shot outline

    b. storyboard

    c. caption

    d. shot sequence

15. _____ refers to the way the various elements within the video screen are arranged.

    a. Head room

    b. Lead room

    c. Continuity

    d. Framing

REFERENCES AND FURTHER READING

Diggs-Brown, B., and J. Glou. *The PR Style Guide: Formats for Public Relations Practice, 2nd ed.* Belmont, CA: Wadsworth, 2007.

Edwards-Tiekert, B. "Writing for Radio: The Basics." Last modified 2005. Accessed October 23, 2010. http://www.kdrt.org/resources/lpfm/writing.

Free Speech Radio News. "Writing for Radio." Last modified 2005. Accessed October 23, 2010. http://www.newscript.com/.

Marsh, C., D. W. Guth, and B. P. Short, B. P. *Strategic Writing: Multimedia Writing for Public Relations, Advertising, Sales and Marketing, and Business Communication.* Boston, MA: Pearson Education, 2005.

Newscript.com. "Newswriting for Radio." Accessed October 23, 2010. http://www.newscript.com.

Telg, R. "Writing for Broadcast." Last modified 2000. Accessed October 23, 2010. http://aee3070.ifas.ufl.edu/Writing.htm.

Telg, R. "Producing an Educational Video." Last modified 2004. Accessed October 23, 2010. http://edis.ifas.ufl.edu/WC024.

Telg, R. "Producing Your Own Video Program." Last modified 2004. Accessed October 23, 2010. http://edis.ifas.ufl.edu/WC022.

Zettl, H. *Television production handbook,* 11th ed. Belmont, CA: Wadsworth, 2012.

11

Writing and Designing for the Web

OBJECTIVES

After completing this chapter, the student will be able to:

- Write and design a basic Web page.
- Develop common navigational and visual elements to help Web users find information.
- Implement the basics of hypertext markup language.
- Describe the elements of a good website.
- Use a Web editor to create a Web page.

INTRODUCTION

*The **World Wide Web**, or **Web**, is one of the major technological developments of the past 20 years. It has affected every aspect of modern life, from the way we work and study to the way we find information and entertain ourselves (Figure 11-1). The Web is one part of the **Internet**, which is the global computer information network that also includes e-mail, listservs, blogs, social media applications such as Facebook and Twitter, and search engines such as Google.*

This chapter is designed to help you understand the basics of how to write and design for the Web. You will learn about the elements of a good website, how graphics work on the Web, and the basics of hypertext markup language (HTML)—the language of the Web—as well as how to use Web editors to create Web pages. This chapter will explain computer graphics formats and screen resolution on the Web, and how common navigational and visual elements are utilized to help Web users find information.

WEB BASICS

What we know of today as the Internet began in the 1970s as a government computer network that people utilized to send information from one computer to another. Unlike the high-speed and wireless connections of today, computers 20 or more years ago were electronically connected to each other

FIGURE 11-1: The Web has impacted every aspect of modern life, and knowing how to write and design for the Web is a valuable communication tool.

© Monkey Business Images/www.Shutterstock.com

only by telephone lines, which allowed computers in different physical locations to "talk" to each other. Since each computer had its own individual **IP (Internet protocol) address**, which is a series of numbers that connects a computer to the Internet, all kinds of information could be shared back and forth.

Originally, the Internet was based on exchanging information as text files, but eventually software engineers discovered a way to send other types of files too, and to display them in a **Web browser**, a graphical interface on your computer that lets you type in a **URL**, which stands for **uniform resource locator,** the formal name for a Web address. A URL looks like this: "http://www.organizationname.com." The three parts of a URL are as follows:

- **http:** which stands for *hypertext transfer protocol.* It is the set of rules for transferring files (text, graphic images, audio, and video) on the Web.

- **www:** which stands for "*World Wide Web.*" Many addresses today do not formally have "www" in the URL, but it is "understood" to be there.
- ***Domain name:*** indicates the type of website.

Domain names can include companies, nonprofit organizations, governmental agencies, educational institutions, and countries. Here are some examples:

- *.com* indicates a company or business.
- *.edu* is an educational organization, such as a school or college (Figure 11-2).
- *.gov* is a governmental website, such as the U.S. Department of Agriculture.
- *.org* indicates a nonprofit organization.
- ***Countries also have their own domains.*** For example, the domain name of Canada is *.ca*, and the domain name of England is *.uk*.

Once you type in the Web address, the site where the page files reside actually loads a copy of the Web page

FIGURE 11-2: This example website has an .edu domain name, because it is for an educational organization. Using multiple columns within a site's layout is a great way to efficiently use space and organize content. This design exhibits a horizontal navigation at the top, with an image banner used to promote the educational organization. The two-column layout is a simple way to organize content with less vertical space, which means that the visitor does not have to scroll down to see the entire story or article. Note that this and some of the other website examples in this chapter use "greeking," which is dummy text, most often using the Greek language, done to show where text will be placed. (*Courtesy of Kevin Kent*)

onto your computer. This simple idea has evolved into a virtual world of possibilities for users, and for many different kinds of services and activities that did not exist before. Companies such as Google, Amazon, YouTube, Facebook, and MySpace are all examples of new ways in which the Web is being put to use to change the way people live and work.

A *website* consists of individual pages that are linked together by some form of navigation. Either inside a navigation bar, or alone, image icons, buttons, or text links, when clicked, take a user to another part of a site. *Web pages* contain text, graphics, and sometimes other forms of media, such as animation or video. All of these elements, put together for some purpose and organized in a specific way, are what is known as the "content" of a website. The technical elements of a website all exist to support and display the content electronically.

Websites are said to be *nonlinear*, because, unlike a book or magazine, users do not typically follow a step-by-step organized path through the content. Instead, they use *hyperlinking*, which embeds a hyperlink (link)—a file name or address to another Web page—into a Web page. Users can choose which links they want to follow, in a pattern of their choosing.

Content on a website can be static and unmoving, or it can be interactive, which is what we call a website or page when it contains something that moves or otherwise engages the user—such as an animation, a video file, a game or a form that asks for user input, a search function, or a quiz. Although it is important to maintain interactive websites, the actual information content contained on a website is text-driven in nature. Because computer screens are hard to stare at for long periods of time, Web writing tends to be short and to the point.

WRITING FOR THE WEB

Although you can transfer the content of your printed documents directly to Web pages, it usually is a good idea to write specifically for the Web. Writing for the Web is not difficult, but it is a little different than writing for a printed document. Following are tips to becoming an effective writer for the Web:

- *Know your audience.* Use the right tone and style for your audience.
- *Keep paragraphs short,* one theme per paragraph.
- *Use headings and subheadings to organize information.*

- *Proofread.* Typographical errors on a Web page cause a site to lose its credibility with an audience.
- *Label all hyperlinks.* People should know where they are going when they click on a hyperlink on your page.
- *Do not overdo the hyperlinks.* Having too many hyperlinks may cause a reader to "link out" to another website. You want readers to stay on your site. Use other links as resources for readers to visit, but try to keep readers on your page.
- *Use bold text and colored text for emphasis.* Do not use light-colored text if readers will print out pages.
- *Use color and graphics to create visual interest and direct the reader's focus from one topic to the next* (Figure 11-3).
- *Use good layout and design techniques, such as white space, bulleted lists, headings, and subheads, to "chunk" information* so that it is easier for your readers to comprehend the information.

WHAT MAKES A GOOD WEBSITE?

Before you begin making a website, take the following questions into consideration. You may see some similarities in these questions with those that you should ask with other communication processes (video, document design, visual communication, and others) that are discussed in this book:

- *What is the purpose of this website?*
- *What needs to be communicated?*
- *Who is the target audience?*
- *What does your audience want to know, and what do you want to convey?*

After you have answered these questions, you are ready to conceptualize the pages for your website. Here are some ideas as you get ready to build your site:

- *Design an organizational chart for the website,* a site plan or model of what goes on each page and how the pages link together.
- *Think about the sites you use on a regular basis.* What do you like or dislike about them? Use these answers to help you incorporate the "likes" and exclude the "dislikes."
- *Decide on the site's theme and style.* Will it be more informational or educational, or will it be entertaining?
- *Determine how many pages you will design.* Small personal sites can be three or four pages,

FIGURE 11-3: This website is a good example of a site that uses attractive color, graphics, and design layout to create a visually appealing page that will attract visitors. (*Courtesy of Kevin Kent*)

whereas large sites can have dozens or hundreds of pages.

Websites follow the conventional rules of good document layout and design. They are, after all, a form of a "page" and are similar to print when it comes to good use of design. Refer to the section later in this chapter titled "Good Design Principles for the Web" and Chapter 6, Document Design, to review elements of design.

The interactive features of a website add some additional factors that should be taken into consideration when determining what makes a good website. For example, users do not read Web pages; they scan them, trying to pick out cues to information for which they are looking. Users do not like long scrolling pages. They prefer shorter pages that are contained within the page window and do not require lots of scrolling. They like white space and visual graphics, but graphics need to be complementary to the rest of the page, and not too big and overpowering, which can slow download times. Good websites share these attributes:

- Accurate, helpful, and useful
- Easy to use and navigate
- Visually interesting, not boring
- Informative
- Not too cluttered or difficult to read

Readers want visually interesting Web pages. You should use these tips to attract attention and maintain readers' interest in your Web pages:

- Give your page a theme and style (consistent color, font, page elements).
- Give readers something interesting to look at from the first page.
- Design your pages to be visual, not text-driven.
- Do not overuse animation and effects.
- Make a site index (also known as a site map), which is a page that includes links to all the other pages in the site in one place.

HTML: THE LANGUAGE OF THE WEB

HTML stands for **hypertext markup language**, which is the programming language or code used to create pages on the Web. Originally, programmers who knew code typed out the basic "tags" into a text editor such as

© Cengage Learning 2012

Erin Freel Best

President, The Market Place

Make every word count. In most cases, visitors are not there (on a web page) long. Put yourself in the shoes or, in this case, at the fingertips of the visitor. What would they be looking for or trying to buy? Make it happen for them, and they will return to you again and again as a resource.

Windows Notepad or Winpad and saved the files with an .htm or .html extension to create a Web page. These days, Web developers use HTML editors (referred to as Web-editing software programs), which automatically generate the HTML code, or they use **cascading style sheets**, also known as **CSS**. CSS styles were developed in the 4.0 version of HTML as a way of defining how to display HTML elements in a Web browser. CSS is beyond the scope of this textbook, but it helps to know that there are multiple ways to "code" a Web page.

HTML codes are sometimes called *tags* because the actual code must be placed inside a "container tag" or "bracket." Spaces and returns do not matter, as HTML will not read them. Commands are not case sensitive, but there are rules that must be followed in order to have the pages display properly. The tags are often nested, so that, for example, all the code for a given page must be placed inside the <html></html> tag. This is read as "open html" and "close html." Anytime you see a slash mark inside the brackets—</>—it means "close the tag." Inside this tag, the head information, which gives the Web page a title that shows up in the browser display, must be placed inside the <head></head> tag, while everything displayed on the page itself must be contained in the <body></body> tag. HTML format tags can express attributes such as font color and style, can be nested inside of each other, and can be used to format how the page looks or to insert a particular sized graphic.

Common HTML Tags

Common HTML tags are as follows:

- *HTML Code Tag:* <html> </html> is placed at the beginning and end of each HTML-coded page.
- *Header:* <head></head> descriptive information.
- *Title:* <title></title> puts title in the top line of the Web page.
- *Body:* <body></body> all text, graphics, and photos must be between these two HTML tags on a page. The *only* tag that comes after the close body tag </body> on the page is the close HTML tag </html>.
- *Paragraph:* <p> puts lines in on a Web page. You do *not* have to have a "close paragraph" tag </p>. This is one of the only tags that does not require a "close" tag.
- *Heading sizes for text:* The smaller the number, the bigger the heading (1−6).

   Example: Heading <h#></h#> (where the # sign denotes a number from 126).

- *Links* are the basic element that makes Web pages interactive. The link tag is <a href>, which stands for *anchor hypertext reference*. You should put quotation marks around the linked page file you are referencing. You also need a "close anchor" </a> to close the hypertext link tag.

   Example: <a href="gallery.htm">Gallery Images </a>

This means that the words "Gallery Images" will be hyperlinked to the page "gallery.htm."

- *Graphics/photos:* You have to use an *image source* <img src> tag to insert graphics.

  Example: <img src="balloon.gif">

This means that the image (graphic) "balloon.gif" will be inserted on your Web page.

- *Linking a graphic:* You can link an image to another page by putting hypertext references around the image. (Do not forget the </a> (close hyperlink) tag.)

  Example: <a href= "index.htm"><img src= "ballon.gif"></a>

This means that the image "balloon.gif" will be hyperlinked to the page "index.htm."

- *Colors:* You can have different colors for text and backgrounds. Certain word colors or numeric combinations can be typed in to represent different colors. Colors are explained in more detail later in this chapter in the "Web Color" section.

  Example: <body bgcolor="white"> <body bgcolor="008080">. This equals dark cyan.

Color can be added to the following types of text on a Web page:
  - <body bgcolor="#colorvalue"> sets the background color for the page as a whole.
  - <body text="#colorvalue"> sets the text color.
  - <body link="#colorvalue"> sets the unvisited link color.
  - <body vlink="#colorvalue"> sets the visited link color.

- *Tiled backgrounds:* Instead of a color, you also can have a tiled background for your Web pages by including the file name of the image as the background "source." Tiled backgrounds are explained in more depth in the section "Background Tiles and Buttons" later in the chapter. The background tag must be typed inside the body tag.

  Example: <body bgcolor= "filename.jpg"> means that the image "filename.jpg" will be repeated ("tiled") as the background of this page.

- *Lists:* There are two types of lists that can be included on Web pages: unordered lists, also known as *bulleted lists*, and ordered, or numbered, lists. *Unordered lists* <ul></ul> have bullets.

Ordered lists <ol></ol> are numbered. You must put <li> before each line of the list.
This is an unordered list:

<ul>
<li> First line
<li> Second line
<li> Third line
<li> Fourth line (add as many as necessary)
</ul>

The unordered list above would look like this on your Web page:

- First line
- Second line
- Third line
- Fourth line (add as many as necessary)

This is an ordered list:

<ol>
<li> First line
<li> Second line
<li> Third line
<li> Fourth line (add as many as necessary)
</ol>

An ordered list, like the one above, would look like this on your Web page:

1. First line
2. Second line
3. Third line
4. Fourth line (add as many as necessary)

- *Other codes:* Although only a few HTML tags are needed to create the basic structure of a page, there are many tags that can be used for specific formatting and alignment of elements on a page (Figure 11-4). You may find the following HTML codes useful when inserting text on your Web pages. These will boldface, italicize, or indent the text on both sides:

<b></b>boldface
<i></i> italics
<blockquote></blockquote> Sets apart text. Indents text on both sides.

GOOD DESIGN PRINCIPLES FOR THE WEB

Good design for the Web includes designing for the right size monitor and screen resolution. The smallest monitor size, 640 by 480 pixels, is seldom used

Quick Tutorial on HTML

By typing in the hypertext markup language (HTML) coding below, you will be able to create a Web page that looks like the screen capture:

Make a file folder on your computer desktop.

Open up **Word or a similar word processing program.** Size the window so it will be on right half of the screen.

Save the page as a **TEXT (.txt) file**, but label this blank page as **example.html.** (Remove the **.txt** ending from the file name ending and use **.html** instead.)

Type in the coding tags and text below:

```
<html>
<head> <title>Example </title></head>
<body>
<p>Everything in here is the body.
</body>
</html>
```

Save the file.

Go to your Web browser and open the file. It should look something like this.

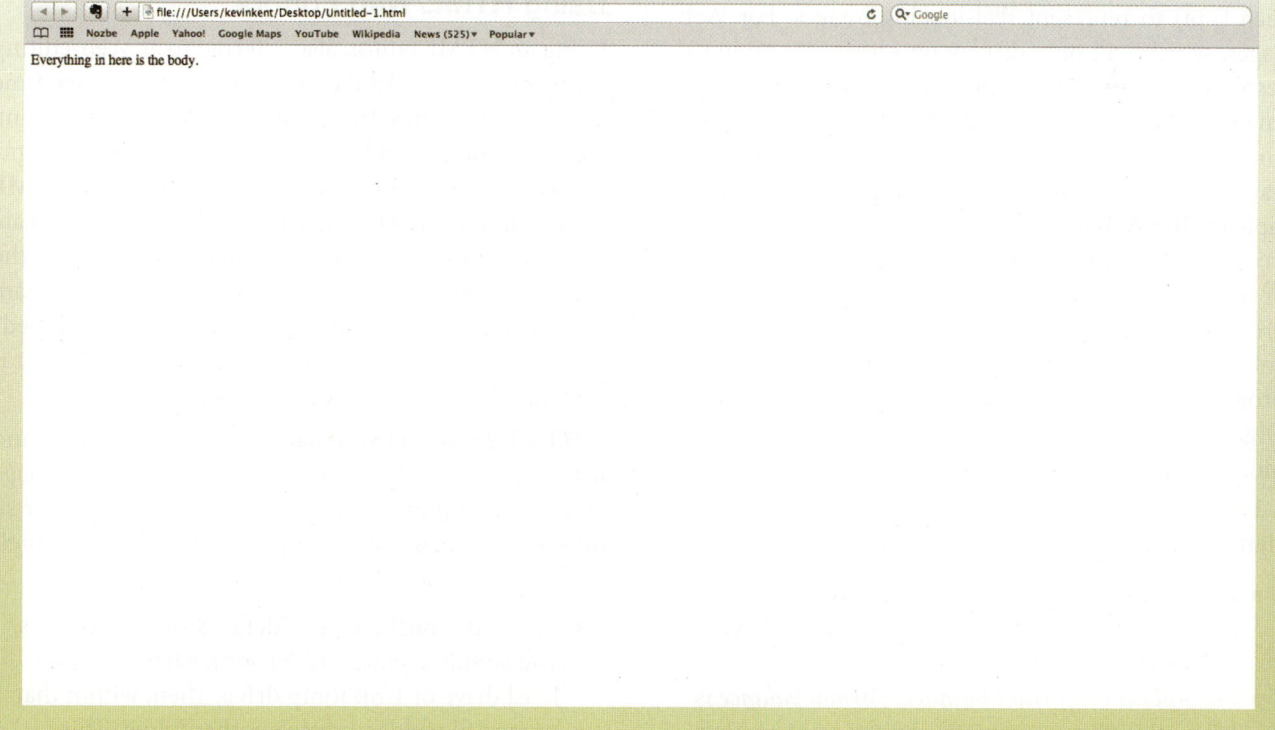

FIGURE 11-4: HTML tags are sometimes called "nested" tags, as all page design tags are nested inside either the head or body tags.

today. Most monitors display at either 800 by 600 pixels or 1024 by 780 and even larger. A **pixel**, which stands for *picture element*, is a single point in a graphic image displayed on a computer screen. The number of *bits*, or elements of information used to represent the pixel on a computer monitor, influences the number of colors that can be displayed, and the number of pixels used to display an image on the screen is called its **resolution**. A pixel is the single point on the screen of an RGB (red, green, blue) computer monitor. Pixels contain color information that, taken together, combines to create the images your eyes see—the more pixels, the more depth to the image. For 8-bit, 24-bit, and 32-bit color monitor systems, each bit represents four channels of color information.

Pixels are composed of a red, green, and blue element, which is why a computer monitor is sometimes called an "RGB" monitor. RGB stands for the red, green, and blue light that combine to create the colors a computer monitor can display. The pixels are so close together in an image that the eye is fooled into seeing the single bits of information as a whole. The finer the resolution, which means the more pixels it takes to display an image, the better the image quality will be. That is most important for printed elements, such as photographs, which need to be at very high resolutions of at least 300 pixels per inch (ppi) to represent real-looking images. For Web pages, which will only display on screen, the monitor resolution default of 72 ppi is usually enough. Larger images with more pixels will take longer to download onto the user's screen. Even though many users now have high-speed Internet access, a good design principle for the Web is to keep graphic file sizes small, and to optimize for the Web. **Optimizing** means resampling or re-creating an image to use fewer colors and pixels in order to compress the file size. As the file size decreases, colors disappear. You can do this in an image editing program such as Adobe Photoshop. Refer to Chapter 9, Digital Photography and Photographic Editing, for more on resizing images for the Web.

In addition to screen resolution, other important design principles include the following:

- *Division of space:* White, or negative, space is used to focus the user on page content and avoid a cluttered look.
- *Formal and informal balance: Formal balance* is when objects are symmetric on the right and left sides of the page. *Informal balance* is when one side is intentionally unbalanced. More and heavier elements draw the eye as a focal point.

- *Visual vectors:* Vertical, horizontal, and diagonal "lines" that are formed by the way objects are arranged on a page. The eye sees the connection of close objects, and the user is, as a result, drawn to read through the page.
- *Repetition:* The same or similar objects appear in a pattern that is attractive, such as a row of navigation buttons.
- *Contrast:* The relationship between light and dark elements on a page. The farther apart the objects are in lightness and darkness, the more contrast there is. Contrast adds interest to the design, attracts the eye, and adds depth to an image or a collection of elements, making them look more three-dimensional. A good example of this is a drop shadow on an image (Figure 11-5).
- *Shape, size, color, texture:* Elements of an object or text that make it stand out and complement the rest of the design.
- *Typography:* The art of choosing type. Typefaces or fonts are styled formats for the characters in the alphabet. Just like in print, Web pages can use different typefaces to emphasize text elements.

Using HTML/Web Editors

Using an HTML editor, also referred to as Web-editing software, such as Adobe Dreamweaver, can save time and produce attractive websites. With Web-editing software, you do not have the control that Web designers who know CSS (cascading style sheets) or HTML coding have, but HTML editors allow you to create Web pages quickly. You can also find free editors on the Web, sometimes called "shareware" because users share the software for free. These editors are often stripped-down versions of a for-sale product, downloadable for free for a limited time—called a "try and buy."

HTML editors let you make pages, insert images and text, create links and tables, add a background image or color, and publish or upload the site to a server so others can access it by typing the Web address in their browser. To make a basic site in an HTML editor:

- First, you must create a "defined site." To do this, you should create a folder somewhere on your hard drive or USB jump drive. Then, within that named file folder, create another folder. This one should be named *images*. Place all of your photographs and graphics in this folder. Once you have done this, you can then use the Web-editing

FIGURE 11-5: This Web home page uses contrast between the vivid colored banner and the white page, as well as drop shadows around the full-color photo frames, to provide a focal point for the viewer. This design features a creative layout that welcomes the visitor with a variety of photos and general information about the company or product. The contrast of the background and placeholders makes it easy for the visitor to locate the important information on the page. The page also features a horizontal navigation at the top of the page, making it easier for visitors to navigate to specific parts of the site. In addition, the right column is designed to contain quick links, or links that are most commonly used by the site's visitors. The top horizontal navigation allows the visitor to see what the site has to offer without having to scroll down the page. Social media icons are also located at the bottom of the page for the visitor to connect directly with the company or product's Facebook or Twitter accounts. (*Courtesy of Kevin Kent*)

software to "define" the site, so that all of the pages you create will be saved to the original file folder. You will insert images on your Web pages from the files that are saved to your *images* folder.

- After you have done this, you can start creating Web pages in the software program. Your main page is your "home page." This is the page your Web viewers will access first. The file name that you use to save your home page should be "index. html" or "index.htm." It should not be "homepage. html" or "firstpage.html" or the topic of the page. Only use "index.html" or "index.htm." Web browsers have been set up to recognize "index" to be a home page. When you visit a website, the home page's actual Web address "hides" the "index.html" portion of the address, so you do not have to type it. For example, if you visited CNN's website, the actual address is "www.cnn. com/index.html," but because the Web browser recognizes "index.html" as the home page, all you have to type is "www.cnn.com." If you save your home page as something other than "index.html," you will have to type the entire

Web address—such as "www.organization.edu/ firstpage.html"—in order to access the home page.
- Make your home page and its navigation first. Spend time designing this page and then use it as a template for other pages on your site.
- Insert buttons and banners already made in an image editor, such as Adobe Photoshop.
- In naming your Web pages, you should never use capital letters or use spaces between letters. You may use underscores if you want to put in two words, but never leave a space between letters.

Most HTML editors have a "publish" or "upload" function, also called **FTP**, or **file transfer protocol**. If you do not see this, use the help index to search for it. Once your pages are all created and linked to each other, you can use this function to upload your site to a server. You must have an account or permission to do this. If a student organization wanted to make a website, for example, you would need to check with your school's Web administrator to get permission and to get access to a folder on the school's server where you could publish your site files.

If you want to develop a website but do not want to learn HTML code or use a Web-editing program, another option is to use a template. A website template gives you an already-coded design with appropriate images, links, and style into which you can insert your own information. Some of the more sophisticated templates also let you manage content and use tools such as blogs, discussion forums, and forms on your website. An example of a more sophisticated template producer is WordPress, available at http://wordpress.org/. WordPress calls itself "a state-of the-art publishing platform with a focus on aesthetics, Web standards, and usability."

An advantage of using a template is that it will have built-in standards, such as what is required to make a website useable by someone with a disability. *Usability* refers to the general ease or difficulty your users have in navigating your website and finding what they need. Creating a website from an existing template, such as WordPress, helps ensure that your website meets usability standards. If you are developing your own site using an HTML editor or by typing in HTML code by hand, you can test your site by showing it to a few of the people who you think will be using the site to see what they think about its ease of use.

Why Learn HTML?

Given the option to use a Web-editing program, you may wonder why anyone needs to learn to use HTML. In fact, understanding the basic structure of a Web page is useful, and when things go wrong, it is often much easier to look at the code to figure it out and adjust. The leading Web editors, such as Adobe Dreamweaver, let you display both the code itself and the actual page at the same time to make this easier. Dreamweaver can also generate CSS, which includes specialized code that allows you to define and create the formatting and design of a Web page with great control.

Images and the Web

Only three specific image file formats can be displayed in a Web browser:

- **GIF (Graphics Interchange Format):** a graphics format that is highly compressible and includes an 8-bit (256 colors) color palette. GIFs are used primarily for line art and for animated graphics. GIFs can be animated, but JPGs (see below) cannot be animated. Photographs should not be saved in GIF format.
- **JPG (Joint Photographic Experts Group):** a format most frequently used for photos. It is not

as good for flat colors and line art drawings. It is the most common photo format for the Web (32 bit, which translates to more than 16 million colors).
- **PNG (Portable Network Graphics):** public domain graphics. This format is not often used.

On the Web, where small file sizes are important, *compression* is used to reduce image files down to manageable sizes. GIFs have *lossless compression*, which means the original information is retained as the file is compressed by an algorithm on a line-by-line basis. Lossless compression is required for simple graphics, text, and data files. JPGs use *lossy compression*, where file information is lost after it is compressed. Lossy compression is used commonly to compress multimedia data (audio, video, photographs). Some designers advise keeping individual graphics under 70 kilobits each. Because your monitor's resolution is set at 72 ppi (pixels per inch), you do not need high-resolution images sizes. Images you want to use in other file formats must first be converted to a Web file format using either a Web editor or an image editor, such as Adobe Photoshop.

Web Colors

Colors on the Web are formed on your computer display monitor using a six-number system called a **hexadecimal color number**. This formula converts the colors your computer monitor displays to mathematical equivalents of other colors. You can use any color you want in an HTML page, but there is a standard 216-color palette that offers a choice of colors that will appear unchanged on almost any computer platform (PC or Macintosh) or browser (Internet Explorer, Firefox, Safari, and others). The system color palettes in Windows for the PC and the Mac are quite different, and this occurs with many other computer systems too. The 216 *"Web safe"* colors display the same way on any monitor system.

The hexadecimal color designation for black is #000000 and for white is #FFFFFF. There are charts that show the hexadecimal values for specific colors. HTML editors let you pick colors visually, so you do not need to remember all of them. If you want to match a specific color on a button to a banner you are adding to a page, you may need to use the specific hexadecimal values, since your eye is not sensitive enough to see the difference. You can insert the hexadecimal color by typing in the values in an HTML editor or inserting the color as part of an HTML tag.

Web Page Design and Tables

One way that designing for the Web is different than designing for print documents is the need to include interactive elements, such as incorporating hyperlinks, on Web pages. Another difference is the **aspect ratio**, which is the ratio of page height to page width. For the most part, a printed page is a vertical space, whereas a Web page visible on a computer monitor is horizontally wider than it is tall. That means it is important to design your elements to fit across the page as well as down.

Until very recently, most designers used invisible tables to establish the spatial relationships and dimensions of objects on a Web page. Now professional Web designers prefer to use CSS and limit the use of tables, but if you are new to making Web pages and using a Web-editing software program, you will probably find tables very helpful to organize the elements on your page. Tables are series of columns and rows that create boxes or cells to hold text, color, and graphics (Figure 11-6). There is an HTML tag that "turns off" the lines around the columns and rows, which is why the resulting table is invisible. Using invisible tables allows Web designers to anchor elements in relationship to each other so they do not move around when the Web page containing them is displayed on different monitors or in different browsers. Using tables creates a page grid that can be a good design element, allowing beginning Web designers to use invisible tables to position elements. If you are using a free shareware Web editor, hand coding, or using a template that you want to adapt, understanding how tables work will help you control your design elements and create an attractive Web page.

Background Tiles and Buttons

Trends in design elements for websites change frequently, and Web pages targeted to specific audiences might differ in how they use color, interactive elements, and background effects. In general, Web pages tend to either have a white background, which follows the design principle of using white or negative space, or they use a color background to convey more of a three-dimensional space. Filling in the background of a Web page can be done either by adding a background color or by making a *background tile*, a small image that, when set as the background, repeats or tiles itself to create a seamless page, like bathroom wall tile (Figure 11-7). To make a background tile, you want to create a small image that has a limited amount of texture or lines; otherwise, these will tend to separate the seamless image into chunks, creating a checkerboard effect.

Buttons are often used as navigation devices. They can stand on their own, but are most often incorporated into the navigation bar or banner on a Web page. By clicking on a button, you know you will be linking to another page (Figure 11-8). *Banners* are like headlines

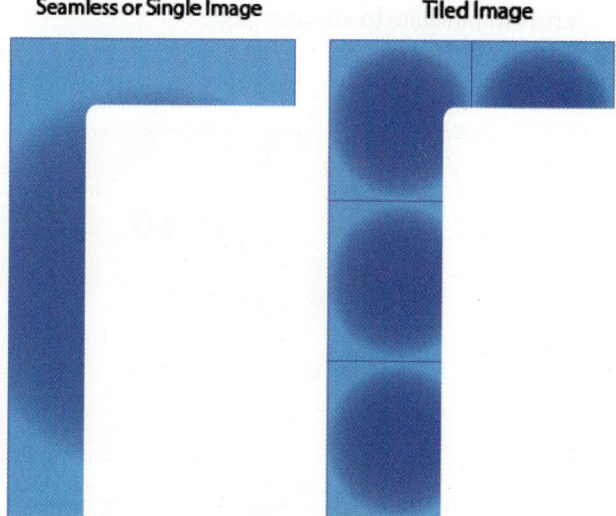

Seamless or Single Image   **Tiled Image**

FIGURE 11-7: Seamless tiles will look like a seamless background, whereas any tile with textural lines will look like a checkerboard when "tiled" (repeated many times across a screen). (*Courtesy of Kevin Kent*)

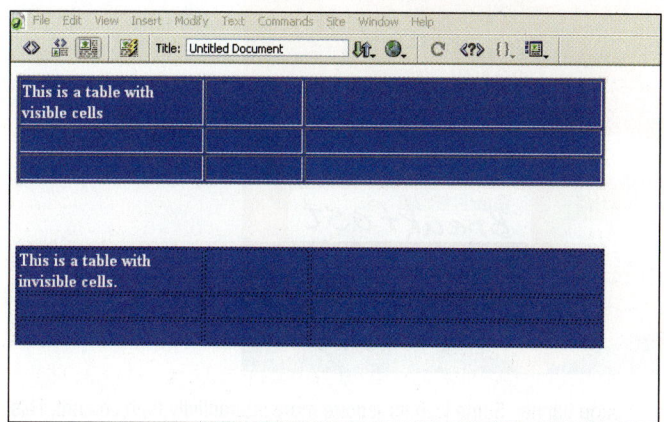

FIGURE 11-6: Tables can be invisible, or they can be drawn with lines you can see. (*Courtesy of Kevin Kent*)

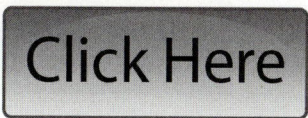

FIGURE 11-8: Buttons are a simple and effective way to indicate to users that something is clickable on a Web page. (*Courtesy of Kevin Kent*)

in print; they contain a head or title for the page, and use graphic elements such as color, texture, and shape to make the head stand out to the user and draw attention to the button or text links on the page.

NAVIGATION AND LINKS

Navigation

Navigation is an important element of a good website. Navigation is how your users get around the pages that make up the site to find information. Over time, certain elements have evolved as efficient techniques for getting users to move from one page to the next. A splash page, for example, is an opening, attention-getting page with graphics that either automatically refreshes to your home page or is clickable. You can use these tips to help people navigate your pages more easily:

- **Button:** Buttons, which can be made as images, can be inserted on your page or into a table as navigation aids.
- **Table cells:** You can set up table cells, and fill the table cell with color and text to create clickable areas to navigate to another page.

- **Icons:** You can use small images, known as *icons*, which you then make clickable as links. Make sure you use BORDER = 0 to remove the link border.
- **Rollovers:** You can create *rollovers*. A rollover is a button that changes to another image—sometimes a different color of the same button—when a mouse rolls over the image. This draws attention to the button and helps the user navigate to another page.

Whatever you use for navigation, it is strongly recommended that your navigation be consistent on all pages. For example, if your navigation buttons are at the top of your home page, your other Web pages should have the navigation buttons at the same location at the top of subsequent pages. You may want to include a list of links to your pages at the bottom of your home page, in very small typeface, called "breadcrumbs." If you do that for your home page, you should do it for all your pages. Most navigation bars are found at the tops or bottoms of Web pages. You should have links to all of your pages, including your home page, on your navigation bars on all pages.

You can link a row of buttons at the top of your page in a navigation bar by putting individual buttons in an invisible table, side by side (Figure 11-9). Another way

FIGURE 11-9: This website uses a navigation bar with buttons in its home page banner. Some layouts require more interactivity than content. This design incorporates several interactive features that include a rotating banner, a text field for asking questions about the product or company, as well as a pull-down menu for specific product and company details. It is important not to overload interactive pages with excessive content, risking that your site's valuable information may not be read. (*Courtesy of Kevin Kent*)

Erin Freel Best

President, The Market Place

For website development, think "Function first. Design second." But you better make sure the design is eye-appealing as well. Presentation is everything. If you have a business, patrons, customers, and members expect you to have a website that is easy to navigate. In today's age, it's bad customer service not to be available 24 hours a day, and with the Web you can accomplish just that.

to use a line of navigation buttons is to create one button that stretches across the page and to make an image map. An image map lets you link parts of an image to different links. So with your button that stretches the width of the screen, you could do an image map of separate parts of your button to link to different pages. If you see a map of the United States and can click on separate states, an image map was used to create that graphic. You can create an image map in an HTML editor or an image editor such as Photoshop.

Links

Web pages can include several types of links. Links connect to other pages, or to other elements in a page. When text or an object is used as a link, the default is to show a highlighted underline or border around the text or image. To remove the border, you can use an HTML tag (BORDER = 0). Common types of links include:

- *Internal links:* from one page to another within a site.
- *External links:* from one page in a site to a page in another site.
- *Anchors and targets:* from one part of a Web page to another part of the same page.
- *Object links:* a graphic object that is also a link.
- *Text links:* text that, when highlighted, indicates a link.
- *Mail-to:* link that opens up an e-mail message window.

Linking is interactive and important in terms of users' evaluations of your site. Here are some ways to use linking:

- Including a "mail-to" link (for e-mail) that a user can click on to request more information
- Linking to an online poll or survey
- Linking to an online form to allow a user to leave feedback or to request information
- Linking to a bulletin board discussion forum

WHAT MAKES A WEBSITE "BAD"?

Websites provide the opportunity to use your creativity to design with graphics, color, white space, photos, and interactive elements in an almost unlimited variety of ways. But because the goal of a website is to communicate, you also need to think about designing your website so that the elements work together harmoniously and communicate in an effective and appealing way. There are some general "rules of thumb" when it comes to designing websites and some things you want to avoid—what not to do if you want your website to be an effective form of communication.

An agricultural website like that depicted in Figure 11-10, for example, uses complementary colors, text broken over three columns rather than running the width of the page, and white space around the graphics to give the page a bright, open look. On the other

FIGURE 11-10: This agricultural website is an example of a well-designed site with harmonious colors and good use of text and white space. This design exhibits a vertical navigation panel along the left side of the page. Vertical navigation is ideal to use when the navigation of a website changes frequently. Vertical navigation panels also serve as a frame or border to the content of the page. More permanent links and navigation could be placed horizontally at the top as displayed in this design. (*Courtesy of Kevin Kent*)

hand, the Web page shown in Figure 11-11 does the opposite—too many contrasting colors, text that runs the width of the page, making it hard to read, and too many elements crammed onto the page. Figure 11-12 is another example of a website with problems; in this case, red text on a black background is difficult to read, there are multiple links on the top left, and the logo graphic has been stretched and distorted.

GETTING PEOPLE TO VISIT YOUR WEB PAGE

Now that you have created your website, it is time to think about how to get users to come to it and use it. Unfortunately, there is no television channel guide for the Web. If you do not let your audience know the

address for your website and what it contains, how can you expect them to look for it?

The closest thing to a channel guide on the Web, though, is a search engine, such as Google or Yahoo. **Search engines** let you type in keywords for something you want to find online. Search engines use powerful programming languages to develop applications that search through all the pages on the Internet and automatically compile a list of site pages that contain your keyword(s). So it is important to title and name your pages something that a user would type into a search engine when looking for your site. In addition to keywords, it is also important to include links to relevant websites and then ask those sites' webmasters (the person who designs and maintains a site) if they would link to you.

FIGURE 11-11: This website is an example of what not to do—too many colors and page elements give it a cramped feeling and the text is hard to read. The colors and design of a website should be appealing but not distracting. Designers should make sure that the layout not only invites the visitor to browse the page, but also make the site's content readable and navigable. This design's background distracts the visitor from the content on the page and makes the text included on the site challenging to read. Quality graphics should always be used to avoid pixilation or distortion. (*Courtesy of Kevin Kent*)

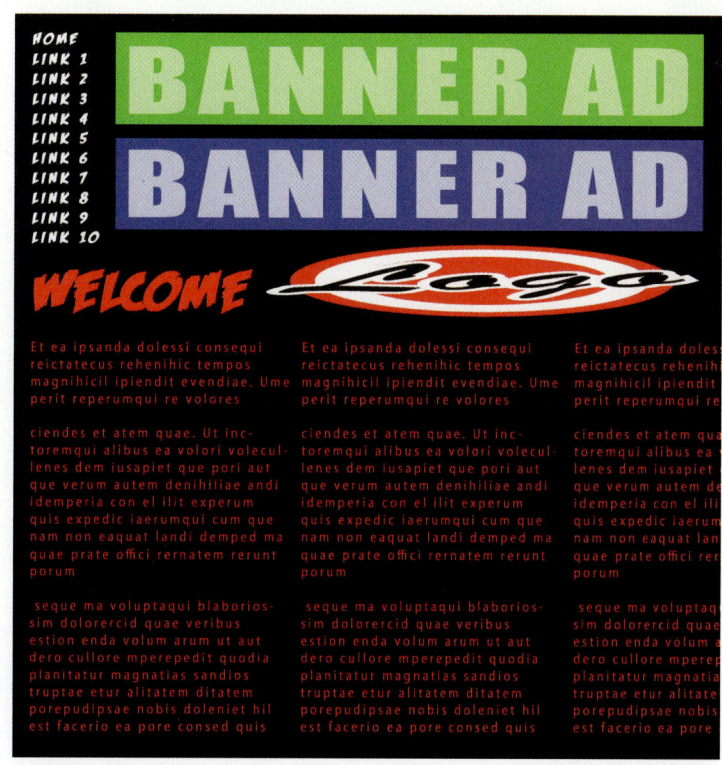

FIGURE 11-12: This is another example of website that could be made more effective by changing some of its page elements. Avoid dark backgrounds when a contrasting text or placeholder color is not used. Do not provide too many links for the visitor in the navigation. Consolidate topic areas and use subheaders and pull-down menus if appropriate. Web advertisements are a great way to create revenue for your website; however, do not let ads consume your page. Visitors may become discouraged before getting to the content of your page. Do not stretch or distort graphics to fill a blank area. The size and shape of the logo should be accommodated in the design of the page. (*Courtesy of Kevin Kent*)

SUMMARY

This chapter covered the major elements in designing for the Web. It included a look at the basics of Web design, including how to write for the Web, good design principles, and how to create and link pages, images, and buttons. Basic HTML tags were explained, as well as site navigation and linking. The chapter ended with a discussion of how to use keywords in file names and links so your website can be more easily found by search engines.

CHAPTER EXERCISES

APPLICATIONS

Following are some ways you can apply what you learned in this chapter:

- Create a personal website for yourself. A personal website allows you to display some of your electronic communication skills. You can provide a potential employer with a Web address, for example, where they can go see your skills and accomplishments for themselves. To create a personal website, you will need to do the following:
  - Make a base Web page that incorporates your name and a header of some type (could be a banner).
  - Decide what links you need to include, and set up your navigation on each page.
  - Include background information: Who you are, what you can do, your families and friends, what job you would like to have someday. Be careful, though, not to include too much personal information. Many stories of identity theft or online stalking have been reported about people who shared too much information on their websites or Facebook pages.
  - Include a link to your resume, if you have one, saved as an HTML document.
  - In addition to a link back to the home page, make sure you include a link to examples of your communication work. There is no need to include all of your work, just a few good pieces. Include material from class, homework, and field trip assignments, and anything you are proud of from other classes, internships, or work. You can scan in things created for print—papers you have written or photos you have taken.
  - Be creative and apply good design principles.
  - Be professional. Just as with social networking sites, such as Facebook and MySpace, you never know who might read your postings. Potential employers have been known to check out job applicants' Web postings to look for inappropriate materials.
- Create a website for an organization at your school. Think about what pages you might want to include, and who can provide the content. You can use a digital camera and post pictures of members, plus biographies. Or you can use your camera to take pictures of activities and post these to their site so everyone can share. How about a calendar for important dates? What links would be handy to have on your site?

CHAPTER QUESTIONS

MATCHING:

Match the component of a business letter below with its definition or characteristics. Not all terms will have a definition associated with them.

_____ 1. _____ The relationship between light and dark elements on a page.

_____ 2. _____ The 512 colors that display the same way on any monitor system.

_____ 3. _____ The format most frequently used for photographs; not as good for flat colors and line art drawings.

_____ 4. _____ A graphics format that is highly compressible and includes an 8-bit (256 colors) color palette.

_____ 5. _____ Objects are symmetrical on the right and left sides of the page.

_____ 6. _____ One side is intentionally unbalanced, with more and heavier elements that draw the eye as a focal point.

a. **GIF**

b. **Web-safe colors**

c. **JPG**

d. **Formal balance**

e. **Informal balance**

f. **Contrast**

MULTIPLE CHOICE:

7. FTP stands for _____.

    a. formatting transfer protocol

    b. formatting transcription protocol

    c. file transfer protocol

    d. file translation protocol

8. The _____ is an opening, attention-getting page with graphics that either automatically refreshes to your home page or is clickable.

    a. intro page

    b. layout page

    c. anchor page

    d. splash page

FILL IN THE BLANK:

9. A(n) _____ links from one page to another within a site.

10. A(n) _____ links from one page in a site to a page in another site.

11. _____ is the formal name for a Web address.

12. A(n) _____ is series of numbers that connects a computer to the Internet.

13. HTML codes are also known as _____.

14. _____ is the way in which HMTL editors upload or publish a Web page or site.

15. _____ is a small image that can be set as a page background by repeating itself seamlessly across the entire page.

REFERENCES AND FURTHER READING

AAAHTML. "HTML Tutorials: Learn HTML Using Our Tutorials." Last modified 2010. Accessed June 1, 2010. http://www.aaahtml.com/.

About.com. "Web Design/HTML." Last modified 2010. Accessed June 1, 2010. http://webdesign.about.com/.

HTMLWriters' Guild. "Web Design Training and Certification." Last modified 2010. Accessed April 1, 2010. http://www.hwg.org/.

Tashian, C. "How Do They Do That with HTML?" Last modified April 16, 2009. Accessed May 31, 2010. http://www.tashian.com/htmlguide/index.html.

The Webby Awards. "The Webby Awards." Last modified 2010. Accessed April 1, 2010. http://www.webbyawards.com/.

WordPress. "WordPress." Last modified 2010. Accessed April 1, 2010. http://wordpress.org/about/

INTRODUCTION

On February 4, 2010, Facebook, originally conceived of as a simple website started by a student in order to keep in touch with friends, celebrated its sixth birthday. In that short time, Facebook has become a multimillion-dollar business and a major global influence on telecommunications. Today, there are more than 500 million active users, more than half of whom log on to Facebook every day (Figure 12-1). According to the Facebook website, the average Facebook user has 130 "friends"—people a user has invited or accepted an invitation from to view their page—and spends over 700 billion minutes a month on Facebook (Facebook, 2010). Facebook was started in 2004 by Mark Zuckerberg, then a 19-year-old student at Harvard University. Currently, it is one of the fastest-growing social networking sites in the world.

FIGURE 12-1: Facebook, one of the fastest-growing social networking sites in the world with significant impact on global telecommunications, has about 500 million active users.

© Diego Cervo/www.Shutterstock.com

Facebook is a prominent example—but it is not the only example—of the rise of new media in the last decade. New media is a term often used to describe Web-based communications technologies based on the concept of file sharing, including such technologies as social networking sites, wikis, podcasts, video podcasts (vodcasts), and even cell phone videos. The term "new media" originally was coined to refer to a small set of online, Internet-based applications—primarily websites, e-mail, and search engines. That small set of Internet-based applications has evolved to include many different kinds of online media experiences, ranging from social networking media, such as Facebook and MySpace; to specialized applications, such as video-sharing (YouTube) and photo-sharing (Flickr) sites; to virtual reality environments, most notably, Second Life.

The Nielsen rating service, which tracks use of all forms of media, recently stated, "Social networking has been the global consumer phenomenon.... Two-thirds of the world's Internet users visit a social network or blogging site and the sector now accounts for almost 10% of all Internet time. 'Member communities' has overtaken personal e-mail to become the world's fourth most popular online sector after search, portals, and PC software applications" (Nielsenwire.com, 2009). According to Nielsen, in May of 2010, Facebook's audience grew 69 percent over the previous year, while the audience of Twitter, another social networking site with 190 million visitors, grew 45 percent (The Nielsen Company, 2010). Interestingly, Facebook is also the number-two-ranked online video site in the United States, with over 58.5 million users, behind Google, which has 146.3 million users (Marshall, 2010).

THE BEGINNINGS OF "NEW MEDIA" RESEARCH

Although many individuals have played major roles in the evolution of new media, the first person to become associated with thinking about the impact of "new media" technology is Canadian Marshall McLuhan. McLuhan was an English professor whose research focused on media analysis. His book *Understanding Media*, published in 1964, focused on how the effects of mass media permeate society.

In the 1960s, McLuhan was known as a "pop" philosopher. However, McLuhan's ideas are current again, as they are being applied to the new media technologies that he predicted. McLuhan's enduring legacy is in connecting the nature of new and evolving media to shifts in society and culture. From McLuhan's

perspective, technological changes in media inevitably lead to changes in audience perceptions, beliefs, attitudes, and behavior. Today, professional communicators can easily see the impact of McLuhan's work when they contemplate the challenges of putting together an integrated communications plan, a crisis communications campaign, or an informal education program. More so now than ever before, communicators must consider multiple media options that are all competing for people's eyes and ears: traditional print and broadcast, the Web, DVDs, social networking media, and more. As a result, the face of American society and professional communications fields, such as agricultural communications, is already vastly different than it was in McLuhan's time.

A NEW AND CONVERGING MEDIA

Since McLuhan's time, some scholars have predicted that technological developments in communications media will eventually lead to the point where all existing media become one medium. This is referred to as converged media. Television and computers will, in this view, merge together into one information device, capable of performing the functions of both. In practical terms, professional journalists are already "converging" their talents: TV reporters are writing print stories that may go on their TV stations' websites, videographers shoot photographs to accompany Web stories, and newspaper reporters are learning to shoot video or edit audio clips that will go on the Web (Figure 12-2).

Another aspect of this discussion is the convergence of media ownership, with fewer major corporations owning more media outlets and distribution channels. This economic element, combined with fewer, converged media distribution channels, has serious implications for "specialized" fields such as agriculture, which have traditionally struggled to gain positive media attention. In a converged world, integrating messages across a variety of media and focusing on ways to add value and build relationships with consumers and stakeholders will be increasingly important.

This new media landscape has implications for society in general, and especially for young adults who have grown up with new media options that have always been available. In 2001, Marc Prensky coined the terms *digital natives* and *digital immigrants* to convey the distinction between those born in the age of digital technology (generally accepted as post-1981, the year that the first personal computer was introduced) and those who lived in the "analog" age

FIGURE 12-2: Media convergence predicts the coming together of television, Web, film, and photography into one information appliance.

FIGURE 12-3: Digital natives, like the boy in this photo, were born in the age of digital technology, whereas digital immigrants, like the grandfather pictured, migrated from the "analog" age into the digital age.

(pre-1981) (Figure 12-3). **Digital natives** are people for whom digital technologies already existed before they were born, whereas **digital immigrants** are those not born into the digital world but who have *migrated* into digital technologies (Prensky, 2001).

Digital immigrants, in particular, retain some of the "accent" of the past. Digital immigrants are more likely to:

- Turn to the Internet second, rather than first, for information.
- Print out their e-mail.
- Print a document from their computer in order to read or edit it.

Digital natives are more likely to:

- Turn to the Internet first, rather than second, for everything from searching for information to reading a newspaper online.
- Text on their cell phones, rather than make a call.
- Own a mobile device or e-reader (iPod, iPad, Kindle; Figure 12-4).
- Download music files.

This shift or migration into digital technologies also is affecting the way traditional professional development training is conducted for journalists. Some mass media companies are retraining their journalists to become better able to use new media. For example, the St. Paul Duluth News Tribune, the St. Paul Pioneer Press, and the University of Minnesota's journalism

FIGURE 12-4: Apple's iPad is a fast-growing hybrid mobile device—a cross between a laptop, a cell phone, and an e-reader.

school received $238,000 from the Minnesota Job Skills Partnership program to retrain newspaper staffs to be effective in the twenty-first century. The goal of the training was to move away from print-based thinking for journalists and ad sales staffs and to retrain them to start thinking of new ways of delivering content and telling stories (Huffingtonpost.com, 2009).

And this new and converging media world is impacting how regular people communicate, as well. According to emarketer.com and reported in a blog by Adam Ostrow (2009), during 2008, more than 82 million Americans created content online, a number expected to grow to nearly 115 million by 2013. In 2008, 71 million people created content on social networks, 21 million posted blogs, and 15 million uploaded videos. There is overlap in the numbers, because many people who created content on social networks also uploaded videos, but the overall number of 82 million is still impressive.

Academic programs that educate students to be agricultural communicators—agricultural journalists, public relations practitioners, videographers, and Web and graphic designers—are also adapting their programs to incorporate new media. At universities across the country with agricultural communications programs, students take design and digital production courses that include print, Web, and video. Although the foundation for these courses remains journalistic writing, various forms of new media, including social media, are being integrated into revamped curricula, with the goal of training future communicators to use new media to communicate about agriculture and natural resources topics.

SOCIAL MEDIA

Social media is a form of new media that includes primarily Internet- and mobile-based tools for sharing and discussing information using two-way communication. Social media most often refers to Web-based activities that integrate technology, telecommunications, and social interaction. Social media applications let users construct and share words, pictures, videos, and audio online.

Current major forms of social media, sometimes called "Web 2.0," include **social networking sites** that let users interact with each other, upload files, and post messages, photos, and videos that visitors can see. Some general social media categories include the following:

- **Blogs (Web logs)**, which are shared online journals maintained by the writer and updated regularly, to which readers can also post comments.

- **Podcasts**, which are series of digital media files (audio or video) that are available for download by users.
- *Video- and photo-sharing sites,* such as **YouTube** and Flickr, which allow users to upload and download media files and share these with others.
- **Wikis**, which are pages or collections of Web pages designed to enable anyone who accesses them to contribute to or modify the content. The word "wiki" comes from the Hawaiian word for "fast." The largest and most famous wiki is Wikipedia, which calls itself "the free encyclopedia that anyone can edit."

Some specific and more popular social media sites include the following:

- *Facebook* is a website that allows users to join networks that are organized around interests, hobbies, and individuals. A person can add friends, post messages on other users' "walls," and maintain a personal profile. Users can upload photos, videos, and lists of personal interests.
- *LinkedIn* is an online network of career professionals from around the world. Users can create personal profiles that focus on their work, career accomplishments, and expertise. A LinkedIn profile allows users to be found by colleagues, clients, and customers.
- **Twitter** is a micro-blog that lets its users send and view other users' 140-character updates, called "tweets."

Social media, like Facebook and Twitter, attract broad audiences of users, from children to senior citizens. Marketers have begun to realize the power of social media in attracting and motivating audiences to respond. From a communications standpoint, a company or organization's use of social media is essential to stay ahead of its audience. General social media sites can be used to publicize an event, deliver a message about a new product or brand, distribute special online coupons or discounts to users, create a group of individuals who follow what you or your organization is doing on a regular basis, or place an advertising or promotional message. More specialized social media sites let you reach a target audience whose members share similar interests and might therefore be interested in a complimentary product, good, or service.

Facebook, currently the most influential social networking site in the world, maintains a page designed to help companies and organizations set up a Facebook page to attract customers. Facebook lets you browse pages by type—everything from products and services

© Cengage Learning 2012

Scott Wallin

Director, Producer Communications, Dairy Management Inc.

Social media has been amazing for us. We have really jumped head-first into social media over the last year or so, and right now if you went on YouTube, for example, you'll see a lot of Florida dairy industry videos on there that tell the story of the farmers. We took a camera crew down to a dairy farm and filmed there all day and we asked every conceivable question we could that might come up from a consumer, and now we have those little 30-second to a minute long podcasts that are on YouTube. So people can go to YouTube, they can click on that and hear a farmer speaking; so that's been big for us. Now we're starting to get into Twitter a little bit and we're even talking about starting a Facebook page for our industry. These are things that weren't even conceivable to us two years ago but now we are taking full advantage of that because we understand, especially for the younger audience, the younger consumer, that is the best way to reach them.

© Cengage Learning 2012

Lisa Lochridge

Director of Public Affairs, Florida Fruit and Vegetable Association

The communication itself is definitely shaping how you word your message. It used to be that you had to make sure you had your one-page news release. Now with Twitter, you're talking about a sentence or two at the most. So you've got to be direct; you've got to be concise. The flip side of that is when you have an issue that requires detail and context and perspective, how do you do that with 140 text characters? It makes our job so much more difficult and so much more challenging, and those are the kinds of things that we're going to have to learn how to do.

to sports teams and celebrities—and shows page hits for the most visited Facebook pages. Facebook ads are free, placed on any page that is accessed by a Facebook user. Facebook also lets you set up a group of users who can follow each other and are related to each other by way of a shared interest or activity.

Agricultural producers are also embracing new media, including smartphones, cell phones, and Twitter, in new and unexpected ways. Farmers use cell phones to monitor irrigation lines, take pictures of their livestock on smartphones, and sell their products on Facebook. In a 2009 CNN article titled "Twittering from the Tractor," Chuck Zimmerman, publisher of the agriculture news blog *agwired.com,* commented, "Most farmers are going to be in their [tractor] cab. You're going all day long, night and day—it can get a little bit boring, you know? So, a lot of them have satellite radio, smartphones, iPhones, BlackBerries. I can't tell you how many farmers are following me who are tweeting from the cab" (Sutter, 2009). To demonstrate how pervasive agricultural blogs are, if you Google "farmer blog," Google returns over 68,000 hits, displaying everything from the somewhat generic "Farmer's Blog" to the esoterically named "Sam's Cold Fusion Blog." The Center for Public Issues Education in Agricultural and Natural Resources (the PIE Center) at the University of Florida invites guest agricultural producers and growers onto its blog to share their insights and experiences with site users. The PIE Center, like many other companies and organizations venturing into new media, links its website, blog, Facebook page, and Twitter account together so that users of one vehicle can feed into the others interactively, which can build total visitors in a multiplier effect.

Wikipedia, itself a social media technology, provides a regularly updated list of social networking sites for people with interests in just about anything. Organizations such as WOMMA, the Word of Mouth Marketing Association (http://womma.org/main/), keep extensive lists of case studies on all aspects of new media, including social networking media.

MANAGING YOUR ONLINE PRESENCE

The following imaginary and cautionary tales about the dangers of disclosing the wrong types of information on social media sites are becoming more real and more commonplace:

- A recent graduate of a small university in the Midwest had just started her first job in her chosen career field when her boss came across her personal blog, where she had been writing about how much she hated her new job. She soon became a former employee.
- A high school senior applying to an exclusive Ivy League college discovered that the school's recruiter had checked out his Facebook page and seen embarrassing pictures of him partying and drinking alcohol even though he was underage. Needless to say, the recruiter did not approve his application for admission.

If you are contemplating getting a job after school or applying to college or graduate school, you should think carefully about what your online presence says about you. As social media continues to grow, managing your online presence is becoming more important than ever. Recruiters and employers now routinely use search engines to screen candidates or check profiles on social networking sites. You may not have thought about this before but everything you have put online stays online, and can and will be used by employers in the workplace to form an opinion about you. What seems like not a big deal to you as a student could be a very big deal to a potential employer. Following are some tips for managing your online presence:

- *Check your blog or Facebook page regularly.* Remove tags that are not appropriate.
- *Be cautious about joining groups and adding applications.*
- *Send private messages instead of wall posts.* Anyone can read wall posts.
- *Be careful about what photos you post.* Use the "mom test"; if you would not show it to your mother, do not post it.
- *Be careful whom you friend on Facebook.* Whom you friend and what their profiles look like make a statement about you, too.
- *Use privacy settings to keep your profile private,* but realize that it is still possible for someone to get to your content.
- *At work and at school, do not visit inappropriate websites,* which is grounds for dismissal. Many companies ban their employees from using the Web and sending personal e-mails on company time.

MAKING NEW MEDIA WORK FOR YOU

Consider the following to make new media work for you:

- *Keep it simple. Start small.* You can always add more complexities after you master the basics. Initially, do not try to "bite off more than you can chew."

© Cengage Learning 2012

Scott Wallin

Director, Producer Communications, Dairy Management Inc.

It almost seems that, at times, just trying to get things into a newspaper is kind of an archaic approach. We understand that the younger generation is not reading newspapers with the regularity that I did as a young person, so we want to be where those young consumers are, which is YouTube, Facebook, and Twitter, and we feel that the more we put out there, the better return and understanding we will get for our industry. If we're not doing it, then the people who oppose our industry, they already are doing it, and they are going to do it even more so. So we have to be where they are on social media, as well, in order to counteract any negativity toward our industry.

- **Be creative.** Think "outside the box." How can you use new media to reach current or new audiences?
- **Consider your audience.** What appeals to your target audience?
- **Assess the available new media tools.** What do you have that you can use or know how to use?
- **Think visually.** Do not use a lot of words, except possibly with blogs, and even then, many blogs are increasingly using visuals. Focus on visuals (graphics, photos, video). The younger your target audience, the more likely they are to be "visual" learners.
- **Think big and small.** Use big visuals because many people will be viewing them on small screens.
- **Develop yourself professionally by learning new media capabilities.** Learn what software and hardware are available and what the positives and negatives are for using them. Question the uses of new media. What would you do with the media? How can you reach audiences? Too many times,

instructional technologies are purchased without a clear understanding of how they can be used.
- **Build in assessment.** If you use new media, build in some type of assessment so you know that what you are doing is working.
- **Start with good communication objectives.** This is the base for everything else. Good communication objectives—knowing what the purpose of your new media project is, for example—allow you to build an effective new media site.

With new social networking media applications being developed almost daily, it is hard to know how new and social media will evolve in the next few years. What is known, though, is that online technologies will continue to evolve and create new ways to interact and communicate with friends, colleagues, peers, and customers online. It will continue to be important for those who want to become professional communicators to stay abreast of these new technologies and seek out ways to innovate and engage online.

SUMMARY

This chapter introduced the concept of new media: Web-based social networking media types such as Facebook and Twitter. The chapter provided statistics on new and social media usage, as well as identified the beginning of research into new media as a form of communication. Media convergence was explained as both the growing ownership of media outlets by fewer and fewer companies as well as the merging of media formats into one medium. The chapter concluded with tips for using new media effectively.

CHAPTER EXERCISES

APPLICATIONS

Following are some ways you can apply what you learned in this chapter:

- Create a wiki. Several websites offer easy-to-use Web-based shareware applications that can be used to create a wiki of your own. To create a wiki:
  - Think of a topic that interests you, and that you can find information about.
  - Go to the Wetpaint site (http://www.wetpaint.com) or another wiki-making site and follow the instructions on how to create a wiki.
  - Invite people (using their e-mail addresses) to join the wiki.
- Create a video in YouTube format and upload it. YouTube (http://www.youtube.com) is the top site for uploading and viewing videos. You can upload videos you have created to the site by first creating a video file in an acceptable format, such as Apple's QuickTime, and then uploading to the YouTube site. To learn more about video production and digital video, see Chapter 10, Video and Audio Production.
- Create an audio podcast and upload it to an audio file-sharing service.
- Interview agricultural company representatives in your area to find out how they are using new media and social media.
- Use Facebook or other social media to create an "interest group" about a particular topic or organization. Keep everyone in your group informed about meetings and special events through this social medium.

CHAPTER QUESTIONS

MATCHING:

_____ **1.** Marshall McLuhan was known as a pop philosopher whose ideas have become new again due to the rise of _____.

_____ **2.** _____ are those for whom digital technologies already existed before they were born.

_____ **3.** _____ are those who *migrated* into digital technologies.

_____ **4.** Photo-sharing sites are a form of _____.

_____ **5.** In order to be effective in a new media environment, journalists are being trained to move away from _____ thinking.

_____ **6.** Facebook is the second-largest _____ site.

_____ **7.** _____ are shared online journals maintained by the writer and updated regularly.

a.   digital natives

b.   digital immigrants

c.   social media

d.   new media

e.   online video

f.   print-based

g.   blogs

MULTIPLE CHOICE:

8. Digital immigrants are more likely to

    a. turn to the Internet second, rather than first, for information.

    b. turn to the Internet first, rather than second, for everything from searching for information to reading a newspaper online.

    c. own a mobile device or e-reader (iPod, iPad, Kindle).

    d. print out their e-mail.

    e. both a and c.

    f. both a and d.

9. Digital natives are more likely to

    a. turn to the Internet second, rather than first, for information.

    b. turn to the Internet first, rather than second, for everything from searching for information to reading a newspaper online.

    c. own a mobile device or e-reader (iPod, iPad, Kindle).

    d. print out their e-mail.

    e. both b and c.

    f. both b and d.

10. _____ is *not* an example of social media.

    a. Photo sharing

    b. A blog

    c. E-mail

    d. A podcast

11. _____ has more than 500 million active users.

    a. Twitter

    b. Second Life

    c. Facebook

    d. MySpace

FILL IN THE BLANK:

12. A _____ is a page or collection of Web pages designed to enable anyone who accesses it to contribute or modify the content.

13. _____ is the point where all existing media become one medium.

14. _____ let users interact with each other, upload files, and post messages, photos, and videos that visitors can see.

15. _____ is an online network of more than 30 million professionals from around the world where users can create personal profiles that summarize their professional accomplishments.

REFERENCES

Facebook. "Facebook Press Room Statistics." Last modified 2010. Accessed November 11, 2010. http://www
.facebook.com/press/info.php?statistics.

Gordon, W. T. *Marshall McLuhan: Escape into Understanding.* Berkeley, CA: Gingko Press, 2002.

Hooper, S. "Facebook Turns 5—But Can It Survive?" Last modified 2009. Accessed January 14, 2010. http://www
.cnn.com/2009/TECH/02/04/facebook.anniversary/.

Huffingtonpost.com. "Minnesota Newspapers to Get State Money to Retrain Staff for the Internet." Last modified
March 3, 2009. Accessed November 4, 2010. http://www.huffingtonpost.com/2009/02/24/minnesota-
newspapers-gett_n_169560.html.

Marshall, J. "August Growth Crowns Facebook Number Two Video Site in U.S." Last modified October 1, 2010.
http://www.clickz.com/print_article/clickz/stats/173678.

The Nielsen Company. "Facebook and Twitter Post Large Year over Year Gains in Unique Users." Last modified
May 4, 2010. Accessed November 12, 2010. http://blog.nielson.com/nielsonwire/global/facebook-andtwitter-
post-large-year-over-year-gains/.

Nielsenwire.com. "Global Faces and Networked Places: A Nielsen Report on Social Networking's New Global
Footprint." Last modified 2009. Accessed December 31, 2009. http://blog.nielsen.com/nielsenwire/wp-content/
uploads/2009/03/nielsen_globalfaces_mar09.pdf.

Ostrow, A. "82 Million User-Generated Content Created and Counting." Last modified 2009. Accessed January 14,
2010. http://mashable.com/2009/02/19/user-generated-content-growth/.

Prensky, M. "Digital Natives, Digital Immigrants." *On the Horizon*, vol. 9, no. 5 (October 2001).

Schonfeld, E. "Costolo: Twitter Now Has 190 Million Users Tweeting 65 Million Times a Day." Last modified
June 8, 2010. Accessed November 15, 2010. http://techcrunch.com/2010/06/08/twitter-10-million-users/.

Sutter, J. "Twittering from the Tractor: Smartphones Sprout on the Farm." Last modified 2009. Accessed
November 11, 2010. http://www.cnn.com/2009/TECH/07/02/twitter.farmer/index.html.

Wikipedia. "About Wikipedia." Last modified 2009. Accessed December 31, 2009. http://en.wikipedia.org/wiki/
Wikipedia:About.

Working with the Media

13

Media Relations

OBJECTIVES

After completing this chapter, the student will be able to:

- Define *media relations*.
- Identify characteristics of an effective media relations strategy.
- Identify the characteristics of a newsworthy news story.
- List effective media interview techniques.
- Write effective news releases.
- Write effective public service announcements.

KEY TERMS

- News media
- Media relations
- Gatekeeper
- Newsworthiness
- News release
- Public service announcement (PSA)
- Communication points

INTRODUCTION

Reporters in your organization have the responsibility of communicating activities and events to outside audiences and other members. One outside audience that should be informed about events is the **news media***—radio, television, newspapers, and magazines. However, many reporters for an organization may not have had experience working with the news media. The news media also may visit events an organization is a part of, such as fairs and farm animal petting zoos, and interview officers and members participating in the event. Those being interviewed may not know how best to answer the reporter's questions. This chapter provides you with the information you need to develop an effective media relations strategy, including how to write effective news releases and public service announcements. The chapter also provides suggestions on how to prepare for news media interviews and how to answer reporters' questions (Figure 13-1).*

FIGURE 13-1: To effectively communicate a message, you need to develop an effective media relations strategy.

© Picsfive/www.Shutterstock.com

Susan Howard, APR (Accredited in Public Relations)

Director, Corporate Communications, DUDA

The news media is the primary way people get their news. The traditional media include newspapers and magazines, television, and radio. Today, most media outlets have a presence on the Web. Learning how the media works—what it considers newsworthy, how it tracks down information, and how it develops news stories for print and broadcast—is important to know so that you can work effectively with reporters, editors, and producers, and help them make their final story accurate.

WHAT MEDIA RELATIONS CAN DO

Media relations is a strategy of working with the news media in order to get out information about an organization's events and activities in local news outlets, such as newspapers, television and radio newscasts, and magazines. A media relations plan consists of such components as contacting reporters directly, writing news releases and public service announcements, knowing how to answer questions that reporters ask, and providing photo or video opportunities for photographers and videographers (video camerapersons) to shoot. The primary goal of any media relations strategy is for you to serve as a credible and trustworthy source of information for the news media. Ultimately, what you want is to present a positive message for the news media to communicate to a specific audience.

Developing an effective news media relations plan for your organization can:

- *Enhance the public's knowledge and understanding of your program, event, or activity.* The public usually does not know about a topic until they hear about it through the news media. Media relations will help get your message out about your program, event, or activity.
- *Build credibility in your program, event, or activity,* because people perceive that what they see in the news media is important.

- *Extend the reach and increase the frequency of your message.* Using the news media may mean your message reaches people in your community, as well as across your state.
- *Possibly recruit new members or support for your program.* Some people will learn about your organization through news stories and may be interested enough in the activity that they want to join. You also may get more support for your organization or for the activity your organization is conducting through the information people see in the news media.

DEVELOPING A MEDIA RELATIONS STRATEGY

You must develop a media relations strategy in order to build an effective relationship with the news media. That is what "media relations" is all about—building relationships. The relationship does not happen just by itself; you have to take the initiative by visiting or calling reporters, instead of having reporters come to you first. Here are suggestions as you map out your media relations plan:

- *Set specific, realistic, and measurable goals.* It is probably unrealistic to expect that every news release you send out or every phone call made to a reporter will result in a front-page story. You should establish realistic goals. It may be that

Lisa Lochridge

Director of Public Affairs, Florida Fruit and Vegetable Association

It really is important to us at Florida Fruit and Vegetable Association to have that trust, and what we do to build that relationship is we work very hard to be a credible, trusted source, not only for our members. Our reputation is what we work on, because our reputation is our brand. It's leadership. It's knowledge. It's helping our members succeed. We just make sure that we engage in activities that are going to continue to build that trust. That's everything for us. In terms of media relations, we work hard to be a trusted source for the media. That may mean things like granting an interview, obviously, being available, being accessible. It also involves just being a help to a reporter. It may be that I don't necessarily do an interview, but I might help a reporter run down some information or supply him with a fact or even just pick up the phone and say, "Hey, did you know this?" Over time, you build that relationship and you become that trusted source so that, especially in the case of a crisis, they know that they can trust what you're saying. They know that they can call you and you will be there. That ensures, too, that your side is going to be told when the story is not so positive.

you set a goal of trying to get one story every quarter in a newspaper. Maybe the goal is to get a notice of an upcoming activity on a television station's community calendar. Be sure that the goal is something that can be measured so that you can evaluate whether you were successful. For example, perhaps the goal will be to increase attendance at an event from the previous year, when you did not have a media relations plan, following the release of a newspaper, television, or radio story.

- ***Decide on your approach to get your goals accomplished.*** Will you communicate through news releases, personal visits to reporters, or on-air television or radio interviews? Which form of news media do you need to use to reach your target audience? Remember that you use the news media to reach the general public, but a "general public" audience is very broad. It usually is better to identify specific audiences to

try to reach. These audiences could be children, teenagers, agriculture producers, teachers, parents, or older adults. Each specific audience gets information differently. For some audiences, they are more likely to watch early morning television programs. Others may hear their news on country music radio stations. Still others may only read a newspaper. Use the news media that your audience uses most.

- ***Decide who is responsible for handling news media requests.*** Who will be the person to answer reporters' questions? It could be your organization's president or reporter or someone else. Whatever the situation, everyone in your organization should know who handles the calls, so that if a news reporter calls, the correct person can answer the questions.

- ***Become a reputable and dependable expert source.*** One of the components to an effective media relations strategy is to become a reputable,

© Cengage Learning 2012

Lisa Lochridge

Director of Public Affairs, Florida Fruit and Vegetable Association

Whether it's mass media, such as a daily newspaper or local television station, or niche media, such as a trade magazine or blogger who covers agriculture, the media reach the people and groups who are important to your business and industry. It's vitally important to understand how media outlets work. What do they consider "news"? Who are their readers or viewers, and what is important to them? Why should they care about your story? If you try to sell, or "pitch," a story to a reporter or editor without a solid understanding of the publication and its audience, you'll have little chance of being heard.

© auremar/www.Shutterstock.com

FIGURE 13-2: Contributing news items about your organization can help you become a reputable expert source.

expert source. This also means that you should contribute news items to let reporters know what is going on in your organization (Figure 13-2).

- *Maintain a directory of reporters in your area.* Find out how the reporters like news releases, photographs, public service announcements, or other materials submitted, and enter that information in your directory. Some places may want all items submitted through electronic mail, whereas other places may want a printed

news release. Update this information at least once a year.

- *On a regular basis, provide informational materials to reporters.* Examples of these types of materials include news releases, public service announcements, brochures, and photographs. Sometimes these items, when they are grouped together, are called *media kits*. Media kits are covered in Chapter 17, Communication Campaigns Development. Many media kits are now posted to organizations' Web pages for reporters' easy access.

- *Get to know the reporters in your area and what topics they cover.* Depending on the story's topic, a story might be covered by an education reporter, a business reporter, or a science reporter. Contact the reporters personally, and follow up with phone calls, e-mails, letters, and personal visits.

- *Explain your need to reporters personally,* especially if you need a good deal of exposure in a short time. However, remember that you are asking for free time or space (in a newspaper or magazine). Any time or space that is given to you is better than none.

- *In addition to contacting reporters, you may wish to send information about your event to*

Scott Wallin

Director, Producer Communications, Dairy Management Inc.

Getting to know reporters and the types of stories that are of interest to them can lead to a positive impact for your organization. Not every news release is intended for every reporter, so avoid a "shotgun" approach of sending everything to everybody all of the time. You'll lose your credibility, and it shows you don't care enough to learn what is important to a reporter you hope to build a relationship with. Read newspapers and watch local television news to become familiar with reporters and their beats, and then put together a tailored media pitch that has a better chance to succeed.

the public affairs director or promotions director at your local newspaper or television and radio stations. Many television and radio stations have a *calendar of events,* which is aired once or several times a day. Newspapers tend to list community events once a week. Stations and newspapers are more likely to publicize the event if it will impact large numbers of people.

■ *Integrate social media.* This is a relatively new process in media relations. Using social media, such as Facebook, blogs, and Twitter, is becoming more popular as organizations try to reach news reporters. Social media also allow your organization's reporter to bypass the news media and go directly to interested audiences who may subscribe to your social media site.

■ *Be ready to go on the air early.* Many television and radio stations invite guests to discuss their upcoming events on the air. However, these interviews usually are early in the day. Be ready and willing to appear during early morning hours if you are asked.

■ *Develop public service announcements (PSAs).* Many television and radio stations air PSAs throughout the day. Radio stations are much better about airing PSAs because they do not take much time to produce. Talk radio stations also

have more time to fill and usually are more likely to air PSAs.

■ *Write news releases.* News releases, explained later in this chapter, are stories written just like a regular news story. Their purpose is to provide reporters with the basic information they need to cover a story.

WORKING WITH THE NEWS MEDIA

The news media—radio and television stations, newspapers, and magazines—pass your information to specific audiences. The news director for a television station or the news editor for a newspaper decides what is important and what is actually reported. In essence, news directors and news editors operate as gatekeepers. A gatekeeper, in the news media, determines what you read, view, and hear about the world around you. Consider this illustration of how a "gatekeeper" operates: Someone operates the gate of a livestock pen, using the gate to keep certain cattle in the pen while also keeping other cattle out. Similarly, a news media gatekeeper opens the "news gate" to allow certain information to reach the general public; the gatekeeper also closes the "news gate" and does not report other aspects, based on a story's content and newsworthiness.

© Cengage Learning 2012

Rod Hemphill, APR *(Accredited in Public Relations)*

Director, Public Relations and Communication (retired), Florida Farm Bureau Federation

The print, broadcast, cable, and online media are neither your friends nor your enemies. They send your message to large numbers of people. Part of their job is to filter those messages; that means they attempt to serve up information that their editors believe is most relevant to their audiences. Knowing the media and their criteria for stories helps you tailor your message to reach your target audience. It is important to build relationships with reporters by helping them serve their audience.

Following are some of the criteria many news directors and news editors use to determine **newsworthiness**. A similar list is provided in Chapter 5, News Media Writing. This list is provided here so that you can determine if your news releases have sufficient news value.

- *Timeliness:* This refers to when your event happened or when it will happen. News reporters do not like old news.
- *Proximity or location:* The event should be in the immediate area so that a local newspaper and television station will want to cover the story.
- *Prominence:* High-profile people, issues, or concerns are more likely to get news media coverage. So if your event has a professional athlete, a well-known governmental official (mayor, state or national representative, governor), or actor, it has a much better chance of being covered.
- *Importance or significance:* The greater the effect and the larger the number of people impacted by your news story, the more likely it is that your story is news.
- *Human interest:* Does the event involve interesting people doing interesting or unusual things?
- *Innovative or unusual:* If the event features something different, unusual, or innovative, it is likely to be covered in a news story.

- *Conflict:* A story that shows a struggle—a person versus the environment, a person versus another person, a person versus a governmental agency—is usually newsworthy.
- *Money:* News stories about financial issues are almost always newsworthy.

With these criteria in mind, you may wonder what story ideas you might have that would be of interest to a news outlet. If you want a reporter to cover a meeting, you first should ask yourself, "Why would a reporter cover this meeting?" If it is a regular meeting and nothing new or exciting is happening, the chance is slim that the reporter would be interested in covering the meeting. Therefore, it probably would be best not to contact a reporter to cover a "non-news" event. If, however, you have invited a special speaker or are doing something unusual, it is very likely a reporter would come. Notice that in this case the focus of the story would be to cover the speaker or innovative event during the meeting, not the meeting itself.

In addition to knowing the criteria used to judge the newsworthiness of possible news story topics, it is also important to understand how the news media cover stories.

Newspaper reporters want lots of quotations, facts, and photo opportunities. Newspapers provide much

FIGURE 13-3: Radio soundbites or actualities need to be short but descriptive and understandable.

more detail because people can reread printed stories. Newspaper reporters will spend considerable time talking with you. Radio reporters want short quotations—also called *soundbites* or *actualities*—of 10 to 20 seconds in length. For radio, you need to be able to describe the situation in a way that can be understood by just hearing the information (Figure 13-3). Television reporters also want soundbites (10 to 20 seconds), but they also need stories that are visual. Television stories cannot be as detailed as newspaper stories, as they are usually limited to 90 seconds or less. TV reporters and their camerapeople will split time between the interview and shooting video footage to illustrate the story.

Also keep in mind that most news media have an online component. Newspapers, television stations, and radio stations likely will provide a version of their stories on the Web. These stories may include text, photographs, and video clips. As a result, people will be able to read, listen to, or watch the story about your event online whenever and wherever they are.

THINKING LIKE REPORTERS

One way to establish successful media relations is to think like a news reporter. Following are some ideas you must keep in mind when working with reporters:

- *Schedule events to maximize news coverage.* Other events are happening at the same time yours is, so make your event count. If you know when one of the most popular and longest-running events in your community is going to happen, do not schedule your activity at the same time as this sure-fire news event.
- *Know the reporter's deadlines.* Arrange your news events so they can be covered well in advance of a reporter's deadline. You should schedule stories with newspaper reporters in the early afternoon, because the deadline for newspaper reporters to complete their stories is early evening to be included in the next morning's paper. Interviews with radio reporters can be scheduled at any time, because radio news programs air many times during the day. Schedule television interviews for early to mid-morning for the noon or early evening newscasts, or early afternoon for the evening newscasts.
- *Reporters are generalists, not specialists.* Reporters know a little bit about a lot of things, but not a lot about any one particular thing. Reporters may not know much about agriculture, your organization, or an event you are sponsoring. Therefore, reporters need a lot of help when developing a story. They need facts presented clearly and concisely.
- *Reporters are good observers.* Anything reporters see or hear is fair game for the story. Be sure that everything a reporter could see is presented in the most positive way. For example, if a reporter is doing a story on preparing for the county fair, make sure all of the livestock pens are presentable.
- *Reporters like to personalize a story.* Submit story ideas that emphasize people or that would interest people on a personal level.
- *Make sure the facts you provide the reporter are correct.* If you do not know if something is right or not, do not guess. Check it out before you give the information to a reporter.
- *Follow trends.* Keep up with the events related to activities you do. Submit story ideas that are "trendy" or timely. For example, if, during the holiday season, giving young farm animals (chicks, rabbits, goats) as presents is the new craze, you may want to submit a story idea to a reporter about what people should consider before giving an animal as a gift.

Last, here are a few suggestions on how you can help reporters do their jobs better. Remember, if you want to develop good media relations, try to accommodate reporters as much as possible (Figure 13-4).

- *Written materials,* such as news releases and brochures, can help reporters tremendously when they write the story.
- *Setting:* Provide tips on where interviews should be conducted. What visuals and audio would improve a television story? Most reporters appreciate any tips to enhance a story.

Scott Wallin

Director, Producer Communications, Dairy Management Inc.

Communication technologies have changed everything in our office. I started with Florida Dairy Farmers Incorporated about eight years ago and we used to fax things. Now I feel almost like a dinosaur having used a fax machine at one point in my life. I've asked reporters, "How do you want to be reached?" And they all say the same thing: "E-mail. Cut back on the phone calls. If I like your pitch over e-mail, I'm going to get back with you and show you some interest and we'll go from there." That's really changed an awful lot in trying to reach the media, that they want things over e-mail. They don't want to look at their phone and see that they've got 40 messages. They don't have time for that. But e-mails are nice and quick and efficient. You've got to make your e-mails obviously very catchy and grab their attention right off the bat, so they're going to look at your e-mail.

FIGURE 13-4: By accommodating reporters and helping them to do their jobs better, you are developing good media relations.

- *Directions/travel:* Provide detailed directions to an event, assistance with camera gear, and help to get reporters and videographers from place to place.
- *Several sources/resources:* Reporters like to have more than one person to interview. If you know someone who would add to a reporter's story, suggest the person's name.
- *Understandable terms:* Do not use unfamiliar words, acronyms, or *jargon* (technical language). Do not assume reporters understand "your language."

NEWS RELEASES AND PUBLIC SERVICE ANNOUNCEMENTS

News Releases

You may be called upon to write a **news release**—also called a *press release*—about your organization's activities, interesting news, or important events. A news release is sent to news media to announce something that has news value. A news release provides reporters with the basics they need to develop a news story (Figure 13-5).

Your news release needs to be local and newsworthy. (Refer to the section earlier in this chapter on the criteria

Model News Release

Organization's Letterhead

NEWS RELEASE

FOR IMMEDIATE RELEASE
Oct. 7, 2012

Contact: *Your name and title*
Address: Address
            City, State, ZIP
Phone: xxx-xxx-xxxx
Fax   xxx-xxx-xxxx
E-mail: name@xxx.com

For Immediate Release (or the date it needs to stop running)
Identifying Headline

Use double spacing. Be sure the contact information is complete, including your phone number and e-mail address. A physical address is optional.

If it is a timed release, delete the "for immediate release" line and insert the proper "ending" information, such as "Run Until Sept. 7, 2012." Be specific.

An identifying headline should summarize the content of the story. Its function is to tell the editors in capsule form what is in the release.

Use at least one-inch margins around the story. Use tab indents for each paragraph.

Paragraphs should be short, preferably no more than three sentences. They should be punctuated correctly. Sentences should average no more than about 20 to 25 words.

One-page releases are more likely to be printed, but you can use additional pages as necessary. If you do, insert "– more –" at the bottom of the page.

At the top of the second page, flush left, write "Add 1."

On the last page, you can use the following notation, centered, on the page, to signify the end of the story: "– 30 –" or "###."

If the release is about a company or organization, typically the last paragraph includes what is called "boilerplate," or general information, about the organization.

###

FIGURE 13-5: News releases, such as this model news release, should be brief, well-written, and newsworthy. News releases should highlight your organization's activities, interesting news, or important events.

for newsworthiness.) Your news release should be brief, no more than two double-spaced pages in length. Most important, your news release must be well written. A news release with bad grammar and punctuation that is not written in Associated Press Style, discussed in Chapter 5, News Media Writing, will likely not be used. It is a good idea to send the release to the person who likely would cover the event; do not just send it to "The Editor." Most news media outlets list reporters, their phone numbers, and e-mail addresses on their websites (Figure 13-6).

FOR IMMEDIATE RELEASE

Contact: Aaron Wockenfuss, Manager
Consumer Communications
Florida Dairy Farmers
407-647-8899, ext. 1113
AaronW@floridamilk.com
166 Lookout Place, Suite 100
Maitland, FL 32751

New Learning Resource Teaches Kids about Dairy Farming in Florida

ORLANDO, Fla. – Children in Florida have a new interactive resource to learn about the state's dairy farming industry. *Dairy 101* is a free, educational kit developed by the Florida Dairy Farmers that uses videos, activity sheets and games to teach children about the health benefits of milk as well as how dairy products get "from the farm to the fridge."

According to the Florida Department of Education, *Dairy 101* materials meet more than 32 Sunshine State Standards for elementary and middle school students, including math, science, language arts and social studies. *Dairy 101* contains materials in both English and Spanish and includes:

- Farm-to-Fridge video
- SunnyBell "Moo"sical video
- Bulletin board elements
- Farming production information packets
- Activity sheets
- Games

Kits are available free of charge to schools, teachers, librarians, daycare centers and other educational groups and can be previewed at www.floridamilk.com/dairy101. To order a kit, contact the Florida Dairy Farmers at 800-516-4443 or kimf@floridamilk.com.

###

FIGURE 13-6: In this news release example, Florida Dairy Farmers informs the media about a new learning resource to teach children about dairy farming in the state. Notice that the news release includes contact information and tells the news reporter that the news release is "for immediate release." (*Courtesy of Florida Dairy Farmers*)

Writing the News Release

A news release is written in just the same way that reporters write news stories. In writing the news release, you should imitate the news writing style described in Chapter 5, News Media Writing. Here are some elements of news writing style to keep in mind:

- *Inverted pyramid structure:* You want to include the most important information first, followed in descending order by less-important information.
- *5 Ws and H:* The most important of the questions (who, what, when, where, why, or how) should be answered in the lead. Others are answered later in the story.
- *Lead:* The first paragraph that is used to grab the reader's attention.
- *Short paragraphs:* Paragraphs usually run one to two sentences in length. Rarely do you see paragraphs of more than three sentences in a news story.
- *Quotations:* Quotations are the exact words of someone talking. It is a good idea to use quotations to bring "life" to your story.
- *Associated Press Style:* The news release must follow AP Style.
- *Proofreading:* The news release should be as free of grammar, punctuation, and spelling errors as possible. Proofread the news release several times.

The *lead*, or first sentence, provides the most important information. The second paragraph should provide any information that will immediately help the reader. For example, for a news release about an upcoming event, the second paragraph may provide the location and exact time of the event. Will there be a registration fee? This part of the news release often explains the "how" question.

Follow the "how" with information that provides context. This may be the "why" of the story. For example, why is it important for people to attend your event? The last part of the story provides useful detail and history. Many times this will be a brief paragraph at the end of the story that explains a little bit about the event. For news releases that feature an organization, the last paragraph usually contains information about the organization, such as the number of members and how the organization supports the community.

In addition to imitating news writing style, news releases also should use the following structure:

- *Spacing and length:* The news release should be double-spaced. A news release should never be handwritten. The news release should be one to no more than two pages in length.
- *Identification:* The name, phone number, and e-mail address of the news release writer should be provided, as well as the name, phone number, and e-mail address of the person(s) interviewed.
- *Release date:* The release date is at the top of the news release and indicates when the news release should be run and when it should not be considered. If a news release can be run or aired as soon as a reporter receives it, the phrase "For Immediate Release" is written. If the release has an ending date, then the release date will include something like this: "Run until Dec. 4."
- News releases about events can either be pre-event or post-event stories. A *pre-event news release* is sent to news media before an event happens, in order to generate coverage of the event in newspapers and on television and radio broadcasts. A *post-event news release* is provided to news media after an event happens. These are used frequently at county fairs to summarize to news media the winners of fair competitions.

Public Service Announcements

A **public service announcement (PSA)** is a free advertisement that radio and television stations air or newspapers and magazines print to highlight information about nonprofit organizations' programs, activities, or events. The message in the PSA should be clear. Include the day and time of the event and any information that the listener may need, such as a Web address or telephone number to learn more.

Keep in mind that you do not have control over when or if PSAs run. On television, PSAs may run during late-night or early morning hours when few paid advertisements run. However, any free airtime is better than nothing. Radio stations can be better about airing PSAs because they do not take much time to produce. Radio announcers also may read the information live on the air.

Writing the PSA

Radio and TV announcers may help you write the PSA. If you have "live" copy for announcers to read on-air, make sure it is complete. It should have the phone number and e-mail address of someone in your organization to contact. Try to make the PSA as brief and as easily readable as possible (Figure 13-7). A 15-second radio PSA, for example, will only be about 38 words,

Rod Hemphill, APR (Accredited in Public Relations)

Director, Public Relations and Communication, Florida Farm Bureau Federation

News writing has some hard-and-fast rules. It is clear, simple, and well organized. A well-written news release can attract media attention to your message. A poorly written release may confuse editors and reporters—or even mislead them.

Example Radio PSA

The Anyville FFA chapter will sponsor a hayride Saturday night, starting at 7. Tickets are $3. Proceeds benefit the student scholarship fund. For more information call here at KUFG, your choice for news.

FIGURE 13-7: Example radio PSA. A public service announcement written for radio, especially to be read "live" on-air, should be brief, but have the important information that listeners need. In this example, the following main points are covered: "who" (Anyville FFA chapter), "what" (sponsor a hayride), "when" (7 p.m. Saturday), "why" (to benefit the student scholarship fund), and "how" (purchase $3 tickets). The PSA also provides information on how the listener can get more information by contacting the radio station.

and a 30-second radio PSA will be approximately 75 words, so your message will need to be to the point. Write the PSA in a conversational style with sentences of 12 to 15 words each. Do not use jargon.

For PSAs found in newspapers and some magazines, you will want to design the PSA so that people's eyes are drawn to it. This usually means making the visual design appealing with an eye-catching headline, graphic, or photograph. Integrate good document design, which you learned about in Chapter 6, Document Design. The words in the PSA should convey the message as briefly as possible.

MEDIA INTERVIEW SKILLS

Being interviewed for the news media is another component of an effective media relations strategy. Many people assume that being interviewed is as simple as walking into an office or television studio and waiting for a reporter to ask questions. However, if you are not fully prepared, both in terms of what you want to say and what you expect during an interview, being interviewed may be a frightening experience. On the other hand, if you know what you want to say and feel confident about your ability and appearance, an interview can be a rewarding and enjoyable experience.

When a reporter calls, you should find out the reporter's name, whom the reporter works for, what the story is about, what the reporter's deadline is, and what the reporter wants from you (Figure 13-8). Determine if you can help. If you cannot answer the reporter's questions, try to direct the reporter to someone who can.

If you are the right person to answer the reporter's questions, you will need to know how best to present yourself in an interview. The following recommendations will help you succeed in most interviews. The interview skills described in this section pertain to all forms of media unless otherwise noted.

FIGURE 13-8: Being able to conduct a proper interview to get information from someone is a good skill to have. News reporters, like the one shown, practice a lot to be able to ask good questions.

■ *Prepare two to three ideas you want to convey.* These are your communication points, also known as "talking points." Your **communication points** are the most important issues or points you hope to address and get across to the reporter during the interview. Make a list of the questions you anticipate that the reporter will ask. Anticipate issues and questions that may arise during the interview, and be prepared to use those issues to launch your communication points. To help you develop your communication points, write three sentences that you could use to get across each point. If they are long sentences, keep refining. Develop at least one answer you can give in 20 seconds or less. Then, practice saying aloud your communication point. You also may wish to prepare three questions that you dread the reporter might ask. Develop responses to each question. The questions in the interview probably will not be as bad as you have imagined, but it is best to be prepared. You may wish to use the blank lines in the Interview Preparation Guide (Figure 13-9) to help you develop communication points and potential interview questions and answers.

■ *Make short, simple, and specific statements.* As you prepare your communication points, practice using simple, everyday language and short answers. Try to keep answers for recorded television or radio interviews to 20 seconds or less. Use real-life examples. Explain your most important point first. Also, summarize and then elaborate, such as in this example: "We

have the best organization in the area because our members really care. Let me explain what I mean…."

■ *Pause after complete statements.* The radio and television reporter will appreciate these breaks during the editing process.

■ *Do not ramble.* When you think you have answered a question adequately, do not feel compelled to keep talking simply because the interviewer has a microphone up to your mouth. If you are satisfied with your answer, stop talking. Rambling leads you to say the wrong thing.

■ *Do not refer to the reporter or to previous answers.* In a radio or television interview, do not say the reporter's name in the middle of a sentence. Also, do not use the phrase "as I explained earlier." The reporter's name and the phrase "as I explained earlier" will be difficult to edit out. Here is an example: "We have the best organization in the area, *as I explained earlier,* because our members really care." If the phrase cannot be edited out, viewers may not know what you and the reporter have discussed previously, and may not understand what you are referring to.

■ *Think before you speak.* Avoid verbal fillers, such as "uh," "ah," "well," "yeah," and "you know," especially for radio and television interviews.

■ *Maintain eye contact with the reporter.* If you are doing a television interview, do not look at the video camera; look the interviewer in the eye the entire time. Maintaining eye contact gives you credibility and implies you believe what you are saying.

Overcoming Voice Problems

For recorded interviews (radio or television), the impact of your spoken message may depend on how you speak. The sound of your voice determines how well you hold the attention of the audience who listens to you. The ability to speak well can be improved through practice. Common voice problems involve pitch, rate, and articulation.

Pitch is the high and low sounds of your voice. The habit of inflecting up (raising the sound) at the ends of sentences and phrases is a pitch problem, because it sounds like you are always asking a question. Making everything you say sound like a question undermines your authority. You will sound more assertive if you lower your pitch and inflect downward.

The speed at which you talk is your speaking *rate.* While sprinting through your message may leave

Interview Preparation Guide

- When a reporter calls you to request an interview later in the day, **write down the topic that is to be covered** on the top line of the *Interview Preparation Guide*.

- Determine your **communication points**. What three ideas do you want to get across to the reporter? Write these on the *Interview Preparation Guide*.

- List **three possible questions** a reporter might ask. Be sure to **prepare answers** to these questions. Write questions and answers on the *Interview Preparation Guide*.

- If time permits before the interview, ask a friend to interview you about the topic in question. Videotape the mock news interview. Analyze it for presentation strengths and areas of improvement.

Topic: _____

Your Communication Points

1. _____

   _____

2. _____

   _____

3. _____

   _____

Possible Question

1. _____

   _____

2. _____

   _____

3. _____

   _____

Your Response (Soundbite)

1. _____

   _____

2. _____

   _____

3. _____

   _____

© Cengage Learning 2012

FIGURE 13-9: Interview Preparation Guide.

listeners behind, talking too slowly may bore them. To find out if you need to slow down or speed up, record yourself talking with someone, preferably in a mock news interview situation. Play the recording and listen to how fast or slow you speak. Practice establishing a rate that is easy for people to understand. Once you have established a good pitch and rate, practice varying them, along with your volume, to add emphasis and expression to your message.

Last, you have to consider how you *articulate*—or distinctly pronounce—words. Do not run your words together. Proper articulation will help your listening audience understand your words.

Mock News Interviews

You can practice before an actual interview by conducting a mock news interview. A *mock news interview* is when someone acts like a reporter and asks you questions that a professional reporter would ask. You may wish to video record the mock interview so you can review and critique your performance (Figure 13-10). Here are some suggestions on how to answer questions or prepare for real or mock news interviews so that you do not succumb to stage fright:

- *Concentrate on the question you are being asked.* Pause before answering a question just long enough to formulate an outline of the answer.
- *Before the interview starts, take a deep breath, get a drink of water, laugh, or yawn.* Why should you yawn? You cannot yawn and be tense at the

FIGURE 13-10: If you know you will be interviewed, practice beforehand. Mock interviews are excellent ways to prepare for the real thing. Video recording the mock interview and then watching it will help you see and hear where you need to improve.

same time. Even a nervous laugh to yourself before the interview will help relieve tension. Any of the practices in this recommendation will help you relax.

- *Remind yourself that you were asked to be interviewed because you know the topic.* You are the expert.
- *Prior to the interview, review previously recorded interviews of yourself to identify presentation strengths and weaknesses.*
- *Try to convince yourself you are having a normal, everyday conversation with someone.*
- *Do not eat certain foods or drink certain beverages before an interview.* Certain foods and beverages coat your throat, causing difficulty in swallowing and speaking. Before the interview, stay away from such items as cola drinks, chocolates, and dairy products. It takes several hours to "uncoat" your throat from these foods and beverages.

Here are some suggestions on how to conduct a mock news interview and how to ask questions like a reporter:

- *Use open-ended questions.* Questions that begin with "who," "what," "where," "when," "why," and "how" are best.
- *Avoid questions that would have "yes" and "no" answers.* Avoid questions with "do you," "have you," or "were you." These are questions that provide only yes/no answers.
- *Avoid compound questions.* A compound question asks two or more questions in one. This is an example of a compound question: "What are your opinions regarding this proposed change, and if you are in favor of it, why? Or if you're opposed to it, what should be done?" Only ask one question at a time.
- *Be a good listener.* Show an interest in the person being interviewed.
- *Maintain good eye contact with the person being interviewed.*
- *Avoid saying "uh-huh" and "I see" while the person is talking.* Acknowledge responses by silently nodding or smiling.

Make a Good Appearance

Television viewers will judge the trustworthiness of your message both on what you say and on your appearance. Following are a few suggestions for dressing for success in a television interview:

- *Clothing in a television studio setting:* If you are not in the official attire of your organization,

© Cengage Learning 2012

Scott Wallin

Director, Producer Communications, Dairy Management Inc.

We do a lot of spokesperson training with our farmers. We bring in professional trainers who are very familiar with our industry, who understand the challenges that farmers face. And we put them through some pretty intensive training, because there are some difficult questions that can come up from a reporter or sometimes when they are just in the grocery store and engage in a conversation with a shopper. We want to make sure that our farmers do go through an awful lot of training where we turn the heat on them a little bit and we ask them some tough questions in a practice scenario just so they can think more clearly on their toes in the event that they are asked a pretty challenging question.

stick to a conservative, professional appearance style. Avoid tight stripes or plaids. On camera, they sometimes cause the colors to "vibrate." This is called a *moire (more-AY) effect.* Wear midtone colors. Extremely dark- or bright-colored clothes can make your face look washed out or dark under television studio lighting. In both television studio interviews and "on-location" interviews, your blouse or shirt should have a place to clip a microphone.

- *Clothing in an "on-location" setting:* If you are not in the official attire of your organization, your clothing should be consistent with your surroundings. For example, you are not expected to wear a suit if you are being interviewed in a peanut field or a citrus grove. Instead, wear jeans and a regular shirt. Avoid hats, but if you *must* wear one, push back the brim so people can see your eyes.
- *Jewelry:* Wear only a few pieces of jewelry. Avoid "clunky" or dangling necklaces or earrings because large gold or high-gloss jewelry can reflect studio lights. Short necklaces are best. In addition, long necklaces can rub against clip-on microphones.

- *Make-up:* A woman's "everyday" make-up should be fine. Use a matte finish to reduce shine. Use a non shiny lipstick, too. Men, most likely, will not have to wear make-up, but be open to the suggestion. The lighting at some television studios may cause men to look washed out; therefore, men may need make-up to highlight their facial features. If any make-up is needed for a man, it will be applied by television producers.
- *Enthusiasm:* Be animated. Use gestures, facial expressions, and body language to add vitality to your words. However, be careful not to overdo it. Smile, because a good first impression can help establish your credibility. Be conversational and deliver your message with confidence.
- *Body language:* Look at the interviewer, not the video camera. Sit still in your chair because rocking or swiveling can take you out of a cameraperson's shot.
- *Gum and pocket change:* Do not chew gum or play with your pocket change or keys while on television.
- *Light-sensitive eyeglasses:* Do not wear light-sensitive eyeglasses. Studio lighting and sunlight will make your glasses darker; viewers will not be able to see your eyes.

Scott Wallin

Director, Producer Communications, Dairy Management Inc.

We do an awful lot of outreach to the media and we feel that we need to tell the farmer's story to the reporters because, like anybody else, they're removed from the farm. It's fascinating when you get a reporter on the phone and you say, "Do you want to go see where milk really comes from? Do you want to go out and talk to a farmer and see a dairy farm?" They're very eager to get out there, and they're learning as much as anybody else would. It's almost like watching a kid be on a dairy farm for the first time, in a sense, where they're asking a lot of the basic questions that a child would who is seeing a dairy farm for the first time.

We feel that every dairy farmer in Florida has a great story to tell and we love the opportunity to bring reporters out and see the farm for themselves because it is something you can't communicate over the telephone or in an e-mail. They've got to get out there and see for themselves. They've got to see the cows. They've got to see the great things that farmers are doing from an environmental standpoint. They've got to see the measures that each farmer goes through to make sure those cows are taken care of.

Checklist before the Interview

In addition to the previous suggestions, you may wish to use the following checklist to make sure you have everything covered *before* the media interview:

- Are you familiar with the television program or publication that you are being interviewed for? If not, it probably would be a good idea to watch a program or two or read the publication before the interview so that you can answer questions that would appeal to the program's viewers or the publication's readers.
- Are you the only person being interviewed? Will many others be interviewed?
- Will this interview be live or recorded? Will there be call-in questions?
- Can you explain your communication points in a concise manner?
- Have you developed a conversational style?

- Have you rehearsed all possible questions and answers with someone else?
- Are you aware of your body language and facial expressions?
- Are you ready to present your message without using jargon?

After the Interview

After you have been interviewed, you should evaluate how well you did before you do another interview. Here are some questions you may wish to ask yourself to evaluate your interview skills:

- Did you cover your communication points?
- Did you create soundbites (short quotations) that were easily understandable?
- Did you listen carefully to the interviewer's questions?
- Did you use short sentences?

- Did you keep good eye contact with the interviewer?
- Were you aware of your body language?
- Did you project a strong, positive image of a person people would trust?

"Be Attitudes"

By following these final *"Be Attitudes,"* you should be successful in any interview setting:

- *Be prepared.* Prepare in advance two or three communication points you wish to get across. Think of questions you believe you might be asked.
- *Be positive.* End every answer on a positive, upbeat note.
- *Be honest.* Always tell the truth. Your credibility is crucial.

- *Be brief.* Crystallize your ideas into a few short phrases that summarize what you are trying to communicate.
- *Be yourself.* Keep your voice at an even pace. Act naturally.
- *Be energetic.* Be animated. Use gestures, facial expressions, and body language to add vitality to your words. Just don't overdo it.
- *Be focused.* Direct your full attention on the interviewer. Look squarely at the person asking the questions. Don't be concerned with distractions.
- *Be comfortable and confident, and take charge.* Relax. You know more about the topic than the interviewer. If not, you would not be interviewed.

SUMMARY

This chapter presented some steps for you to take in order to develop effective media relations with the news media: Get to know reporters in your community, write news releases and PSAs on a regular basis, and know how best to answer a reporter's questions. Most important, become a dependable and reputable source.

CHAPTER EXERCISES

APPLICATIONS

Following are some ways you can apply what you learned in this chapter:

- Develop and implement a media relations plan, based on the suggestions listed in this chapter. Areas you should include in the plan are as follows:
  - Determine news media goals.
  - Determine the approaches you will need to take to accomplish your goals.
  - Assign responsibility for handling news media inquiries.
  - Develop a news media directory with reporters' names, business addresses, phone numbers, and e-mail addresses.
- Prepare as if a reporter were going to interview you on the most important topic you are working on at the current time. Focus your message so that you can describe your communication points quickly and smoothly. After 5 to 10 minutes, ask for volunteers to present their messages as if they were being interviewed.
- Maintain a scrapbook of current news stories from newspapers and magazines, so that you can stay up to date on issues related to your area of interest. This reference may be in the form of an actual scrapbook of cutout clippings or folders with clippings in them. Keep this information handy for quick reference if a reporter should call.
- Determine how contacting the media would promote your organization and its programs—how would working with the news media help?
- Brainstorm several ideas for stories that you believe reporters would be interested in covering. Maintain this list. Try to "pitch" to reporters at least one story idea a month generated from the list.
- Using one of the ideas generated in the previous activity, determine the following:
  - Does the idea meet "news value" criteria?
  - How can you make it "different" so the news media will want to cover the story?
  - How can you "pitch" the same idea to different media (newspaper, television, and radio)?
- Invite a panel of news reporters, representing different media (newspaper, television, and radio), to discuss the characteristics they like to see in a news story idea, the different approaches reporters take in covering stories, and ways you can better work with reporters to get your stories covered.
- Use the Interview Preparation Guide as you complete this suggested activity:
  - Brainstorm a potential topic that a reporter may call about.
  - Determine your communication points. What three ideas do you want to get across to the reporter? Write these on the Interview Preparation Guide.
  - List three possible questions a reporter might ask. Be sure to prepare answers to these questions. Write questions and answers on the Interview Preparation Guide.
  - If time permits, conduct a mock news interview about the topic in question. Video record the mock news interview. Analyze it for presentation strengths and areas of improvement. Evaluation may be done individually or in a small group. An interview should last no longer than five minutes.
- Conduct a mock news interview with someone else serving as a reporter. Analyze your mock news interview for pitch, rate, and articulation. Do you vary your pitch? Do you speak too fast or too slow? Do you speak distinctly? Work to improve any weak areas of speech that you determine. Analyze your mock news interview for "nervousness." Do you appear to feel comfortable in front of a camera? Do you look nervous? Do you look "natural"?

CHAPTER QUESTIONS

MULTIPLE CHOICE

1. The two or three ideas you want to convey during a media interview are called your _____.

   a. communication points

   b. communication guidelines

   c. educational tips

   d. complete statements

2. Which of the following is NOT a part of the development of an effective media relations strategy?

   a. Setting realistic news coverage goals

   b. Getting to know the reporters in your geographic area

   c. Taking a deep breath before an interview starts

   d. Developing a news directory of reporters in your area

3. Tight stripes or plaids on a person's clothes will cause which of the following effects on television?

   a. The "plaid wave"

   b. Giraffe stretch marks

   c. The Moire effect

   d. C & W line dancing

4. Which of the following is a way to overcome stage fright for a news interview?

   a. Yawn.

   b. Go into the interview unprepared.

   c. Convince yourself you are being interviewed by a reporter from *60 Minutes*.

   d. Dress inappropriately.

5. Answers to questions that will be used in a recorded interview (television or radio)—on average—should be no longer than which of the following?

   a. 5 seconds

   b. 20 seconds

   c. 40 seconds

   d. 1 minute

6. The best time of day to schedule an interview with a newspaper reporter is _____.

   a. early afternoon

   b. early morning

   c. early evening

   d. midnight to 6 a.m.

7. The style of writing used in news stories and news releases is _____.

    a. Associated Press Style

    b. Modern Language Style

    c. Associated Writing Style

    d. News Press Style

8. _____ is the high and low sounds of your voice.

    a. Rate

    b. Articulation

    c. Pitch

    d. Communication point

9. A _____ provides reporters with the basics they need to develop a news story.

    a. public service announcement

    b. news release

    c. caption

    d. storyboard

10. The _____ is at the top of the news release and indicates when the news release should be run and when it should not be considered.

    a. masthead

    b. lead

    c. release date

    d. contact information

11. _____ refers to a story that has the most important information first, followed in descending order by less-important information.

    a. Inverted pyramid

    b. Pyramid

    c. Triangle of facts

    d. Composition

12. _____ refers to high-profile people, issues, or concerns that are more likely to get news media coverage.

    a. Conflict

    b. Importance or significance

    c. Proximity or location

    d. Prominence

13. The speed at which you talk is your speaking _____.

    a. diction

    b. articulation

    c. pitch

    d. rate

14. A _____ determines what you read, view, and hear about the world around you.

    a. news anchor

    b. media specialist

    c. gatekeeper

    d. shepherd

15. _____ refers to an event that is in the immediate area, so that a local newspaper and television station will want to cover the story.

    a. Conflict

    b. Importance or significance

    c. Proximity or location

    d. Prominence

REFERENCES AND FURTHER READING

Diggs-Brown, B., and J. Glou. *The PR Style Guide: Formats for Public Relations Practice.* Belmont, CA: Wadsworth, 2004.

Marsh, C., D. W. Guth, and B. P. Short. *Strategic Writing: Multimedia Writing for Public Relations, Advertising, Sales and Marketing, and Business Communication.* Boston, MA: Pearson Education, 2005.

Telg, R. "Getting out the News." Last modified 2000. Accessed October 23, 2010. http://mediarelations.ifas.ufl .edu/2effectivemediarelations.htm.

Risk and Crisis Communication: When Things Go Wrong

OBJECTIVES

After completing this chapter, the student will be able to:

- Create an outline for a crisis communication plan.
- Apply the concepts of risk communication and crisis communication.

INTRODUCTION

Organizations probably do not want to think about it. Organizations might even wish to deny it. But the sad fact is that something could go wrong at almost any organization. It could be just a matter of when. *It could be a death or injury at the job, a chemical spill, a food recall. You name the "bad luck" scenario, and it could happen. Think about some crisis situations that could happen in an agricultural setting. Did these come to mind?*

- *Bacterial pathogen resulting in a food recall*
- *Food contamination*
- *Chemical run-off*
- *Pesticide poisoning*
- *Soil erosion*
- *Weather*
- *Disease*

Why are organizations not prepared for these situations "when things go wrong"? Being prepared for a crisis—ahead of time—will help your organization get through the rough times when things do go wrong. And part of that overall crisis plan should be the integration of **crisis communication** (Figure 14-1). In this chapter, we will examine an extremely important aspect of communication practice: crisis and risk communication.

FIGURE 14-1: Being prepared for a crisis means also having a plan for crisis communication.

© Michelangelo Gratton/www.Shutterstock.com

Edward R. Albanesi, APR (Accredited in Public Relations)

Associate Director of Public Relations, Florida Farm Bureau Federation

Although it is not possible to anticipate every conceivable crisis that your organization might face, you can prepare yourself for the most common ones. In agriculture we have to be concerned about weather-related disasters, outbreaks of plant and insect pests, and diseases like bovine spongiform encephalopathy (BSE, "mad cow" disease). It's also important to set up communication channels and staff responsibilities that will kick in regardless of the crisis being faced. Managing the flow of communications during a crisis can be the most crucial step taken to help your organization and its constituencies emerge from the crisis intact. Quite simply, have a plan. Anticipate what will happen when a crisis presents itself, and then develop a strategy for dealing with it. Branch out the consequences by detailing action that will be taken under varying circumstances. Assign specific jobs to specific staff. Stage mock crisis events and walk through the process. Then, evaluate how you performed.

RISK COMMUNICATION

Risk communication refers to the knowledge and skills used by professional communicators to inform people about hazards to their environment or their health, to manage issues, to disseminate information, and to communicate effectively about potential crisis and emergency situations. Here are some examples of health and safety issues that qualify as risk communication topics:

- How safe is the water we drink?
- How polluted is the air we breathe?
- What risk does the landfill down the street pose to my family and my community?
- Is it safe to eat beef?

Risk communication skills and techniques are used to handle both risk and crisis situations. With risk communication, communicators try to lay the groundwork for the trust that needs to be established between the community and the organization dealing with the risks involved. However, bad risk communication could cause a crisis communication episode to develop. Communication experts generally agree that three elements exist when communicating a risk:

- **Message:** Messages are the overall information an organization wants its audience to comprehend,

even if the audience forgets the details. The message should inform and persuade; it should help audiences understand the message and get them to take certain action.

- **Medium:** The medium for the message depends on the specifics of the situation. It could be a brochure, a billboard, or a television commercial, for example.
- **Audience:** Risk may vary dramatically in different populations. Targeting a specific audience is extremely important.

Successful risk communicators must know how the public perceives risk and how to distinguish between objective risk and subjective risk. **Objective risk** is calculated by scientists based on research. **Subjective risk** is the risk that the public perceives to be hazardous and is affected by issues of familiarity ("I knew someone whom this happened to."), dread, and personal control (U.S. Environmental Protection Agency, 2007, p. 5). For example, it is much safer statistically to fly than to drive, and the chances of getting bitten by a shark are small compared to being attacked by a dog. These statistics are *objective risks*. However, people fear flying and shark attacks much more than they fear driving or dog attacks. It is the *subjective risk* that plays into people's fears.

Vincent T. Covello, director of the Center for Risk Communication, and Frederick W. Allen, counselor for the National Center for Environmental Innovation, Office of Policy, Economics, and Innovation with the U.S. Environmental Protection Agency (EPA), produced a pamphlet titled the "Seven Cardinal Rules of Risk Communication" (Covello & Allen, 1988). Any organization communicating risk issues should follow these rules summarized here:

- *Accept and involve the public as a partner.* Involving the public early, preferably before any decisions are made, helps establish an atmosphere of trust and sincerity. The release of information from your organization to the public goes more smoothly if you are seen as trustworthy with nothing to hide. If you fail to involve the public at an early stage, the community may become angry and overestimate risks.
- *Plan carefully and evaluate your efforts.* Communication strategies and plans must be developed as early as possible. Begin with specific objectives, such as providing information to the public, motivating individuals to act, or contributing to the response of a conflict. Recruit spokespersons who are good at presentations and interactions.
- *Listen to the specific concerns of community members.* Risk communication is a two-way exchange. You must listen; do not make assumptions. Recognize people's emotions.
- *Be honest, frank, and open.* Trust and credibility are your most important assets when communicating risk information. If trust and credibility are lost, they are almost impossible to regain. To maintain trust and credibility, you may want to state your credentials, admit mistakes, and disclose risk information quickly.
- *Work with other credible sources.* Develop partnerships, when applicable, with other organizations or governmental agencies. Try to partner with other trustworthy experts, such as university scientists, trusted local officials, or doctors.
- *Meet the needs of the media.* The news media transfer risk communication to the public, so you must meet reporters' needs. Be open and accessible to reporters. Before a risk situation occurs, build a relationship with reporters. This is covered in Chapter 13, Media Relations. "Banking goodwill" ahead of time with reporters may help you in a risk situation.
- *Speak/communicate clearly and with compassion.* Do not use technical language. Be simple. Be sensitive to people's emotions. Promise only what you can do, and be sure you can do what you promise.

Distinction between Risk Communication and Crisis Communication

Risk communication differs from crisis communication. Crisis communication usually deals with bad things—crisis situations—things that go wrong, such as a severe injury or death that happened at an event sponsored by your organization. Risk communication is not restricted to major disasters; risk communication is about things that might go wrong, the accident that could happen. It is any event that could cause public concern and could create media attention that will be focused on an organization. Here is an example of how the distinction between risk communication and crisis communication could play out:

- *Risk communication:* You own a food processing facility. A food product that a competitor sells has been found to have salmonella. The competing company issues a food recall. Your product, though, does not have salmonella. You are in "risk communication" mode. Although your product is not tainted with salmonella, you initiate a toll-free telephone hotline, informational websites, and distribution of information through various media to inform consumers that your product is safe. You are being proactive and are responding to and listening to the public. Because you acted quickly, the public's concern about your product—even though your food product was not tainted—was alleviated because you brought in the public as a partner. You suffer some economic loss at this time because sales decreased, but because you have responded in a way that enhances the public's trust, you are seen as a responsible company.
- *Crisis communication:* Your food processing company unknowingly shipped out salmonella-tainted food. Within a short time, people around the country are getting sick, and the cause has been traced to your company. You are in "crisis communication" mode. In this scenario, you must respond quickly to the media and the public's food safety concerns. If you respond in a way that addresses their concerns, you can maintain credibility and trust.

© Cengage Learning 2012

Jim Handley

Executive Vice President, Florida Cattlemen's Association

It is really important that we create and maintain a comprehensive crisis communication plan because we need an outline to follow when a crisis comes up. Whether it is a storm event—such as a hurricane—an animal disease outbreak, or a food safety issue, a plan helps the entire industry—the businesses and individuals within the industry—with a plan of action to respond to the crisis. The plan ensures a set of action steps is outlined, whom to communicate with, and how to manage the crisis for the best possible outcome. If it is a food safety crisis, then a good plan will maintain consumer confidence; if it is a storm event, the plan helps the affected parties deal with the aftermath of the storm and resume business; if it is an animal disease issue, a good plan may save the industry from extreme economic loss through proper control of the disease and utilizing the media through proper communication to inform the public and people in the business being impacted.

It is vital that proper advance preparation be done to be ready to respond when a crisis occurs. This might include spokesperson media training. This ensures that a unified, organized message is being communicated to the media when the crisis happens. It is important that a communication chain is established that includes all the key people, agencies, organizations, professionals, and media outlets to be included in the execution of a crisis management plan. An organization can anticipate the type of questions and scenarios that may unfold in a crisis and have talking points and factual information ready to disseminate when a crisis is at hand. One idea may be to create a website that is held "dark" or inactive but can be turned on or activated if a crisis occurs. This site may include all information about a given animal disease, a food safety issue, or management ideas for a storm. It will provide media with a place to go to get the facts and ensures a uniform message is distributed to the media and the public once a crisis occurs.

CRISIS COMMUNICATION

The term *crisis communication* is most often used to describe an organization facing a crisis and the need to communicate about that crisis to decision makers and the public. All crises have common characteristics:

- *They are nearly always negative.* They cast shadows of doubt about the credibility of an organization in the eyes of the public.
- *A crisis can create improper or distorted perceptions.* A crisis may involve allegations that tell only part of the story and stimulate negative impressions by the public about the organization. Unfortunately, perception is too

often reality. An organization, therefore, must be prepared to deal with incorrect viewpoints and comments.
- *Crisis situations are almost always disruptive to the organization.* Work is placed on hold until the crisis is resolved.
- *A crisis generally takes the organization by surprise.* The organization is placed in a "reaction" mode, where it reacts to the situation, rumors, comments, and potentially hostile interviews. The communicating organization is experiencing an unexpected crisis and must respond.
- *Crisis also implies lack of control.*

Preparing for the Crisis

The best way to handle a crisis communication situation is to *plan*—in advance—for a crisis situation. Of course, you will not know what specific crisis might occur, but having a contingency plan in place—so that your organization knows who will talk with the media, the "chain of command" for decisionmaking, and how communication will be handled overall—is extremely important. Your organization's overall crisis plan should devote significant time and effort to the **crisis communication plan**, especially if the crisis affects a large part of the public. The more people a crisis impacts, the more important it is to communicate to the public. The plan should address these key issues:

- *Organize a "what-if" brainstorming session with others.* Come up with "what-if" scenarios about potential crisis situations. Determine steps on how you would respond to the "what-if" crises (Figure 14-2).
- *Select crisis management and crisis communication teams.* Who is responsible for communicating with the media during a crisis? Who fields telephone calls? Who makes decisions about what to say to the media? Everyone in your organization should know who is on the crisis communication and crisis management teams.

Your crisis communication plan should be developed to fit your organization. There are many more actions that could be taken, but for most crises, the following will be the ones most needed:

- *Identify key audiences.* Determine to whom you want to communicate. It could be parents, students, local officials, or the general public.

FIGURE 14-2: Good crisis management includes brainstorming potential crisis situations and determining appropriate responses—ahead of time.

- *Designate a spokesperson.* One person should answer all questions and make all presentations. This ensures that information comes from one source.
- *Provide guidance to the public.* In a crisis that entails physical harm or a health risk to the public, give the public the information they need quickly. If it is an evacuation, for example, have a plan to communicate information so that the public evacuates in a safe and timely manner.
- *Develop messages and then communicate the messages and the facts.* Develop a few clear, simple messages for the media. These messages should be delivered repeatedly and clearly and by one person. The content of the messages should communicate concern about what is happening and explain what is being done to alleviate the crisis.
- *Anticipate the tough questions.* Make a list of potential tough questions and be ready to respond to them.
- *Control the message.* Stick to the message and the facts. Control the information that is disseminated. If bad news is to be released to the reporters, be up front about it and get it out at once, instead of letting the bad news trickle out a little at a time. This way all of the negative news gets out at once, instead of having smaller "negative news" reports throughout the day.
- *Control the flow of information.* Hold regularly scheduled news conferences or reports so that the information "flows" at certain times of the day.
- *Keep track of media calls and requests.* You will use this information later as you evaluate how well your crisis communication plan was handled.
- *Respond to the news media quickly and fairly.* Reporters want to get the message to the public. If you try to avoid the news media, reporters will not paint a good picture of your organization. Therefore, cooperate with reporters, be sensitive to deadlines, and provide all reporters with the same information.

Several good examples of crisis communication plans can be found on the Web. Use a Web search engine and type in "crisis communication plan." Also, from the Canadian Centre for Emergency Preparedness (2010), you can download a draft crisis communication plan that you can tailor to your organization.

Scott Wallin

Director, Producer Communications, Dairy Management Inc.

Why is a crisis communication plan important for your organization (or any organization, for that matter)?

During a crisis, your first goal is to maintain consumer confidence in your product. If your industry appears to be unorganized, unsure, or dishonest, you run the risk of losing consumer trust and hurting your farmers' livelihood. Having a crisis plan in place is the best insurance policy you can have for your industry's immediate credibility and future welfare.

The Florida dairy industry has a comprehensive crisis communications plan in place that deals with most every conceivable emergency. We practice "real-life" scenarios to get a feel for how we would react and think during an actual crisis. Hopefully, we'll never use this training, but we feel we are prepared. Another key goal is to build relationships with credible allies who can represent you effectively should a crisis occur. There is no such thing as having too many experts on your side during a crisis, but they'll be better able to assist you and your industry's needs if you built the relationship ahead of time. Our industry has made contacts, for example, with state veterinarians, department of health officials, and various university experts and researchers.

Communicating during the Crisis

Following are some pointers on how to communicate to the news media during a crisis situation:

- *Get the facts.* Miscommunication heightens during a crisis and can be exaggerated by half-truths, distortions, or negative perceptions. Get to the heart of the real story and tell it.
- *Be active, not reactive. Tell it all; tell it fast.* Take the offensive when a serious matter occurs.
- *Gather and classify information into categories, such as "facts" and "rumors."* Facts should be routinely updated; rumors should be verified or exposed as myths.
- *Deal with rumors swiftly.* Tell only the truth about what you know to be fact. Do not repeat others' opinions, hearsay, or assumptions.
- *Centralize information.* Designate one spokesperson. A central spokesperson provides

a singular "face" for the reporters. Viewers begin to become familiar with a central spokesperson, so this is one way to begin building credibility with the organization, if the person comes across as trustworthy. Centralized information also will minimize miscommunication.

- *Do not get mad.* Keep your cool in an interview or news conference with reporters. Some of their questions may be hostile, and some questions and comments may seem to be a personal attack to you, but remember that they are trying to get information on a crisis-oriented story that may have widespread impact for their audiences. So do not get mad when you are asked the "hard" questions.
- *Stay "on the record" in all interviews.* "On the record" means information that you provide to a news reporter can be attributed to you. "Off the

© Cengage Learning 2012

Jim Handley

Executive Vice President, Florida Cattlemen's Association

Do you have a "best case scenario" of when your organization faced a crisis communication situation?

The beef cattle industry managed a crisis situation in the media that started December 23, 2003, when a cow in the United States was diagnosed with BSE (bovine spongiform encephalopathy), also known as "mad cow" disease. It was the first-ever discovery of a cow with this disease in the United States and there was tremendous interest from the media and potential fear in the consumers of beef. We activated a "dark" website, conducted a series of information-sharing conference calls, mobilized a network of spokespersons to respond to media inquiries, and conducted interviews. Fact sheets were distributed to the media, as well as within the industry, to make sure factual information was distributed in a timely manner nationwide. Daily conference calls were conducted to educate the media, answer questions, and keep all the facts front and center for the media to use in the reporting of this news story. This was a very good example of managing the crisis with a plan, maintaining consumer confidence, and serving the needs of the media to a positive outcome. The American public was more confident about the safety of beef following this incident than it had been prior to it. The result of the crisis management plan worked with tremendous efficiency to get science-based facts out to the public. Having a plan was vital to managing this situation. I did four television interviews within three hours of the story breaking, and did interviews all over the state with all types of media. In Florida, alone, representatives of the Florida Beef Council and the Florida Cattlemen's Association did 82 interviews in three weeks. This crisis management has been used as a model in planning and preparation of new crisis management plans within the U.S. beef industry and in other industries.

record" means you are providing information to a news reporter that you do not want attributed to you. Do not go "off the record." Any comment worth saying should be said "on the record." If you go "off the record," be ready to read it in print the next day. Anything can, may, and will be done to advance a story. You should not be lured into going "off the record" under any circumstance.

- **No "no comments."** Try to have an answer for reporters' questions. But if you do not have an answer, do not be afraid to say, "I don't know, but I'll find out." Saying "no comment" appears to television news viewers and newspaper readers that you have something to hide.

- **In any crisis situation, follow every order, direction, or suggestion from emergency officials.**
- **Write everything down.** Maintain a crisis communication inventory of what was said by whom and at what time. This way, you will have a record of the event and how it was communicated. You can evaluate your responses so you will be better prepared if another crisis happens in the future.

After the Crisis

After the crisis is over and all communication with the news media has ended, do not just sit back and do

© Cengage Learning 2012

Scott Wallin

Director, Producer Communications, Dairy Management Inc.

Do you have a "best case scenario" of when your organization faced a crisis communication situation?

The hurricane season of 2004 devastated many Florida dairy farms. Facilities were destroyed or were without power and, as a result, cows were unable to be milked as scheduled (Figure 14-3). For the farms that could milk cows, a fuel shortage kept many tanker trucks idle, and those that could travel often were challenged to reach farms by debris that blocked roadways. Our crisis plan included a section on dealing with natural disasters. We fielded calls from reporters who wanted to know how badly the supply of milk to stores had been affected, not to mention what the extent of damage was and how quickly the industry could recover. Our plan had identified ahead of time which key industry leaders would handle questions concerning the milk supply to the stores, in addition to who could best address the "state of the industry."

© Cengage Learning 2012

FIGURE 14-3: This photograph shows some of the damage at a Florida dairy following the 2004 hurricane season.

nothing. It is time to evaluate how you handled the crisis. Your review should include the following:

- *A review of why the crisis occurred.* Could you have done anything to prevent the crisis?
- *An evaluation of how the crisis was handled.* Was information disseminated through one spokesperson? Did miscommunication occur?

- *An examination of similar scenarios.* What would you do in a similar situation in the future? What did others do in similar situations?

A crisis will happen in the life of most organizations. Taking time now to prepare for a crisis—even if you think it will never occur—and planning how to communicate to the news media during a crisis is your best defense.

SUMMARY

This chapter covered the distinction between risk communication and crisis communication. The best way to handle a crisis situation is to develop a crisis communication plan before a crisis occurs.

CHAPTER EXERCISES

APPLICATIONS

Following are some ways you can apply what you learned in this chapter:

- What are potential subjective risks that an agricultural organization or company should be prepared to address?
- After reading this chapter, what components of crisis communication planning do you need to adopt or practice more often? What elements of risk communication do you want to implement?
- Develop a crisis communication plan for a real or imaginary agricultural organization or company. Some questions you might want to consider as you develop your plan are as follows:
  - What happens when your issue becomes a "hot" topic?
  - What response should you give as a member of your organization?
  - Whom could you suggest to the news media as extra informational resources?
- Record news programs in which spokespeople are interviewed during a crisis situation. Evaluate how well you believe they responded to reporters' questions. You also may wish to show these videotaped segments to people in your organization as a way to generate discussion about crisis communication.
- Stage a "mock" crisis communication situation. Choose a hypothetical, yet possible, crisis situation. Some possible mock crisis communication examples follow. Have participants work in small groups to determine what message(s) should be presented to the media. Have the groups conduct a mock news conference in which group participants would provide information to the media. Participants in other groups should serve as reporters. How did the participants react during the mock crisis? Lead a discussion on how to properly provide information to the media during a crisis.

Mock crisis communication 1: Beef recall: Five-hundred-thousand pounds of ground beef have been recalled because of possible contamination with *E. coli*. Two hundred people nationwide have gotten sick, but so far, no deaths. You, as a local beef producer, have been called by a local reporter to offer your perspective on the story.

QUESTIONS YOU MAY BE ASKED TO ANSWER:

- The company that is recalling the ground beef reused materials from the previous day's processing and placed it in today's meat. Shouldn't this be against the law? Are you comfortable with this process?
- Do you ship cattle with *E. coli* to processing plants?
- Do you find *E. coli* or other bacteria in other types or cuts of meat?
- Aren't you frightened that the hamburger you eat can kill you?
- Can you guarantee that I won't get sick from eating hamburger?
- Are you at risk from eating ground beef at a restaurant? What about at home?

Mock crisis communication 2: Groundwater contamination: A recent study by a local water management district shows increased levels of nitrates in the groundwater. Fingers begin to point to your cattle operation and others—cattle ranches and dairies—as the source of the groundwater contamination. You have been called by a reporter to offer your perspective on the story.

QUESTIONS YOU MAY BE ASKED TO ANSWER:

- Why isn't something being done about the pollution you and other cattle ranches are causing to local groundwater?
- Can't you do something about run-off? Are there preventative measures?
- Shouldn't you and other cattle ranches/dairies be charged for cleaning up the water?
- Aren't you concerned that people will get sick from drinking water contaminated by your ranch/dairy?
- Should homeowners pay for cleaning up something you're messing up?

CHAPTER QUESTIONS

MATCHING:

Match each of the following definitions with the correct term.

_____ **1.** Calculated by scientists based on research.

2. Used to describe an organization facing a bad situation and the need to communicate about that situation to decision makers and the public.

_____ **3.** What the public perceives to be hazardous; affected by issues of familiarity, dread, and personal control.

_____ **4.** The process of informing people about hazards to their environment or their health.

a. Crisis communication

b. Risk communication

c. Objective risk

d. Subjective risk

e. Persuasion

Match each of the following definitions with the element used when communicating a risk.

_____ **5.** The overall information an organization wants its audience to comprehend; it should inform and persuade.

_____ **6.** A group of people targeted for the risk communication information.

_____ **7.** How something is communicated—it could be a brochure, a billboard, or a television commercial, for example.

a. Message

b. Medium

c. Audience

d. Communication

MULTIPLE CHOICE:

8. Which of the following is NOT one of the seven cardinal rules of risk communication?

 a. Accept and involve the public as a partner.

 b. Plan carefully and evaluate your efforts.

 c. Takes the organization by surprise.

 d. Be honest, frank, and open.

9. Which of the following is NOT one of the key factors in a crisis communication plan?

 a. Designate more than one spokesperson so that the media have greater access to information.

 b. Anticipate the tough questions.

 c. Keep track of media calls and requests.

 d. Respond to the media quickly and fairly.

10. Which of the following is NOT a characteristic of a crisis?

   a. Nearly always negative

   b. Controlled flow of information

   c. Disruptive to the organization

   d. Takes the organization by surprise

SHORT ANSWER

11. List five of the steps of communicating *during* a crisis.

REFERENCES AND FURTHER READING

Canadian Centre for Emergency Preparedness. Last modified 2010. Accessed August 30, 2010. http://www.ccep.ca/ccepweb.asp?m=94&ap=3.

Covello, V.T., and F. W. Allen. "Seven Cardinal Rules of Risk Communication." OPA-87-020. Washington, DC: U.S. Environmental Protection Agency, April 1988. Accessed August 30, 2010. http://www.kfa-juelich.de/mut/rc/covall88.html.

Fearn-Banks, K. *Crisis Communications: A Casebook Approach.* Mahwah, NJ: Lawrence Earlbaum Associates, 2002.

Groth, E. "Risk Communication in the Context of Consumer Perceptions of Risk." Consumers Union. Last modified 1998. Accessed August 30, 2010. http://www.consumersunion.org/food/riskcomny598.htm.

Hogue, J. "Avoiding Disaster: The Importance of Having a Crisis Plan. Last modified 2001. Accessed August 30, 2010. http://iml.jou.ufl.edu/projects/Spring01/Hogue/index.html.

Joint Institute for Food Safety and Applied Nutrition. FoodRisk.org. Last modified 2010. Accessed August 30, 2010. http://www.foodrisk.org.

Lundgren, R., and A. McMakin. *Risk Communication: A Handbook for Communicating Environmental, Safety, and Health Risks.* Columbus, OH: Battelle Press, 2008.

U.S. Environmental Protection Agency. "Risk Communication in Action: The Risk Communication Workbook." Last modified August 2007. Accessed August 30, 2010. http://www.epa.gov/nrmrl/pubs/625r05003/625r05003.pdf.

Putting it Together

Persuasion and Persuasive Informational and Educational Campaigns

OBJECTIVES

After completing this chapter, the student will be able to:

- Describe the psychology of persuasion.
- Define what persuasion is and explain how it can be used to influence others.
- Describe the three types of argument strategies.
- Describe and provide an example of how persuasion is used in the media.
- Prepare a persuasive paper or speech using the five-part argument strategy.
- Develop an informational/educational campaign plan.

INTRODUCTION

It can be argued that all types of communication involve some form of persuasion. **Persuasion** *attempts to influence or convince others to take a specific action or to reach a certain conclusion about an issue. When we argue for our point of view, we are engaging in persuasion. When we present information as part of a claim or statement, we are being persuasive.*

What is the difference between information and persuasion? **Information** *can be neutral, or it can be biased, but it is generally intended to show evidence, facts, and details about something. Persuasion uses information as part of an attempt to make claims or arguments designed to reinforce or change beliefs. Persuasion is what we employ when we want someone to undertake a specific action or behavior. Think about a time when you wanted your parents to do something for you, such as buy you a special birthday gift, let you stay up late, or go out with your friends. Your attempts to persuade your parents probably focused on communicating specific reasons why your parents should take the action you wanted (Figure 15-1). The ability to persuade others depends on good communications skills. Persuasion is an activity that we engage in to convince others, but it is also the basis for communications campaigns that attempt to move an entire audience to a desired action.*

This chapter is designed to help you develop communications skills that can be used to persuade and influence others. The chapter will cover strategies for developing rational arguments that can be used to communicate orally and in written form, such as the five-part argument, which is the basis for a persuasive essay. In addition, you will learn about logic and reasoning, as well as what false arguments are and how to avoid them. The chapter will also focus on using persuasion as a specific approach in communications campaigns aimed at informing and educating others.

FIGURE 15-1: Persuasion is an activity that we engage in to convince others.

PERSUASION STRATEGIES

Have you ever given a persuasive speech or written an essay for a class where you were asked to take a position on something? Persuasion is a type of communication with specific characteristics. When communicating persuasively to an audience, as in a speech or written document, the objectives of persuasion are:

- to reinforce a belief an audience holds,
- to change a belief an audience holds, or
- to move an audience to action.

Good persuasive techniques include the following:

- Make claims or assertions.
- Present evidence (statistics, facts, observations, visuals).
- Present a specific line of reasoning.
- Address skeptical readers or listeners by refuting arguments that oppose the communicator's main proposition or claim (Figure 15-2).

Persuasive communication includes the use of strategies that are designed to create a convincing argument.

FIGURE 15-2: An effective speaker addresses the audience with evidence and facts to support the claim being made.

Sometimes called **argument strategies**, these include the following:

- Proposition-fact-evidence
- Common ground
- Logical reasoning (induction, deduction, comparison)

Proposition-Fact-Evidence

In this strategy, you start with your basic proposition, followed by supporting facts, and then forms of evidence. A **proposition** is a claim indicating your stance or position on an issue. It can also be your proposed solution to a problem. Let us say you are arguing for your school to allow more students to park cars on campus. Currently only seniors can park their cars on school grounds. Your proposition is that by letting both juniors and seniors park their cars at school, it will alleviate the congestion caused by parents dropping off their children in front of the school each morning.

Facts are statements of what is known to be true in a given situation, whereas **evidence** includes any proof you have that helps you argue your main points. A fact statement about school parking might be:

"Residents who live on the street where the school's main entrance is have been complaining about the traffic."

Although this a fact that supports your case by describing an aspect of the problem, additional facts that lend credibility or believability to your main argument are equally important, such as:

"There are enough parking spaces available to accommodate the number of juniors who would like to park on campus if allowed to do so."

This fact statement adds to the value of your argument because it shows that your proposed solution is valid.

Evidence may include testimonials and statements that support your claim, cited documents, literature, and statistics. In the previous school parking example, supporting evidence might be a signed petition from fellow students, letters of support from teachers and parents, and references from articles that show that this solution has worked for other schools in the same circumstances.

Common Ground

Common ground refers to the act of taking polarizing viewpoints and showing where they agree. The idea is to find enough common areas of agreement that the two sides can be shown to be more similar than different and thus stimulate a compromise where a consensus can be reached. This strategy is particularly effective when there are two opposing sides to an issue that has caused controversy. In the school parking example, perhaps there is a group that is very opposed to more students parking because it will add more cars on campus, whereas the other side sees the proposal as a way to eliminate congestion caused by parents driving their children to school. Both sides could find common ground on the need to reduce congestion.

Logical Reasoning

Logical reasoning can be defined as a way of thinking. We do not always use logic in our thinking processes. Sometimes we let our emotions influence our thinking. When making a persuasive argument, the writer or speaker draws on logic and reasoning to make and support points and refute or pre-empt any objections. There are three types of logical reasoning processes:

- Induction
- Deduction
- Comparison

Induction is a method of logical reasoning that is associated with the sciences. *Inductive reasoning* involves the writer or speaker moving from particular facts to general conclusions. Induction occurs when a person looks at a set of facts and makes an educated guess to explain them. Induction is used in science, where the assumption or induction a scientist makes is called a **hypothesis**. The scientific method is based on inductive reasoning. For example, a plant scientist might decide to test which variety of a plant can withstand drought-like conditions better than other varieties. Drawing on the characteristics of the plant, soil, and

FIGURE 15-3: Plant scientists use the process of inductive reasoning to form a hypothesis about which plants might grow best.

climate conditions that exist, the scientist makes a hypothesis as to which plant might work best and then conducts an experiment (Figure 15-3). The results of that experiment allow the scientist to form a general conclusion as to which plant worked best.

Scientists use inductive reasoning to discover new knowledge that might be applied to solve real-world problems, such as growing more viable plants.

Deduction is when the writer or speaker moves from the general to the particular. *Deductive reasoning* starts with a general principle, then applies the principle to a fact, and finally draws a conclusion concerning the fact. How the deduction process works can be best illustrated with an often-used syllogism:

- *Major premise:* All professional golfers are good athletes.
- *Minor premise:* Judy is a professional golfer.
- *Conclusion:* Therefore, Judy is a good athlete.

A **syllogism** is a logical argument with three propositions or claims: a major premise, a minor premise, and a conclusion. The conclusion is a deduction of the major and minor premises. A good deduction requires careful analysis of the premises upon which it is being made. In the previous syllogism, for example, the conclusion is only true if being a good golfer automatically makes you a good athlete.

Comparison is the form of logical reasoning in which writers and speakers choose between or among best alternatives based on a set of standards or criteria. To effectively use comparison, you must show how what you are comparing is similar or different in some way. Discussing alternatives or making comparisons can also be used when making a recommendation or proposing an action or solution to a problem. Advertisers and marketers tend to use comparison in order to persuade consumers to buy their product over someone else's:

Product A	Product B
■ The most popular-selling brand with consumers	■ Not the best-selling brand
■ Extra features	■ Less features
■ Comes in many colors	■ Comes in few colors
■ Easy to clean	■ Many parts make it harder to clean

The one to buy: Product A!

After making a "head-to-head" comparison like this one, the advertiser ends with a "*call to action*" that prompts consumers to conclude, based on the comparison, that they should purchase the advertiser's product. Good comparisons compare objects or alternatives on similar characteristics and conclude by recommending a specific choice or course of action.

PREPARING YOUR PERSUASIVE PAPER OR SPEECH

When writing a persuasive paper or preparing for a persuasive speech, your goal should be to focus on an issue or problem that people disagree about. Try to avoid problems where the solution can be determined immediately by available factual material. Some of the hardest subjects to write about are controversial topics that are based on personal judgments and those for which no conclusion can be reached. These types of topics are usually tied to feelings. The most effective persuasive argument appeals to reason, not emotion.

When writing or speaking persuasively, you should directly state what is good or bad, and why you think so, near the beginning of the persuasive argument. This is your thesis statement, or main point, that you want to make. Because your purpose is to persuade using logic and reasoning, this communicates to readers or listeners that you want to convince them of your point of view. Remember that you are trying to influence. To communicate persuasively, you must try to influence the beliefs and attitudes of your readers or listeners.

One way to structure a persuasive argument in written or spoken form is to use the five-part argument.

The **five-part argument** breaks a persuasive essay or speech into five parts:

- The *introduction* attracts the attention of the audience, sets the tone, and describes what the argument will be about. The introduction usually includes the thesis statement, the specific sentence that explains what the main point of the argument will be.
- The *background* provides the context and details needed for a reader or listener to understand the situation being described, as well as the problem or opportunity being addressed.
- **Lines of argument** make up the body of the essay or speech. Here is where you include all the claims, reasons, and supporting evidence you have that help you make your points effectively.
- **Refuting objections** is an often overlooked part of an effective persuasive argument. "Refuting" means disproving, ruling out, and countering any potential objections before the readers or listeners can think of reasons not to be persuaded.
- The *conclusion* is where you present your closing arguments. To be effective, the conclusion should restate your thesis statement and summarize the main points of your argument. If you are advocating a particular solution to a problem or a decision to be made, you should close by asking your reader or listeners to adopt your point of view.

PERSUASION IN THE MEDIA

Persuasion is an element of what we see, hear, and read in the media. Advertisers and marketers want to persuade audience members to buy their products. Politicians want to persuade voters to support them. Policymakers and activists hope to persuade citizens to agree with their views on the issues. One way they do this is through a concept called "framing" (Figure 15-4).

Framing refers to how a message conveyed in the media provides contextual cues that can indicate to receivers how to think about what is being communicated. Although journalists strive for objectivity in their reporting, the way in which they describe what they report helps to "frame" a story's content. This could be done in ways that promote one idea over another, meaning a conflicting idea might be left out or portrayed negatively. When a topic is framed or connected to an existing way of describing something, the topic's meaning can be greatly influenced. The way information is framed is the way people come to

© Monkey Business Images/www.Shutterstock.com

FIGURE 15-4: Framing is one technique used in the media to persuade individuals to buy products, support candidates or policies, or take some desired action.

understand that issue. Journalists use media frames to write stories that help uncover meaning from related stories and events.

Even the selection of given words can affect the consideration of information and the audience's reaction to it. Framing can significantly affect the perception of an issue and the evaluation of alternative options. The use of the word "Frankenfood" in a story about genetically modified food is an example of a frame that adds a negative element to the description. Framing can elevate certain aspects of an issue in the perceptions of individuals. Frames can be included in newspaper or magazine articles, TV stories and programs, and even movies. Framing focuses on this question: What message is portrayed, and how is it portrayed?

Framing also can be a valuable tool for communications professionals in advertising, marketing, and public relations. Ultimately, communicators want their product, service, event, or company "framed" in as positive of a light as possible. Understanding framing helps media professionals describe their topics more favorably when they seek media coverage.

Positioning refers to describing or framing your subject in a way that influences how others "see" or perceive it. Professional communicators frame, or position, their message based on their knowledge of the audience they are trying to influence. When framing a message, ask yourself the following:

- What idea or proposal do I want accepted?
- How will my readers relate to this idea or proposal? Will they benefit in any way?
- Can I construct a detailed, complete argument from this positioning?

The most common ways to frame or position a message include the following:

- A *positive frame* describes rewards, value, or benefits.
  - *Example:* Farmers are "stewards of the land."
  - This is a positive frame, because the selection of words conveys a positive image of farmers in relationship to the land and the environment.
- A *negative frame* focuses on the consequences or risks.
  - *Example:* Agricultural production is based on "factory farming."
  - This is a negative frame, because the term "factory farm" is a word choice that associates farming with a mechanized, non-natural manufacturing process.
- **Fear appeals** raises fear and anxiety.
  - *Example:* Describing genetically modified food as "Frankenfood."
  - This is a fear appeal because it associates a food technology with a well-known cultural icon (the *Frankenstein* novel and numerous "Frankenstein" movies) involving the creation of a monster and misuse of science.

THE PSYCHOLOGY OF PERSUASION

Persuasion attempts to use convincing arguments to alter attitudes or behavior. Effective persuasion motivates people to yield to influence. Persuasion can be either **pro-social**, in which a person stands to gain something positive or be rewarded by agreeing, or **anti-social**, in which people are warned about a potential punishment or negative outcome that could happen if they do something. Public health and safety messages in the media are good examples of the negative approach to persuasion, such as:

- "Don't drink and drive."
- "This is your brain on drugs."
- "Only you can prevent a wildfire."

From a psychological standpoint, persuasion is most effective when the person doing the persuading is viewed as credible and trustworthy, and the message being conveyed is perceived to be truthful, accurate, and aligned with an individual's or audience's values and interests. According to psychologist Robert Cialdini in his book *Influence: The Psychology of Persuasion* (1984), you can be more persuasive if you follow one or more of these elements:

- **Reciprocate:** Do something for someone else. Have you ever told friends or siblings they could borrow something of yours if they let you borrow something of theirs? When you negotiate by reciprocating, that is a form of persuasion.
- **Commitment and consistency:** Getting people to say they will commit to doing something is more effective than just providing them with the information they need to adopt the behavior you want. An example of a *commitment* is a contract or petition. Being consistent is necessary if you want to establish trust. Saying one thing and doing another is not consistent and can lead to **cognitive dissonance**, which is the basic incompatibility of holding two or more beliefs simultaneously (dissonance). When we hold a belief or are told something by someone that does not correspond to what we already know, we experience cognitive dissonance. Dissonance is not a comfortable feeling. When this happens, we either try to ignore the new information or we do not believe what we have just heard or seen. Imagine believing that your best friend liked you for yourself, only to find that person saying bad things about you behind your back. That is dissonance.
- **Social proof:** This means that people will do what they see others doing, as long as it seems to be socially acceptable. That is what advertisers do when they want to persuade people to buy a new product. They run an ad showing it being used by a lot of people; therefore, it makes using the product seems like the popular thing to do.
- **Liking and authority:** People are perceived to be more persuasive when they are likable and perceived to be in a position of power.

In addition to these elements, effective persuasion is influenced by the believability of the person who is communicating, as well as by the truthfulness of the message being communicated.

Message Characteristics and Effective Persuasion

Message characteristics are also related to effective persuasion. **Message "sidedness"** is the degree to which communications messages present only one side, both sides, or both sides plus an evaluation of the arguments or claims. The most effective messages are usually those that present both sides of an argument as well as refute the other side's position. **Message ordering**, on the other hand, involves the placement of elements of the persuasive argument. For example, is it more persuasive to place the main persuasive point early or at the end? Although there is no universal law to follow on this, it is generally a good idea to "get to the point" early and repeat it at the end of your communication for emphasis. Finally, a **"call to action"** is a specific directive designed to get listeners to act on the advice of the communicator. If you influence an audience's attitudes but do not include a call to action, you do not give audience members a way to take action or show support.

False Arguments

False arguments are a form of persuasion based on illogical or poor reasoning. We can say that this kind of persuasion lacks argument quality because the argument being made, rather than being supported by facts and logic, uses faulty reasoning and generalizations. Common false arguments include:

- **Bandwagon:** A bandwagon argument is one in which you try to persuade someone to do something because everyone else is doing it. ("Get on board, because everyone else is on board.")
  - **Example:** "But everyone else is going to the concert, why can't I?"
- **Ad hominen:** An ad hominen argument is basically trying to persuade by attacking the other side.
  - **Example:** "That kid is a jerk; he can't be right about what student government candidate to vote for."
- **Red herring:** A red herring is when you deliberately try to distract someone from the facts in order to persuade the audience to your way of thinking.
  - **Example:** Your mother asks who forgot to take the trash out. You know you were supposed to do it, but to distract her, you say, "Did you see the 'A' I got on my math paper?"
- **Begging the question:** When you beg the question, you refuse to acknowledge that there is any question and claim that your view is the only right one.
  - **Example:** "It is an undisputable fact that the only way to reduce traffic in the student parking lot is to limit parking spaces to seniors only."
- **The either-or fallacy** sets up the perception that if you do not agree with what is being conveyed,

there will be a direct negative result. ("If you don't go along, things will be bad.")

- *Example:* "Global warming will not stop unless everyone stops using all fossil fuels."

■ *Post hoc, ergo propter hoc:* This argument plays on logic. It assumes that if one thing is true, then the other thing must be true as well, not leaving any room for any other factors that may be influencing an outcome. ("If A, then B.")

- *Example:* "I wore my lucky shirt and got an A on the test. I must have gotten the A because I wore the shirt." In fact, other factors, such as the amount of studying you did, had much more to do with your A grade than your shirt did.

■ *Hasty generalization:* Hasty generalizations—jumping to the wrong conclusion—are made when people do not think about possible conclusions and when they do not anticipate a possible or probable result.

- *Example:* In the late nineteenth century in the Florida Everglades, developers began to build dams on Lake Okeechobee in order to clear the land so that development of real estate projects could occur. They did not think through what might happen if a major storm system came through the area, bringing with it large amounts of rain. In 1928, a Category 5 hurricane hit the area, killing thousands. The catastrophe was caused by flooding from a storm surge when winds drove water over the mud dike that developers had built to circle the lake. As a result, major flooding in the Everglades wetlands areas occurred (Figure 15-5).

FIGURE 15-5: The Everglades are a subtropical wetlands area in southern Florida that is susceptible to flooding. In 1928, this led to a major disaster.

INFORMATIONAL AND EDUCATIONAL CAMPAIGNS

An *informational campaign* is a special type of communications campaign in which one of your chief goals is to communicate information, using appropriate tools packaged for specifically targeted audiences and strategically framed messages, which is sometimes called a "message strategy." An example of an informational campaign might be informing students at your school about what your student organization is doing, such as achievements and recognitions of its members or activities and events that have been planned.

An *educational campaign* is very similar; the goal is usually to communicate information, but with the added purpose of *educating* a targeted audience. An effort aimed at educating your community about the importance of local agriculture would be an example of an educational campaign.

A *social marketing* campaign is a special type of educational campaign that uses persuasion, communicated via the mass media, to educate and inform audiences to engage in socially desirable behaviors, such as recycling, conserving water, or improving one's health and fitness levels. Although most educational campaigns are aimed at raising awareness and knowledge levels, social marketing campaigns focus on trying to change people's behavior. Social marketing campaigns have been used to "stop AIDS" through promotion of sexually responsible behaviors, to get people to use seat belts, and to promote providing young girls with opportunities to develop "girl power" in developing nations.

Both informational and educational campaigns use a combination of communication techniques, ranging from interpersonal to mass communication, and both kinds of campaigns typically start with developing a plan that identifies the problem and provides background and rationale for the suggested campaign elements. Campaign elements are next developed, then presented, either to internal or external decision makers, and then implemented. An evaluation method of some type is usually suggested as a way to determine or evaluate the success of the methods used.

Planning an Informational/ Educational Campaign

The following is a comprehensive outline that can be used to develop an informational/educational campaign that focuses on an issue or problem to be

Jennifer Nelis

Director of Public Relations and Marketing, Florida Nursery, Growers and Landscape Association

At Florida Nursery, Growers and Landscape Association (FNGLA), we use an educational platform to craft persuasive messages designed to change consumer behaviors related to their lawns and landscapes. For one such project, we developed a public service announcement (PSA) that featured Mother Nature as an attention-getting character with an educational message. Drawing on Mother Nature's natural credibility and inherent persuasion, we set out to educate Florida's consumers on gardening basics with the underlying message that you can still buy plants during a dry spell.

solved. Persuasion strategies are utilized in such a campaign to develop the goals and refine the messages you want to get across effectively. An informational or educational campaign's plan will consist of goals and objectives, strategies, rationale, tactics, and evaluation.

Informational/Educational Campaign Elements

The *goal* is the overall purpose of the campaign. For example, who will benefit and why? What are the potential outcomes, if this plan is implemented?

An *objective* provides clear guidance that permits the orderly presentation of content leading to some effect in the identified audience. Objectives consist of the following:

- *Behavior:* What the audience will be expected to do as a result of the effort to be undertaken.
- *Conditions:* Circumstances under which the audience members will be expected to perform the behavior.
- *Standards:* How well the audience members will be expected to perform the behavior.

As an example, let's take the case of trying to educate teenage girls about the dangers of websites dedicated to promoting anorexia, called "pro-anorexia" sites. Here is how goals and objectives could be developed for a campaign to provide information to restrict the spread of pro-anorexia websites:

- *Example goal:* Have legislation passed restricting underage admittance into pro-anorexic websites.
- *Example goal:* Make Web hosting services responsible for restricting admittance into harmful sites.
- *Example objective:* To increase awareness by 20 percent of the potential harm that unrestricted access by minors to pro-anorexia websites can cause among middle and high school students and their parents.

Strategies are statements as to how tactics will be employed to achieve the stated objectives and goals. Strategies are decisions based on what you know of communication processes, your target audience, and other situations that have similar aspects that you can draw on to fit your current circumstances. Strategic decisions indicate how you will target your audience and the rationale for doing so. Some strategies to consider implementing are as follows:

- Develop a strategy for the type of mass media communication(s) you plan to use, such as print, television, radio, and billboards.
- Develop a message strategy of how you will frame the informational or educational message.

See Chapter 2, Effective Communication and Message Development, for more on how to develop effective messages.

- Develop a strategy for the non-mass-media communication(s) you plan to use, such as special events, public relations, or speakers.

Rationale means to draw on what you know of communication processes and explain why this package of strategies is the best possible way to achieve your objectives and goals. *Tactics* are specific elements that you will use in your campaign, such as a news release or a print advertisement. Identify a set of tactics you plan to use. Make sure you describe what each tactic is and what message or information it will contain. Following is a partial listing of some specific communications-oriented elements you might use for informational/educational campaigns:

- Brochures
- Print advertisements
- Radio and television commercials
- Posters
- Billboards
- Fliers

In addition to these tactics that you might use in your campaigns to communicate to a mass audience, do not forget the following more educationally oriented tactics for an educational campaign:

- Lesson plans
- Curriculum guides
- Computer slide presentations

- Videos
- Websites
- Newsletters
- Training programs
- Books
- CD-ROMs and DVDs

Evaluation

Evaluation consists of measuring the success of your campaign. Typically, evaluation consists of measurement of *intangibles*—using surveys, questionnaires, and focus groups to assess new opinions that the audience formed, changed behaviors, or knowledge that was gained. Refer to Chapter 3, Research Methods, for more information on surveys, questionnaires, and focus groups. Evaluation also can consist of measurement of *tangibles*, such as measures designed to calculate exposure, or *impressions*, including numbers of audience members reached, the number of times the campaign was mentioned in the news media, the number of brochures or other elements produced and disseminated, or the number of hits on a website. Some questions you may want to consider in your evaluation are as follows:

- Did your audience become more aware of the issue or problem?
- Did the audience come to understand the issue or problem more fully?
- Did the audience's attitudes change?
- Did audience members decide to adapt or change their behavior?

SUMMARY

This chapter covered persuasion, including what persuasion is, the psychology behind the concept of persuasion, and how persuasion is used in the media. Tips for preparing your persuasive paper or speech were given, along with strategies used to persuade. The chapter concluded with a detailed description of the elements of informational and educational campaigns that attempt to persuade.

CHAPTER EXERCISES

APPLICATIONS

Following are some ways you can apply what you learned in this chapter:

- Take a look at a recent newspaper or magazine story to see if you can tell how it was "framed." Were the facts attributed to a source? Did the sentences use leading words?
- **Persuasion Activity:** Work in groups of four to brainstorm a controversial issue that has more than one side or position. In your group, agree on which side of the argument you will take. Develop and outline your five-part argument:
  - Introduction
  - Background
  - Lines of argument
  - Refute objections
  - Conclusion

  Identify your persuasion strategies. Be sure to refute objections. Base your arguments on the evidence. Is there any common ground?
- **Informational/Educational Campaign Project**: Using the plan outlined in the chapter, your goal—as a member of an in-class group—is to develop an informational or educational campaign for some real group or organization associated with an issue of importance to agriculture and natural resources audiences. Determine the following:
  - How best can you reach your audience?
  - How should you package or frame your message?
  - Should your message be one-sided or two-sided to be most effective?

CHAPTER QUESTIONS

MATCHING:

_____ **1.** Refers to describing your subject in a way that influences how others "see" or perceive it.

_____ **2.** A statement of what is known to be true in a given situation.

_____ **3.** Attracts the attention of the audience, sets the tone, and describes what the argument will be about.

_____ **4.** The degree to which communications messages present only one side, both sides, or both sides plus an evaluation of the arguments or claims.

a.   Introduction

b.   Message sidedness

c.   Positioning

d.   Facts

MULTIPLE CHOICE:

5. The most common ways to frame or position a message include which of the following?

    a. Positive frame (describes rewards, value, or benefits)

    b. Negative frame (focuses on the consequences or risks)

    c. Fear appeals (raise fear and anxiety)

    d. All of the above

6. "All professional golfers are good athletes" is an example of a _____.

    a. major premise

    b. minor premise

    c. conclusion

    d. syllogism

FILL IN THE BLANK

7. _____ is the form of logical reasoning that writers and speakers use to choose between or among best alternatives based on a set of standards or criteria.

8. _____ attempts to influence or convince others to take a specific action or to reach a certain conclusion about an issue.

9. _____ involves the writer or speaker moving from particular facts to general conclusions.

10. _____ processes start with a general principle, then apply it to a fact and draw a conclusion concerning the fact.

11. _____ means disproving, ruling out, and countering any potential objections before readers or listeners can think of reasons not to be persuaded.

12. _____ are decisions based on what you know of communication processes, your target audience, and other situations that have similar aspects that you can draw on to fit your current circumstances.

13. _____ are specific elements that you will use in your campaign, such as a news release or a print advertisement.

14. _____ refers to the act of taking polarizing viewpoints and showing where they agree.

15. _____ refers to how a message conveyed in the media provides contextual cues that can indicate to receivers how to think about what is being communicated.

REFERENCES AND FURTHER READING

Cialdini, R. *Influence: The Psychology of Persuasion.* New York: William Morrow and Company, 1984.

Griffin, E. *A First Look at Communication Theory.* New York: McGraw-Hill, 1997.

O'Hair, D., and M. Wiemann. *Real Communication.* Boston, MA: Bedford/St. Martin's, 2009.

Petty, R. E., J. T. Cacioppo, and A. Strathman. "To Think or Not to Think? Exploring Two Routes to Persuasion." In *Persuasion: Psychological Insights and Perspectives,* 2nd ed., edited by T. C. Brock & M. C. Green, 81–116, 2005.

Stone, G., M. Singletary, and V. Richmond. *Clarifying Communication Theories.* Ames: Iowa State University Press, 1999.

Vyncke, P. "Lifestyle Segmentation: From Attitudes, Interests and Opinions, to Values, Aesthetic Styles, Life Visions and Media Preferences." *European Journal of Communication,* vol. 17, no. 4(2002): 445–463.

Wimmer, R. D., and J. R. Dominick. *Mass Media Research: An Introduction,* 7th ed. Belmont, CA: Thomson Wadsworth, 2003.

16 Planning Special Events

OBJECTIVES

After completing this chapter, the student will be able to:

- Create, plan, and evaluate special events.
- Implement the logistics of a special event.
- Describe the elements of a successful special event.
- Develop a special events budget.
- Describe event marketing and public relations as they are used in special events.

INTRODUCTION

If you have ever planned a meeting, put together a party for someone, or worked on a fundraiser for a nonprofit organization or a bazaar at your church or school, you have experienced the planning of a special event. A **special event** *is a planned gathering of people who get together to participate in some type of shared activity. Whether it is a potluck supper shared by neighbors or a national conference with hundreds or thousands of attendees, special events do not just "happen" (Figure 16-1). Events are special occasions that need to be planned in order to be successful. Planning an event takes special skills in communication, management of details, creativity, budgeting, teamwork, and leadership. Learning how to plan special events is an important skill that can benefit you personally, as well as lead to a potential future career.*

FIGURE 16-1: Special events, both large and small, require careful planning in order to be successful.

Erin Freel Best

President, The Market Place

Nothing lends credibility to a group or organization more than a well-run meeting or event. Good speakers, succinct agendas, and programs can make or break a meeting or event. You can set yourself apart in the "real world" by running an effective meeting and putting on an organized special event.

This chapter is designed to help you understand what special events are and how to conduct them. You will learn what professional event planners do, as well as all the elements that go into planning a successful event. This chapter will explain how event sites are chosen, what criteria are used to plan events, and how to handle site permits, contracts, and budgets. You will learn the five critical stages in planning a special event, and you will also learn how to evaluate an event.

MAKING AN EVENT "SPECIAL"

In one sense, you could say special events have been around since the first gatherings of cavemen and cavewomen, when they gathered to share stories. Most of us have a natural desire to get together with others and share an activity. Special events can range from a small meeting or get-together of members of a group or organization to large-scale conventions, concerts, and parades. Many credit Bob Jani with inventing the term "special event." Bob Jani was Walt Disney's first director of public relations when Disneyland opened in 1955. He said Disneyland's Main Street Electric Parade was a new and different type of programming, calling it a "special event." Jani later went on to produce other major events, such as the Radio City Music Hall Christmas Spectacular.

Since Jani's time, the people responsible for organizing and conducting a special event have become known as **event managers**. Event managers create, design, plan, coordinate, and manage events of all types. To be successful, an event manager needs to be a person of vision and have a great deal of initiative, energy, and self-motivation. As managers, all elements of the event are their responsibility, so event managers must be good leaders and team members, capable of inspiring others to do the best job they can. Event managers must be able to communicate effectively and with a diverse range of people and personalities. They also need to be organized and able to take on a variety of roles, as well as being able to remain calm despite what may be happening around them (Figure 16-2).

FIGURE 16-2: Event managers must be good leaders and team members, able to work effectively with different people and personalities while taking on a variety of roles.

Not every event needs the services of a professional special events manager, of course. Anyone can put on a meeting or event. Volunteers who are members of a group or organization can find themselves being asked to manage a special event for their organization. Conducting a meeting or organizing a school fundraiser uses the same skill set that professional event managers use, and it is good training for anyone interested in undertaking a valuable learning experience, even individuals who are not looking at event planning as a potential career.

STAGES OF A SPECIAL EVENT

Imagine you are the event planner for a fundraising event at your school. To keep the event on track so that it runs smoothly requires a great deal of preparation. Successful events are those that are produced in the following stages:

- *Planning the event:* At the start, you must determine your goals and objectives, identify your audience, and conduct research to determine if the event is viable.
- *Event logistics:* In this stage, you work on putting the planning elements in place, often with a team of other people who are responsible for specific tasks.
- *Event-day coordination:* On the day of the event, you coordinate all the details and the people working on the event with you.
- *Event evaluation:* At the end of the event, you wrap up and evaluate how it went, making sure to note ideas for improvement in the future.

Planning the Event

All successful events start with a plan. A special event is an organized activity, incorporating many elements, so a plan helps you stay organized and includes all the details necessary to make your event a success. The elements that go into a plan for a special event of any type include determining your goals and objectives, identifying your audience, and research:

- *Goals and objectives:* To have a successful fundraising event, you need to balance the needs of your audience with your *goals and objectives* for the event. Do you primarily want to raise money? Build school spirit? Make members of your community more aware of what goes on at your school? The answers to these questions will tell you who your primary audience is, and that will help you decide what kind of event you want to have.

- *Audience:* What is the audience for your event? If it is a fundraising event to be held at your school, for example, are you trying to attract fellow students, faculty and staff, parents of students, and/or members of your community to come and participate? Sometimes your audience will consist of more than one group, but there should be a primary audience that you want to build the event around. A school carnival, banquet, or festival might attract several different audiences. The type of activities you plan will depend on which audience you are most interested in; students might be attracted to a dance, for example, whereas parents might be more interested in a banquet or a speaker (Figure 16-3).

- *Research:* Doing some research before you plan and select the elements of an event helps you avoid and reduce risks. Having people not come to an event that you have spent a lot of time and money to develop is a big risk. **Primary research**, such as talking to people who might come to a meeting to find out the best day and time for them or sending out a survey in the case of a large event, can help you determine your likelihood of success before planning begins. Refer to Chapter 3, Research Methods, for suggestions on how to write surveys to gather primary research information. **Secondary research**, which involves reviewing published sources of information, is

FIGURE 16-3: It is important to target your event to the audience you are trying to reach. This photograph shows the chef for a gala New Year's Eve banquet, an event that will appeal more to mature adults than young people.

© erwinova/www.Shutterstock.com

Scott Wallin

Director, Producer Communications, Dairy Management Inc.

We build public relations through a variety of ways. We have a lot of outreach to schools, for example. We want to make sure that educators are getting our materials and our information that tell the story of the Florida dairy farmer. We do an awful lot of outreach to health professionals throughout the state. We talk to doctors and dieticians and pediatricians and school nurses. We want to make sure that they're getting the proper information on the benefits of dairy foods that need to be a part of children's diets and all their patients' diets, as well, because milk isn't just a drink for children. We feel that it is a nutritional choice for all ages. We do an awful lot of consumer marketing events. For example, we do some great things with the athletics programs with some of the universities in Florida. We search for events that have a high population of consumers. We can kind of get in there and tell them our story about the Florida dairy farmers and let them sample our products and really get to know us a little. It's really trying to focus on the major markets in the state of Florida, finding out where the population is going to be, and then hitting that segment as hard as we possibly can.

also helpful. Looking up the dates for potentially conflicting organizational activities in a published calendar or on a website is a form of secondary research. **Benchmarking** is a special kind of research that involves looking at what other organizations have done to plan events similar to yours. You might be able to get ideas from benchmarking other organizations' events, and you might also be able to find out how things went, so you can avoid planning activities that did not go well and perhaps adopt elements that did go well for your own event.

Logistics

Logistics refers to working out everything involved in selecting and developing the right type of event to achieve your objectives. Even if you are organizing a meeting, for example, the selection of the location, timing and date, speakers, and activities involves crucial decisions that influence the ultimate outcome.

Generally speaking, event logistics includes all of the following considerations:

Food

Will food be available? If so, what kind? Food is one of the biggest costs budgeted for an event. If there is going to be food, how much will there be and what type: finger foods, buffet, or a sit-down meal? If you have enough money budgeted—as you might for a wedding or a conference—you might be able to work with a caterer, someone who is a professional and can order, cook, and serve all the food for your event. Otherwise, you might need to consider having a "pot-luck," where attendees or members all bring a food dish. Your organization might contribute the main dish, to make sure there will be enough food to go around.

Entertainment

Event entertainment can range from background music playing on a computer with speakers to full-scale entertainment on a stage with performers or a live speaker.

Entertainment can also take the form of games and activities for attendees, which can be informal or very organized. It is important to match the entertainment to the audience and to the time allotted for the event. Even the most interesting of speakers, if he or she goes on too long, can cause the audience to tune out and be bored.

Theme and Decorations

Events often have *themes* to make it easier to decorate the site location and promote the event. Good themes add to participants' enjoyment and memories of the event (Figure 16-4).

Think about a school dance or prom. How did the theme add to the overall mood of the event? Coming up with an effective theme takes some thought. Effective themes should be short, should easily relate to the event, and should be visually appealing. All print and other promotional materials, such as tickets, posters, and fliers, should be designed around the event theme. All decorations—including costumes, props, sets, and signs—also should be consistent with the theme in some way, such as the use of the theme as a logo or the use of consistent colors.

Date and Time

Date and time are elements that must be communicated to attendees as well as those who will be helping out, so they must be set early. It is important to think

FIGURE 16-4: A wedding is an event that is traditionally focused around a special theme that extends to the colors of the bridesmaids' dresses to the table arrangements and printed invitations.

through the best time to hold your event, so that there are not any conflicts with other events that might attract the same audience. Are there any dates that will not work because of a conflict, or are too near a holiday or vacation period? Many event planners can tell stories about setting the date for an event only to find out it conflicted with something else. Even a big sporting event can be a conflict. If many in your audience could be expected to be attending or watching another event, they will be unlikely to go to your event. Get out the calendar and cross out dates and times that do not work before picking some that do. Check the date and times with a few of those who might be attending first before committing to it.

Location

Selecting the location for an event is usually up to the event manger to decide. Picking a good location involves knowing the size of your potential audience, and then scouting potential locations to make sure they can accommodate the number of people you are expecting. If you have ever been in an overcrowded meeting room or school assembly auditorium, you know how quickly overcrowding can influence your appreciation of the event you are attending. Always allow for a few more people than you expect might come, just in case more show up. This is a good policy when figuring out how much food to serve, too. Caterers will usually add a few more "places" than what was ordered for this reason. Running out of food can be as bad as not having enough room.

To work out how much room you might need, you might consider using a site diagram. **Site diagrams** are sketches of the facility, usually with dimensions included. Items you should sketch into your site diagram include tables, chairs, audiovisuals, stage/podium, entrance/exit, decorations, displays/exhibits, and trash/recycling (Figure 16-5). Once you have made your site diagram, you can estimate how much room there will be available for attendees.

Sponsorship

Getting sponsors to financially support your event is a way of ensuring that you do not lose money. It is also a way of developing linkages and adding to the credibility of your goals. Sponsorship is extremely time-consuming. It is not the same as a goodwill donation, where donors contribute without any expectation of commercial benefit. Sponsorship is a commercial transaction between the involved parties. Sponsors

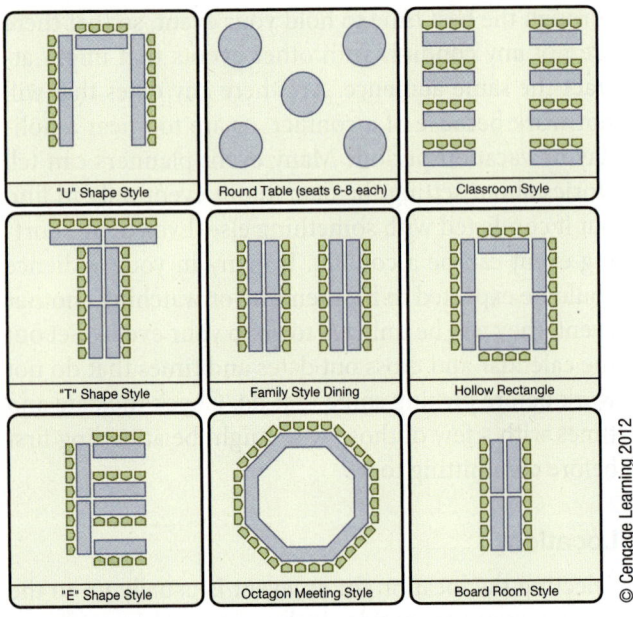

© Cengage Learning 2012

FIGURE 16-5: A site diagram is a useful tool in planning an event. These site diagrams show some of the most common ways tables and chairs can be arranged for special events.

want to see a return on their investment. For example, they may see that participating in a special event that targets their customers may be a way of generating positive publicity and favorable consumer attitudes. In an indirect way, such a sponsorship may even help drive sales, if attendees at the special event recognize the sponsor and subsequently patronize the company. Some broad objectives companies have for sponsorships are as follows:

- Enhance image and consumer attitudes about the company.
- Drive sales.
- Obtain positive publicity and visibility.
- Promote goodwill by positioning the sponsor as a community "good citizen."
- Contribute to community development.

Some examples of special events and prospective sponsors are as follows:

- *Fair:* soft drink distributor, grocer, automotive dealer, bank
- *Festival:* department store, restaurants
- *Sports:* athletic wear, soft drink distributor, hospital
- *School program:* children's toy store, children's clothing store, amusement park
- *Meeting/conference:* bank, commercial printer, insurance broker

There are several types of sponsorships that can be used for a special event or activity. A **title sponsor** pays a premium fee to have its name become a part of the event itself.

An event can have more than one sponsor, called **co-sponsorship**, in which more than one organization shares sponsorship of the event. An **exclusive sponsor** is one who pays extra to have an exclusive role, that is, to be the only sponsor of the event. If the sponsor is a media outlet, such as a local newspaper or radio station, the **media sponsor** may not provide funds, but often does provide a predetermined amount of advertising for the event. An in-kind sponsor is similar. **In-kind sponsors** do not financially support an event, but they do donate products or services. For example, a school club may decide to have a plant sale to raise funds for a trip. If club members are able to get the plants donated by a local nursery, the nursery would be an in-kind sponsor of the sale.

Cross promotion is the term used for an event that is co-sponsored by more than one organization, or an organization and another type of sponsor, such as a media outlet. It is called a cross promotion because the different organizations are each contributing something and are each benefiting from the association or partnership. An example of this is when a local restaurant chain agrees to publicize a charity's fundraising event in its restaurants and has a booth set up at the event to serve food. The charity and the restaurant chain are co-sponsors and partners in the event. They both benefit, and working together they can make the event more successful than it would have been if only one of them sponsored it.

Planning Committees

Sometimes events, especially those that are conducted on behalf of charities or nonprofits, are coordinated by a *planning or administrative committee*. The key to success for a planning committee is having the right balance. The size of the committee may vary, but it will be most effective if the number of members is related to the complexity of the event. For large events with many types of coordination roles and responsibilities, the committee might split into smaller subcommittees, each charged with a specific task or set of tasks.

Planning committees benefit from having people with imagination, persistence, patience, and a sense of humor. Well-balanced committees are diverse, in terms of personalities, demographics, experience, and interests. A committee planning an event needs structure and organization, and it needs leadership. Often, if the event to be planned is an annual one, there will be leadership slots that are filled with new members each year. Assistant chairs of subcommittees may move up to subcommittee chairs, and then to leadership for the entire committee. Some planning committee tasks may include the following:

- Event coordination
- Accounting
- Communications
- Decorations
- Entertainment
- Facilities, equipment, and supplies
- Maintenance
- Marketing
- Security
- Signage and banners
- Transportation
- Vendors
- Membership

Permits allow you to conduct activities at your event that are regulated and for which you need permission, such as having a fireworks display or serving food to the public. Permits are issued by local, state, and federal government agencies. If you are holding an event in a public place, it is very likely that you will need to secure a permit. Permits provide legal permission to hold your event and to conduct activities at your site (Figure 16-6). Usually there will be a fee to obtain a permit. Here are some examples of types of permits and which local government department issues them:

- *Bingo or lottery:* lottery or gaming department
- *Food handling:* health department
- *Occupancy:* fire department
- *Parking:* transportation department
- *Parks usage:* parks department
- *Pyrotechnics (fireworks):* fire department
- *Sales tax:* tax collector's office
- *Sign and banners:* zoning department
- *Street closing:* transportation department

Budget

An **event budget** represents all of the income and expenses of the event. A budget is based on income and expenses you reasonably believe you can expect with the resources available. Setting up a budget for an event involves taking into consideration all of the costs you anticipate and coming up with a bottom-line figure that falls within what has been allocated for the event. Things to consider that may cost money include—in addition to food costs or fees for a meeting room—speaker's fees, printed materials, audiovisual equipment, prizes or raffle items, giveaways, decorations, lighting, costumes and props, publicity, invitations, and entertainment. In addition to the costs involved, a budget includes income or revenue that results from any part of the event. Income can include ticket sales, silent auctions, raffles, pledges or donations, sales, and gifts. To create a budget:

- Estimate the audience that will attend.
- Estimate all costs.
- Add 20 percent as a **contingency**, in case costs are more than anticipated.
- Calculate the total expected income, for example, through ticket sales and donations.
- Calculate the profit or loss.

Here is a *quick budget guide* for a food-focused event and for a non-food event:

Food-focused event

- Site fee = 20–35 percent of total budget
- Food costs = 60–75 percent of total budget
- Miscellaneous costs = 10–15 percent of total budget

Non-food event

- Site fee = 50–75 percent of total budget
- Miscellaneous costs = 25–50 percent of total budget

FIGURE 16-6: A trade show, like this one featuring agricultural equipment, is a special event where it is important to make sure you have secured the appropriate permits and assurances so that everything occurs as planned.

© Robert Kylio/www.Shutterstock.com

Event-Day Coordination

Now you are ready to execute the plan you have developed. Event-day coordination involves executing the special event plan and making sure everything comes off perfectly. As coordinator, it is the event manager's job to be the point person responsible for the coordination of all activities. That could be as simple as arriving at the meeting room early and making sure officers are there and ready to give officer reports, or as complex as orchestrating the details for a crowd of attendees at a banquet. Event managers need to be good team leaders, good listeners, and good communicators. Other elements to consider are discussed next.

Programming

In addition to personal leadership qualities, successful event managers also rely on setting up a solid program or script to coordinate event timing and activities. A program usually includes titles for activities to be held, as well as days, times, and locations for everything that is planned to take place. For events that include performances and introductions, scripts are often used that include everything that will be said and by whom, as well as sound effects, music cues, and special effects.

Atmosphere

Every little detail contributes to the appropriate atmosphere for an event. At a wedding, for example, little details—the bride's dress and veil, her bouquet, the church service, the table linens at the reception, the music being played, whether it is a daytime or evening wedding—all add to the atmosphere and the enjoyment of the event. At a conference or trade show, the location of the meeting rooms, how the chairs are set up, the type of speakers, handouts and other materials, and even what kind of food is served, all contribute to what attendees will take away from their experience. Think about how you could add to the atmosphere of a meeting or school fundraiser. When considering atmosphere, try to involve all of your audience members' five senses (Figure 16-7):

- *Sound:* Design the sound system and sound effects to capture attention.
- *Sight:* Strong visual elements are needed (signs, video, symbols, logos).
- *Touch:* Textures of napkins and the printed program help in this area.

FIGURE 16-7: A lighting display is an example of a strong visual element that can enhance an event's atmosphere by appealing to the audience's sense of sight *(Photo by Kevin Kent).*

- *Smell:* Perfume and food odors add to any event, but avoid bad odors.
- *Taste:* For any food-focused event, food and drink are powerful atmosphere elements.

Staffing

Who will be working at the event to handle registration, greet attendees, coordinate programming duties, and serve food and drinks, if appropriate? Successful event managers make sure they know everyone who is on the team. They also make sure team members know their responsibilities and understand how to communicate with each other. If you are running an event, it is always important to thank your team for their hard work and to give team members credit. For example, an organization's president who is coordinating a meeting might want to thank members for attending, and then acknowledge officers who helped set up the event.

Event Evaluation

What causes poor events? Events not turning out as planned is sometimes due to chance—such as an outdoor event being rained out by an unexpected thunderstorm. However, there are some common reasons why events fail:

- Not being willing to try something new to update an event that has been held for many years
- Lack of creativity and innovation in developing the event so that attendees are engaged and not bored

- Lack of marketing and publicity for the event
- Poorly selected and trained personnel
- Not enough money to put on the event
- Poor timing
- Poor site conditions

All of these factors are typically related to decisions made by the event manager or other team members. Good **evaluation** takes into account what cannot be helped or anticipated while trying to identify areas that can be improved upon.

Event evaluation is the process of determining how effective an event was. Evaluation is done to help understand what worked and what did not work in order to learn from the experience and plan even better events in the future. You also want to know what impacts and outcomes an event had. Looking at how much money was raised or how many people attended is one part of evaluation, but, by itself, it does not explain enough about how the different parts of the event contribute to this outcome. It also does not explain what people thought about the event, or how the team members who worked the event think they could improve it in the future.

You can evaluate an event along the way or comprehensively at the end. Event managers might do an evaluation "in-house"—by themselves or with the help of others on the team, also known as a *self-evaluation*— or with the use of "monitors" who observe the event and provide feedback to the event manager. Evaluation may also use surveys of attendees.

Evaluation is done to improve future events. To evaluate an event, you must first decide what you want to know, and then come up with a set of questions to ask, such as the following:

- Did attendees enjoy the event?
- Were they made to feel welcome?
- What was the best part of the event?
- Who was the best speaker?
- Did they like the location?
- Did they get their "money's worth"?
- Was the event held at a good time of year, month, week?
- Would they recommend the event to others?

In a self-evaluation, the event manager and team members evaluate themselves and how they think they did. Here are some personal questions an event manager may need to answer:

- Was I a good leader?
- Was the special event worth it?

- What parts of the job did I like or dislike?
- Was the event a success or failure?
- Which people did I work well with, and are there people I will never work with again?
- Do I want to manage another event?
- Did I keep complete notes and fill out the paperwork?

EVENT MARKETING AND PUBLIC RELATIONS

Marketing your event so that people know about it and are motivated to come is extremely important. The **five P's of marketing**—product, promotion, price, place (location), and public relations—must be taken into consideration in order to have an event that is well attended:

- **Product:** Your event is the product you want to sell. Whether it is a free event or one for which you are charging admission, your goal is to create an event that will be appealing to your audience.
- **Place** is the location where your event is held. In addition to finding a location that is the right size, you also want to find a location for your event that will be relatively easy for your attendees to get to. If it is a publicly held event, is public transportation (buses) available? How about parking? If it is a private event, for example, a festival held at your school or church, parking may be available, but you might need to have team members on hand to help direct people where to park.
- **Price:** If you are going to charge admission to your event, it is important to decide on a price that not only covers your expenses but that is reasonable and affordable for those you wish to attend. If you have not held this particular event previously, you can ask people beforehand what they would be willing to pay or you can adjust your price to that of other similar events being held in your area.
- **Promotion** involves all of the ways you publicize and make people aware of your event, from word of mouth through advertising and public relations. Putting up a flier around school to let people know about a special meeting is a form of promotion. Another form of promotion is designing a T-shirt that features your event's special logo and theme

Scott Wallin

Director, Producer Communications, Dairy Management Inc.

Defining "Public Relations"

There's a lot of different ways to define public relations. We see ourselves as the voice of the farmer. We carry that voice out to the public because it's not realistic that a farmer is going to be able to leave the farm. They're so busy, they're not going to be able to get off the farm and tell their story to the public, so we learn as much as we possibly can about their business, about the industry, and we convey that message out in a very positive way to consumers, to health professionals, schools, and the media. So we're the voice of the farmers, and our goal is to make people aware and appreciate what goes on behind the scenes. Milk doesn't just magically show up in the grocery store. There's a long, great story behind it. Public relations for us is making sure that people do have an appreciation for what goes on before that glass of milk hits their table.

FIGURE 16-8: Designing a logo for your special event and featuring it on a T-shirt can be an effective form of event promotion (*Design by Kevin Kent*).

(Figure 16-8). Taking out an ad in the school newspaper, or sending a press release to the newspaper's editor to provide information about the meeting and perhaps get a reporter to come to the meeting and write a story about it are ways to use advertising and public relations to promote your event.

■ **Public relations** is one of the most effective ways to promote a special event. It has been said that advertising is what you say about an event, whereas public relations includes what you say about an event as well as what others say about it. Advertisers pay to run their ads in some form of mass media—newspapers, radio, television, and magazines. With advertising, you control the message you want to communicate in your ad, and also when it runs. This costs money, both for placement and for any production that is done to create the ad. But advertising can sometimes be viewed as not credible because ads are basically the claims of the advertiser. If you do decide to advertise your event, the costs to create and run your ad are part of the event budget.

Public relations involves using the news media to promote your event. If a newspaper or a television or radio station decides to cover your event, it will write a news story about it. You do not have control over what is said in the story, or over when it runs, but the resulting publicity is free. In addition, a news story might be seen as more credible and objective than advertising, and it is certainly more affordable.

© Cengage Learning 2012

Lisa Lochridge

Director of Public Affairs, Florida Fruit & Vegetable Association

Importance of Public Relations for Agriculture

Agricultural organizations are no different than any other company or any other organization. They have publics that they need to be communicating with on a regular basis. They have messages that they need to be communicating, and so it's important that agricultural organizations be involved in ongoing public relations to those groups that are important to their success.

If you decide to use public relations, you may have some costs to create press materials. Generally, these costs would include paying someone to create the materials and any costs to produce them, such as printing costs. For example, you might send a printed news release to a newspaper or magazine. A **news release** is a story written by someone in your organization and sent to the media in the hope they will publicize it. For a radio or television station, you could send a news release or, if it is a nonprofit or charitable event, a **public service announcement (PSA)**. PSAs run during the time the station allots for commercials, so they are usually the same length as commercials, either 30 or 60 seconds. If the station runs your PSA, the air time is free, as the station is doing a "public service" for the community by letting the public know about your event. For this reason, public relations exposure can be more valuable than advertising. See Chapter 13, Media Relations, for more information about PSAs and news releases.

TYPES OF EVENTS

Retail Events

Special events are often held to raise money for an organization. Fundraising events, also called retail events, are those where your goal is to raise money in some way. This can be done through selling tickets to the event itself, by selling or auctioning off items at the event, or perhaps doing both. Here are some retail event ideas that are commonly used to raise funds and the audiences that these ideas appeal to:

- *Arts/crafts show:* women and senior adults
- *Circus/carnival/petting zoo:* young families
- *Computer show:* men
- *Cooking demonstration:* women
- *Magician/puppeteer:* young children
- *Health fair:* senior adults
- *Fashion show:* women, teenage girls
- *Sports celebrity:* men and boys
- *Music concert:* teenagers

Retail events require a special focus on looking at the **return on investment**, or ROI, from the event itself. Based on the actual expenses, how much income was generated? How many tickets were sold and/or donations made? Looking at the ROI of a fundraising event can help the event manager determine the success of the event, and make decisions about what might need to change to insure the fundraising elements result in adequate revenue for the sponsoring organization in return for the expenditures that were made. If an event didn't make money, or the revenue generated was smaller than expected, ticket prices might need to be increased, or expenses to hold the event might need to be reduced.

© Cengage Learning 2012

Lisa Lochridge

*Director of Public Affairs, Florida Fruit &
Vegetable Association*

Defining "Public Relations"

I define public relations as outreach to those key groups that are important to the success of your organization. Those groups and those publics can evolve over time depending on your situation, where your company is, and what scenarios that they're dealing with.

SUMMARY

This chapter provided an overview of special events. Topics covered included defining a special event, the role of the event manager, stages of planning and conducting a special event, and the evaluation process. In addition, the concept of event marketing was introduced, as well as how to use public relations to promote a special event. Various types of special events were discussed, along with planning and logistics considerations.

CHAPTER EXERCISES

APPLICATIONS

Following are some ways you can apply what you learned:

- Working in a team of three or four students, brainstorm a special event. Work out a site diagram and potential budget. Identify three ways to promote the event around your school and in your community.
- Develop a one-page news release or 60-second public service announcement to promote a special event on behalf of your school. Ask another student to review it and make suggestions for improvement.
- Think about an event you participated in. Analyze and evaluate it, drawing on the following key considerations:
  - **Planning of the event**: How well was this done? Were there any issues or problems that were not anticipated?
  - **Audience for the event**: How was attendance? If the event was a fundraiser, was the money that was raised close to what was expected? What could be done to increase attendance or raise more money if the event was held again?
  - **Promotion of the event**: How was the event promoted? Was advertising or public relations used? What did you think of the information that was communicated? How could promotion of the event have been improved?

CHAPTER QUESTIONS

MATCHING:

_____ **1.** Held when your goal is to raise money in some way.

_____ **2.** The process of determining how effective an event was.

_____ **3.** For a non-food event, this is typically 20–35 percent of the overall budget.

_____ **4.** Includes sketches of the dimensions of the facility.

a. Evaluation

b. Site diagram

c. Site fee

d. Retail events

MULTIPLE CHOICE:

5. A(n) _____ sponsor is one who pays extra to be the sole sponsor.

  a. media

  b. exclusive

  c. co-

  d. in-kind

6. A _____ is a story written by someone in your organization and sent to the media in the hope they will publicize it.

    a. public service announcement

    b. feature

    c. cross promotion

    d. news release

7. Which of the following is NOT a form of event research?

    a. Primary

    b. Secondary

    c. Logistics

    d. Benchmarking

8. The stages of a special event include all of the following, EXCEPT _____.

    a. planning

    b. logistics

    c. sponsorship

    d. evaluation

9. _____ refers to working out everything involved in selecting and developing the right type of event to achieve your objectives.

    a. Logistics

    b. Benchmarking

    c. Contingency

    d. Five Ps

10. A(n) _____ sponsor provides a predetermined amount of advertising for the event.

    a. media

    b. exclusive

    c. co-

    d. in-kind

11. _____ refers to adding an additional amount in case costs are more than anticipated.

    a. Logistics

    b. Benchmarking

    c. Contingency

    d. Five Ps

12. _____ involves using the news media to promote your event.

    a. Public relations

    b. Promotion

    c. Product

    d. Placement

FILL IN THE BLANK

13. _____ involves all the ways you publicize and make people aware of your event, from word of mouth through advertising and public relations.

14. _____, _____, _____ and _____ are four examples of retail special events.

15. _____ is run as a public service during the time the station allots for commercials, so they are usually the same length as commercials, either 30 or 60 seconds.

REFERENCES AND FURTHER READING

Allen, J. *Event Planning: The Ultimate Guide to Successful Meetings, Corporate Events, Fundraising Galas, Conferences and Conventions, Incentives, and Other Special Events,* 2nd ed. Mississauga, ON: John Wiley & Sons Canada, 2009.

Craven, R. E., and L. J. Golabowski. *The Complete Idiot's Guide to Meeting & Event Planning,* 2nd ed. New York: Alpha Books, 2006.

Devney, D. C. *Organizing Special Events and Conferences: A Practical Guide for Busy Volunteers and Staff.* Sarasota, FL: Pineapple Press, 2001.

Friedman, S. *Meeting & Event Planning for Dummies.* Hoboken, NY: Wiley Publishing, 2003.

Jago, L., A. J. Veal, J. Allen, and R. Harris. "Events beyond 2000: Setting the Agenda." Australian Centre for Event Management. Last modified 2000. Accessed May 18, 2010. http://utsescholarship.lib.uts.edu.au/dspace/handle/2100/430.

17 Communications Campaign Development

OBJECTIVES

After completing this chapter, the student will be able to:

- Explain what a communications campaign is and identify the elements of a campaign.

- Define *integrated marketing communications* and explain its role in marketing.

- Identify the elements of the marketing mix.

- Define *reach, frequency*, and *cost per thousand* (CPM) and be able to apply them in a campaign.

- Develop an integrated marketing communications campaign for a company, organization, product, or issue.

INTRODUCTION

When you see a television commercial for a major brand such as Nike, Pepsi, or McDonald's, you are experiencing the results of the intensive planning and integration of many elements—graphic design, message development, video production, and media placement. Good communications campaigns are essentially strategic plans that successfully target relevant audiences with communications messages and media that have been carefully designed to achieve specific objectives. Developing communications campaigns that are targeted and likely to be well received requires a combination of careful planning and a strong understanding of the attitudes and behaviors of the audience members you are trying to reach. For that reason, the process of developing a campaign is fairly structured, and can be undertaken in a step-by-step approach. Once you know the steps to developing an effective communications campaign, whether you are trying to promote your student organization's spring banquet to ensure good attendance, developing a campaign proposal, or marketing a major new product, the campaign planning process is essentially the same (Figure 17-1).

FIGURE 17-1: The process for developing an effective communications campaign is essentially the same, regardless of the objective.

© auremar/www.Shutterstock.com

This chapter is designed to explain the communications campaign planning process in detail. You will learn what a campaign is, what the major elements that go into a campaign consist of, and the sequence of steps involved in developing an effective campaign for a product, group, issue, or organization. This chapter will give you the tools needed to write a good communications campaign plan, and to brainstorm effective techniques and tactics to make your campaign plan a success.

COMMUNICATIONS CAMPAIGNS

A **communications campaign** is a strategic, structured plan consisting of a mix of media and message strategies and tactics with a consistent, unified theme. Communications campaigns are based on analyzing the client, the competition, the potential target audience(s), and the marketing/media mix as a prelude to developing the plan. Campaigns can be developed for an individual (political campaign), a product (brand), an organization (corporation), an association (commodity group), a public service (charitable), an issue (voter ballot initiative), or an institution (university or government agency).

Communications campaigns have changed over the past few decades, due to the consolidation, globalization, and evolution of new types of communications media. Less than 50 years ago, there were only three major television networks from which to choose. Newspapers published both morning and afternoon editions. There was no such thing as cable television, social media did not exist, and the Web would have seemed like something from a science fiction TV show or movie. Today, there are more media outlets available than ever before, creating **media "clutter"**—the situation that exists when there are so many media channels and messages out there that it becomes difficult for your message to break through. Major brands are often marketed globally, and competition and regulation have increased. Consumers have more choices than ever before as well, and are often less loyal to a particular brand (called **brand loyalty**) than their

parents were. These shifts have created the need for more integrated marketing communications, which focus on using a variety of marketing and media elements in a consistent way.

INTEGRATED MARKETING COMMUNICATIONS

Integrated marketing communications, or **IMC**, is the term used for the process of developing and implementing a set of integrated campaign elements that utilize common and consistent themes across multiple elements. IMC developed in response to the expanding number of communications channels (print, broadcast, Web) and the need to unify the theme of a campaign across all of these channels in order to ensure that audiences exposed to a campaign via one type of communications would associate it with the same campaign in another medium. The **marketing mix**, sometimes called the "*marcom*" *mix*, for "marketing communications," includes the following ways to communicate a campaign message (see Figure 17-2):

- Advertising
- Public relations
- Brand marketing
- Sales promotion
- Direct marketing
- Internet marketing

FIGURE 17-2: The marketing mix includes all of the relevant ways to communicate your message via some form of media.

An **integrated marketing campaign** is one in which your chief goal is to communicate a message, using appropriate tools packaged for specifically targeted audiences and a strategically framed message—what is known as the **message strategy**. It is important to identify the specific audience for your campaign, because not everybody is interested in the same things. For example, you might like hunting, whereas your best friend may be into sports. Carpentry might be a hobby of your father's, while your mother really likes to garden. Targeting a group that shares an interest, belief, attitude, or value that is related to what you are trying to communicate makes it easier to get your message across. See Chapter 2, Effective Communication and Message Development, to learn how to analyze audience characteristics.

An IMC campaign uses a combination of communications techniques, ranging from interpersonal to mass communication, and typically starts with developing goals and objectives and conducting a **situation analysis**, which identifies the problem and provides a background and rationale for the suggested campaign elements. Conducting the situation analysis helps identify a potential target audience or audiences for the campaign. Once the target audience has been established, strategies and tactics are developed to create the **campaign message**, or *theme*, as well as to determine the ways in which the message theme will be delivered to most effectively reach the targeted audience. Because more than one dissemination tactic is typically used in a campaign, integrating the messages so they look and feel consistent across different types of media is important. Finally, some type of evaluation is usually suggested to determine the success of the campaign methods used and to refine the campaign elements in order to improve their effectiveness.

Advertising

Three criteria must be met for a communication to be classified as *advertising*:

- The communication must be *paid* for.
- The communication must be delivered through *mass media*.
- The communication must be attempting to *persuade*.

Advertising involves integrating a campaign theme or message and delivering it in a format suitable for a specific type of media. The major forms of advertising media include newspapers, magazines, billboards, transit, radio, television, and the Web. Advertising may be done on a local, regional, or national basis. The benefits of advertising are that the advertiser has control over

the message and where it is "placed" in the media. This control is an important benefit, because advertisers want to display their advertising messages in forms of media that reach their desired target audience. An advertising message targeted to men, for example, might be placed during a sporting event on television that attracts a large male audience, whereas an advertiser hoping to target young women might run a print ad in a magazine such as *Glamour* or *Cosmopolitan*. Another advantage is that, in addition to the placement, an advertising message runs in print, on the air, or online exactly as the advertiser intended it without editing or other adaptation.

Although advertising has many advantages from a control standpoint, disadvantages include cost, since the advertiser must pay for the ad to be developed as well as pay for the media to be placed, and less credibility than if the information was covered as a news story.

Ads can be used to promote products, goods, and services, as well as promotional activities and events and even political campaigns. The magazine print advertisement shown in Figure 17-3A is for a campaign asking voters to vote against a state ballot initiative known as "Amendment 4." Print ads used in a political campaign are designed to develop a connection with the reader and the issue of the campaign. The ad in Figure 17-3B is designed to appeal to college voters. Notice the message used in order to connect with the target audience and how the message differs from the campaign logo used with other audiences.

Public Relations

Public relations is the attempt to promote goodwill and garner publicity on behalf of a company, organization, or individual through **earned (non-paid) media**. The practice of public relations uses many tactics and tools, but the fundamental purpose of public relations is to get news outlets to cover events related to an organization's products or activities in a news story. This is the main advantage of public relations. If a press release gets picked up and covered as a news story, this "earned" media is free to the company or organization that initiated it. In addition, the information has more credibility than if it had been conveyed in an advertisement. The disadvantage, of course, is that the organization does not have control as to how it will be covered by the news media.

The fundamental tools of public relations are the news release, the media kit, and the public service announcement. News releases and media kits are tools to "package" an organization's story in ways that journalists will find useful and interesting. News releases are

FIGURE 17-3A The logo for this campaign is designed to appeal to a more general audience.

FIGURE 17-3B This magazine print ad for a political campaign uses a strong visual image and message to appeal to college-aged voters.

stories written just like a regular news story. Their purpose is to provide reporters with the basic information they need to cover a story. News releases, photos, facts sheets, DVDs, and other media that are related to a specific topic or event are often put together in a folder or binder with a specially printed cover called a **media kit**, because the contents are like a kit of elements that reporters can use to develop their news story. Media kits package news releases, fact sheets, photos, brochures, and other items in an attractive, logo-identified package for news editors and journalists. Public service announcements (PSAs) are free advertisements that media outlets agree to run as a public service. In order to have an ad run as a free PSA, it usually must be on behalf of a nonprofit or charitable organization. See Chapter 13, Media Relations, for descriptions of these tools.

Public relations strategies include:

- Generating publicity through attracting media interest in covering a story you promote to them via a news release or media kit.
- Raising awareness and attracting attention of target publics (special events, event sponsorships, announcements, spokespersons).
- Creating audience attention, favorable attitudes, and goodwill through partnerships (cross promotions, cause-related marketing).

Brands

A **brand** is a product, good, or service that is given an identifiable name and image. Companies and organizations try to develop their products, goods, or services into brands, because a brand is something their audiences can easily recognize and understand when they see it on a product, on a company office door, or in the media. A brand name is the specific name given to a product or service, which is usually represented in a visually designed identifier called a logo. When consumers buy brand-name items on a regular basis, they are said to have brand loyalty. Branding, the process of creating and marketing a brand, is done through a communications campaign that consistently uses elements of the brand—the brand name and logo—to create a desirable image of the brand in the consumer's mind.

Products marketed as brands are called "packaged goods"—things that are labeled and sold on shelves by retailers. Examples include:

- Food (Tyson's chicken parts; Sara Lee snacks)
- Health and beauty aids (Cover Girl cosmetics, Old Spice deodorant)

Susan Howard, APR *(Accredited in Public Relations)*

Director, Corporate Communications, DUDA

Successful public relations campaigns do not happen by accident. Setting clear objectives and determining who you want to reach and how to best communicate your messages to them are keys to achieving your public relations goals.

© Cengage Learning 2012

- Brands can also consist of the retailers that sell the goods (Gap, Home Depot, Clinique)

Services can be brands as well:

- Entertainment (professional sports teams)
- Airlines (Delta, United)
- Travel/recreation (Disney)

Many agricultural products have been historically treated as **commodities**, such as corn, wheat, cotton, or citrus, which are products that are not sold on the basis of a brand or product name. As a result, many consumers do not realize how many brand items include elements derived from agriculture and natural resources industries (see Figure 17-4).

© Kletr/www.Shutterstock.com

FIGURE 17-4: Agricultural commodities, like the wheat being harvested in this picture, are a main ingredient in many branded food and fiber products.

Sales Promotion

Sales promotion involves a variety of tools and techniques to develop incentives to create a perception of greater brand value, while encouraging immediate acceptance or adoption of a product, good, or service by a consumer or businessperson. It includes everything from sweepstakes, contests, and premiums (free items given away, such as pens or cups, usually with the organizational logo on them) to coupons, frequency discounts (buy more than one and get it for less, such as a buy-one-get-one-free offer, or BOGO), gift cards, and free samples or trial offers.

Sales promotion can target business audiences, but people are probably most familiar with sales promotions aimed at consumers. A sales promotion that includes two or more organizations partnering together to conduct a promotion is called a **cross promotion**, which is described in more detail in Chapter 16, Special Events. When one of the partners is a nonprofit or charitable organization, it is called **cause-related marketing** because the nonprofit or charity is the "cause" that is being promoted.

One special form of sales promotion is the **media promotion**. In a media sales promotion, an organization partners with a media outlet to jointly promote something. The media outlet uses its channel to convey information to its audience, while the organization includes the media outlet as a sponsor on its materials.

Direct Marketing

Direct marketing is marketing that focuses directly on the end users. It includes use of direct offers to consumers, often through either a form of advertising media, such as a television infomercial, or through postal mail. The distinction is that the offer is made available directly to consumers without their having to go to a retail outlet or location to purchase it. When companies use the postal system to deliver a direct offer, it is called direct mail, which includes such items as catalogs, mailed offers sent directly to the home, and forms of shared mail, where information materials from several different organizations might be included in one mailing. Direct mail can be expensive. Because it relies on the mail service, postage must be included, along with the cost of producing the informational materials, in the total budget.

Web Marketing

Before the Web came into existence, the Internet was primarily a communications and collaboration network for government employees and academic researchers. The advent of the Web browser, however, ushered in a new age, where it became possible to market products, goods, and services online and even to advertise on Web pages. Today, the Web is an important element of the marketing mix. Every company, organization, and brand needs a website, at a minimum, and Web banner advertising, e-mail newsletters, and marketing on search engines and social networking sites such as Facebook are all fast becoming "must do's" for any communications campaign. That is because Web marketing is relatively cost effective, especially in comparison to other types of marketing. Other than banner ads, the Web is not considered to be a paid form of media, and the cost to produce a website can be fairly inexpensive.

MEDIA USAGE

Communications campaigns use mass media to efficiently deliver the message as effectively as possible to the targeted audience. Mass media includes all forms of media that reach a mass audience, from newspapers and magazines to TV, radio, and the Internet.

As individuals, we choose to be exposed to specific types of media because the information that is delivered is considered to be useful, and we find it entertaining and gratifying. As you can tell, mass media is a very important factor in our lives, and we are exposed to many different media vehicles in a day. A media vehicle is another term to describe media outlets or channels (e.g., Web, television, print); the term "media vehicle" is frequently used in the development

of a communications campaign. Communications campaigns seek to use a combination of message and media designed to be most effective at reaching the target audience of the campaign. That means picking forms of media that attract enough members of the target audience and doing so frequently enough that audience members are motivated to process the information directed at them and to respond in a favorable manner.

Media Usage Survey

Having members of your intended target audience answer the following questions can give you an idea of how much and the kind of media your target may use. This should be done at the beginning and throughout the IMC process:

- How much media have you been exposed to in the last seven days?
- Have you seen a movie? Which one(s)?
- How much television did you watch last night?
- Which television network, cable channel, or local station did you watch? What times of day? What programs?
- Did you read a magazine? A newspaper? National, regional, or local? What type of magazine?
- How many hours did you listen to the radio? What times of day? How many minutes?
- Did you drive past a billboard or other form of transit (bus signs)?
- Did you go online? How many hours? Did you use a search engine? Did you buy anything?
- What about non-media—CDs, cell phones, video games, other?

Reach and Frequency

Reach and frequency are important concepts in any communications campaign. Reach is the number of members of your target audience who are reached one time by the media vehicles you have chosen, whereas frequency is the number of times your targeted audience members are exposed to your message. In an ideal campaign, you would be able to have adequate amounts of both reach and frequency, but that is not always the case, depending on your budget. High-reach media vehicles such as broadcast (television and radio) cost more than high-frequency media, such as newspaper and the Web. Research tells us that it takes three to five exposures to a message to stimulate action on the part of the receiver. So reaching a lot of audience members one time may not be enough to make your campaign a success.

Let us say you are working on a campaign for a local feed store. You know the store's marketing budget is limited, which means you have to decide between reaching a lot of potential store customers once, or fewer numbers of customers several times. What would you do, and why?

Although there are no easy answers here, it would probably be more effective to target loyal consumers who already shop at the store or similar stores with messages and sales promotion activities designed to increase their purchases and the frequency of their purchases than to target potential customers who have not bought from the store before. If you have unlimited funds at your disposal, targeting nontraditional customers makes more sense and will pay off because you are increasing the base of those who might potentially become customers at some point. Otherwise, trial-generating promotions, such as discounts, special events, and media promotions, make the most sense in this kind of situation.

Cost per Thousand

Because media costs differ according to how different types of media are bought, and whether or not they include production of materials as a cost, there is no simple way to judge the effectiveness of one medium over the other in a campaign plan other than with cost per thousand. **Cost per thousand**, or **CPM**, is a measure that looks at the total cost divided by the total estimated audience for using a particular media vehicle. It is a way of making all media equivalent in some way. Calculating CPM is a good way to determine how much reach you will achieve in a campaign, based on a budgeted cost, and it can be used with all media that generate some kind of audience figures, even Web. CPM can be looked at on the basis of total audience as well as targeted audience, which is known as CPM-TM. If you look on the Web, you will find many free CPM calculators that explain how to use CPM and calculate it for you. A great website that does CPM, reach and frequency, and CPM-TM calculations is the SportsMedia.net site, http://sportsmedia .net/MediaKit/calculator.htm. Online cost per thousand (CPM) calculators calculate CPM as well as other metrics, such as reach and frequency, that help you determine the overall effectiveness of the media you are proposing to use.

DEVELOPING AN EFFECTIVE COMMUNICATIONS CAMPAIGN, STEP-BY-STEP
The Campaign Development Process

The **campaign development process** consists of specific steps and tasks that help you develop a plan that utilizes elements of the marketing mix, a consistent theme, and carefully developed strategies and tactics in order to achieve a communications or marketing objective. Here is the process:

- Identify the problem, issue, or opportunity and set objectives for your campaign.
- Identify who the decision makers are and who the potential target audience is by doing an audience analysis.
- Develop a situation analysis so that you have all the information you need on the client, the issue, and the goal. For a definition of situation analysis, see earlier in this chapter.
- Do a SWOT analysis based on what you know, and use it to develop strategies and tactics. The *SWOT analysis*, which stands for *strengths, weaknesses, opportunities,* and *threats*, is a fundamental part of a communications campaign. Your client's identified strengths and weakness are internal, whereas opportunities and threats are external. Weaknesses can be countered by advertising or promotion, and threats can be mitigated by public relations. Maximizing opportunities and minimizing threats is where campaign efforts should be concentrated. (See Step 5 for how to conduct a SWOT analysis.)
- Develop strategies that extend from your SWOT analysis. *Strategies* are how you go about achieving your objective. You should have a strategy for the message, elements of the marketing mix, and media/campaign scheduling.
- *Tactics* are very specific elements, such as a news release or a specific print ad. Do not confuse these with strategies.

Following is a step-by-step guide designed to help you and members of your team think through developing a communications campaign plan. These can be used as planning tools alone, or in combination with the plan outline in Chapter 15, Persuasion and Persuasive Informational and Educational Campaigns. These tools can be also used to develop a campaign for other purposes as well, such as a campaign for your chapter, your school, your church, or all the way to a big national brand. No matter the size of the "client," the campaign development process is the same.

Rod Hemphill, APR *(Accredited in Public Relations)*

Director, Public Relations and Communication (retired), Florida Farm Bureau Federation

Your campaign strategies are your roadmap to success. If you do not know where you are going, how will you know when you have reached your goal? Decide what you want to accomplish and develop objectives: What audience do you need to reach? How will you reach that audience and what response do you wish to draw from that audience? Adopting strategies that meet your measurable objectives is essential to your success.

Step 1: Team and Client Profile

Team: Most campaigns are developed in a group or team of individuals with special skills, duties, and responsibilities. These roles and responsibilities are based on positions that exist within a marketing communications agency in a company or organization that hires communicators internally to perform these duties. At the beginning of the campaign development process, it is a good idea to divide the responsibilities and have each team member pick his or her role (Figure 17-5). Of course, roles may overlap from time to time, and team members may back each other up or overlap a bit, but clearly identifying team roles is part of a successful campaign planning process. Some of the campaign roles and responsibilities include:

- *Account executive/team leader:* leads the team and ensures that all members are communicating, doing their jobs, and working together to solve problems.
- *Recorder:* records meeting minutes, takes notes, works with the account executive to lead group meetings and to ensure work is completed for assignments and major projects, and coordinates and communicates with team members.
- *Technical support:* backs up the recorder. This role is responsible for ensuring the technical details.
- *Deadline coordinator:* responsible for managing deadlines and making sure teamwork is getting done and assignments are not being missed.

Client profile: In addition to determining team roles and responsibilities, one of your first tasks in the campaign planning process is to develop a profile for your client. A "*client*" is the person or persons in charge of communications for the organization, brand, or company with whom you are working. This person is called the client in recognition of the fact that your team is working on his or her behalf.

FIGURE 17-5: Identifying team roles is an important initial step in the campaign planning process.

In order to understand your clients and their problems/opportunities, it is first necessary to develop the *profile*—who they are, what publics/audiences they target, and what communications contact points they utilize, among other things. To do this, you will need to get some background information on your client. Following are some questions you may want to include in the client profile:

- What kind of issue, key message, or organization are you working with?
- Is it a national, regional, or local campaign?
- What do you know about the company or organization's background/history/competition?
- Who are the audiences involved?
- What is the client currently doing for the message and media?
- How effective do you think the client's communication is?
- Does the client currently have a specific logo, theme, or spokesperson?

Step 2: Analyzing the Audience: Needs and Motivations

In order to thoroughly understand the audience you want to target and why, your next step should be to conduct a *needs assessment*, which includes an *audience analysis* as described in Chapter 2, Effective Communication and Message Development. A needs assessment involves collecting data in the form of surveys or answers to questions from members of your target audience. Refer to Chapter 3, Research Methods, to learn how to develop good survey questions. You can use this data to help form message and media strategy and tactics. Here are the basic components of an audience needs assessment:

- *What are the needs?* The need determines why the communication is required.
- *What is the purpose?*
  - *Motivate:* Generates interest or stimulates the audience to action.
  - *Inform:* Serves as an introduction, overview, or background.
  - *Instruct:* Involves the audience in activities so that learning takes place.
- *Who is the audience?*
  - You need to identify as much as you can about your intended audience. See Chapter 2, Effective Communication and Message Development, for more on identifying the audience.

- *What do members of the audience think?*
  - What is their attitude (positive, negative, or neutral) and level of involvement? For example, are they "pro-animal rights," "health-food conscious," or "anti-biotech"?
  - What are their relevant behaviors and intent?
- *What do audience members do (and why)?*
  - Where do they go for information?
  - What is their media usage?
  - What do they already know about the issue or product?
  - Is this new information you are presenting?
- *How do you motivate them?*
  - What motivation do they already have to be exposed to your message?
- *What is the setting the audience will be in?*
  - What will audience members be doing when the product is provided to them? For example, will they go into a store to purchase or will they attend an event?

Step 3: Develop Campaign Objectives

An *objective* provides clear guidance that permits the orderly presentation of content leading to some effect (e.g., acceptance, purchase, awareness, learning) in the identified audience. When writing an objective for your campaign, think about the outcome you want to influence. The outcomes of an objective can be grouped into three categories:

- *Psychomotor:* actions. Do you want your audience to engage in some behavior, such as purchasing your product?
- *Cognitive:* thinking. Do you want your audience members to learn, understand, or have knowledge about something they did not know before?
- *Affective:* emotions. Do you want your audience members to hold an attitude or belief about your product?

Objectives should have the following components:

- *Behavior:* What the audience will be expected to do as a result of the effort to be undertaken.
- *Conditions:* Circumstances under which the audience members will be expected to perform the behavior.
- *Standards:* How well the audience members will be expected to perform the behavior.

Avoid imprecise verbs that are difficult to measure, such as "understand," "know," and "appreciate." Instead when writing your objectives, make sure you are using specific action verbs, such as the following, to describe what you want the audience to do:

- Explain
- Generate
- Identify
- Establish
- Build
- Maintain
- Raise
- Enhance
- Grow

Two types of objectives are used in communications campaign planning. *Marketing objectives* focus on sales, trends, and growth. *Communications objectives* focus on audience perceptions and behaviors, including their awareness, attitude, and intent.

Examples of objectives

- "Increase market share and overall sales growth by 5 percent in our target markets within one year" is an example of a *marketing* objective.
- "Increase brand awareness among non-customers by 5 percent in a 60-day period" is an example of a *communications* objective.

Step 4: The Situation Analysis

The situation analysis is the starting point for a campaign. A situation analysis provides in-depth background on all facets of your client's business or all perspectives of an issue. It is used to bring a clear understanding of the situation that prompts the development of the campaign objectives and plan. Most situation analyses are designed to give background on the situation, as well as to identify the problem or opportunity at hand. You will need to use secondary information sources (books, trade publications, Web, newspapers), as well as primary sources (annual report, media kits, interviews, brochures) to develop the information that should be included in your situation analysis. Use this outline as a guide to get you started, and to set up your campaign proposal.

Analysis of client's company/organization:

- History
- Mission
- Sales, profits, and trends
- Sources of revenue/customers
- Current and past campaign efforts
- Who are the main competitors?

Consumer/target market:

- Who are they? You can glean this information from the audience analysis.
- How do they spend money? What do they buy?
- How do they spend their time?
- What interests them?
- How do they use media?

Market/product analysis:

- What is the current key message of your client and its main competition?
- How is the client covered in the news media, and to what extent does the client use paid media to advertise or promote its products/services?
- Product distribution: Where is your client located? Where does your client do business or distribute products or provide services for sale?

Step 5: The SWOT Analysis

The purpose of a SWOT analysis is to help your team develop a perspective on your client and the issue you are focusing on for your campaign. A SWOT analysis is usually done after the situation analysis, as a brainstorming exercise that will lead to goals and objectives and eventual strategy development (Figure 17-6).

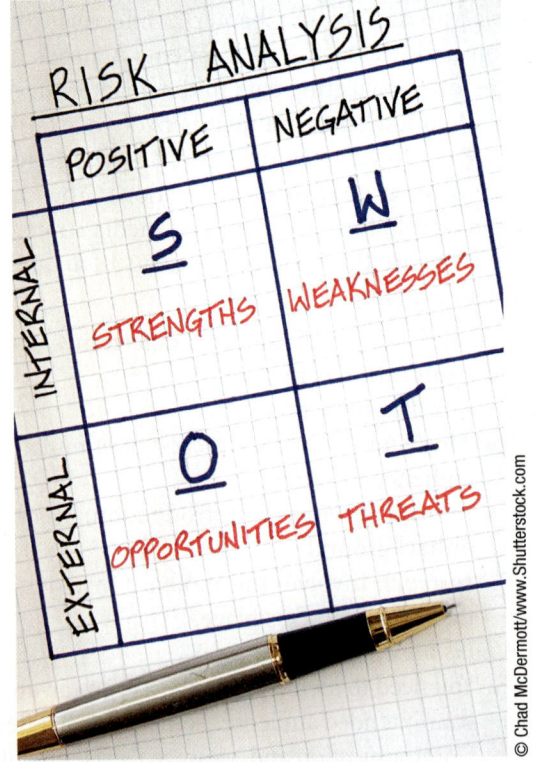

FIGURE 17-6: Example of a SWOT analysis.

To do the SWOT, appoint one person on your team as discussion leader/recorder. Make a grid with four quadrants as shown below on a flip chart or chalkboard. Have team members brainstorm their ideas and record what they come up with in the boxes.

Strengths	Weaknesses
Opportunities	Threats

Transfer the ideas from the SWOT analysis grid to a worksheet, like the one below, and use it for brainstorming strategies and tactics.

Internal Analysis

Strengths:

1.

2.

3.

Weaknesses:

1.

2.

3.

External Analysis

Opportunities

1.

2.

3.

Threats

1.

2.

3.

Step 6: Strategy Development

From your SWOT analysis, develop two or three key *communications campaign strategies. Strategies* can be thought of as how you will go about achieving your campaign objectives. You should develop strategies for how you plan to use elements of the marketing mix, as well as the key message you want to communicate. Provide your rationale by drawing on what you know of communications processes and explaining why this package of strategies is the best possible way to achieve your objective. Do not confuse strategies and tactics. *Tactics* are specific tools or techniques you will use to carry out your strategy.

- *Strategy:* Use advertising to promote your youth group's annual banquet.
- *Tactic:* Run a half-page ad in your school's newspaper.

Step 7: Creative Tactics

People hired in communications campaign companies are highly creative, but everyone has the capacity to be creative to some extent. An effective communications campaign requires creative use of elements that get the attention and sustain the interest of your target audience. At this step in the process, utilize some creativity to develop tactics. You may want to brainstorm some ideas and use these questions:

- *What elements of the marketing mix do you think you will use for your campaign?* List and rank these in order of importance. Pick one of these elements, and attempt to construct your message theme (something that will be consistent with the rest of the elements you use). Develop your key message theme in one sentence. Now rewrite the sentence using an active verb at the beginning and focus it on your target audience. Try to cut out any extra words.
- Then, think of your theme visually. *What is the central image or visual that mostly represents your message theme? What are the logos/identifiers necessary to explain who is involved and what they do?* Make a sketch of your idea.

Step 8: Media Objectives, Strategies, and Tactics

Media is, essentially, how you go about delivering your message to the audience you have selected using some form of mass media (newspaper, magazine, radio or television commercial, outdoor billboard). Because the media portion of your campaign includes choosing the media you want to use, scheduling when and where and how much you want to use, and developing a budget for placement of media, your campaign plan should include media objectives, strategies, and tactics based on your media mix.

Media objectives include reaching your target audience(s); generating a certain number of *"impressions,"* or times an audience member is exposed to your message; and distributing impressions so as to cover a specific area or region. *Media mix* is your choice as to the specific media (also called "media vehicles") you want to use, and media strategy is how you deploy audience delivery, scheduling, size, and placement characteristics to achieve an objective. Keep in mind that depending on the client and your overall objectives, you may or may not use paid media. But if you do, you will need to work out your objectives, strategies, and tactics, and include media costs in your

Media Objectives, Strategies, and Tactics Worksheet

Use this worksheet to help you create a media objective, strategies, and tactics.

Write a media objective, such as the following example:

- Reach my target audience of_____ using a media mix of _____ with a frequency of _____ number of exposures.

Based on the objective, write a media scheduling strategy:

What are the start and end dates for your campaign?

What media vehicles do you plan to use, and why?

What non-mass media vehicles do you plan to use (public relations, sales promotion, direct marketing, Web)?

© Cengage Learning 2012

FIGURE 17-7: Use this media objectives, strategies, and tactics worksheet to determine media buying, timing, and scheduling for your campaign.

campaign budget. Media planners develop the objectives and strategies, using a media plan to determine when, where, and how media will be placed given a specific budget. Media buyers negotiate the best deal for the media vehicles in the media plan.

Media strategies include reach and frequency, as well as timing and scheduling of media. As mentioned earlier in this chapter, media reach refers to the size of the audience you are able to reach with a specific media schedule, and frequency is the number of times your message is exposed to an audience (see Figure 17-7).

Media timing and scheduling strategies include the following:

- *Consumer aperture:* Running your advertising when consumers are most likely to be motivated to pay attention. An example would be to promote pizza at the dinner hour, when consumers are most likely to be hungry.
- *Continuity:* Ads run over a specific period of time, with no gaps.

- *Seasonality:* Scheduling that occurs during specific time of year, such as Christmas, spring, or Independence Day.
- *Flighting:* Ads run intermittently, with some media vehicles running at times different from others.
- *Pulsing:* Ads run more or less consistently, but with periods of intense advertising, then backing off to very little.
- *Roadblocking:* Scheduling the same ad on several networks or local broadcast stations to run at the same time.
- *Saturation:* Running as many ads on as many media vehicles as possible, seemingly "saturating" the market. For example, if you run a radio commercial on all five radio station in a market at the same time, that would be an example of saturation.

Step 9: Public Relations

Using public relations, or earned media, to gain publicity for your client is an effective way to stretch your campaign budget. Even if you are using some paid media, you can promote your campaign message by sending out a news release or a media kit, or hold a special event in order to get media coverage. Following are some issues to consider as you create a public relations component for your communications campaign plan:

- What is the key message you want to communicate in your public relations activities?
- How can you make it newsworthy? Because you will be trying to get news media coverage, you need to consider the "hook" or "slant" that makes your campaign newsworthy. Newsworthiness is the degree to which a news media outlet thinks your story is news, usually determined by how much it matters to the media outlet's audience.
- If you will use a news release, describe what you will include and where you will send it.
- If you will use a media kit, describe what you will include and where you will send it.
- If you will use public service announcements—which can only be done if your client is a nonprofit or charitable organization—describe what you will communicate, how long (30 or 60 seconds) or what size (full page, half page) it will be, and where you will send it.
- If you will hold a special event, describe what you are doing and how you will get your target audience to come to it. More detail about special events can be found in Chapter 16, Special Events.

Step 10: Media Cost Planner

How much does paid media cost? Media vehicle costs are determined by the size of the audience the media will attract, which differs from market to market. National or network media vehicles, which reach a large audience, cost much more than local media outlets do. Examples of national media include a television network, such as CBS, or a national magazine, such as *Time* or *People*. Examples of local media include your local FM radio station, television cable channel, and local newspaper.

Each different type of media has characteristics that determine the cost. Print—newspapers and magazines—is bought by the size of the ad, for example, whereas radio and television are bought on the basis of "air time," time slots between programs that are sold as advertising.

Following are some rough "ballpark" average cost estimates for different types of media. Because costs change frequently, a list of potential media vehicles is provided for you to consider as you integrate them into your communications campaign plan. You will want to contact your local, regional, or possibly national media sales representatives to determine how much costs will actually be.

Rough estimates of local media costs (will vary by market)

Newspaper advertisement (daily paper in small town or rural community)
Full-page ad: $2,000–$5,000
Small ad (3 columns × 10 in.): $200–$300

Magazine advertisement (city magazine)
Full page, back cover: $20,000
Full page, color, inside: $10,000
Full page, black and white: $5,000–$8,000

Solo direct mail piece: $50/thousand
Shared mail piece: $20/thousand
Media kit, full color: $50/thousand
Billboard (small painted): $3,000/month
Transit (bus panel): $300/month

Television (local or cable) (run at least five or six commercials per day; run a mix of 30- and 60-second commercials): $5,000/week
Radio (run at least 20 commercials per day. The run might be a mix of 60- and 30-second ads, a mix of 30-second ads and a few 10-second promos, or all 30-second ads): $1,000/week

Sales promotion
Coupons: $10/thousand
Press conference or local tour: $2,000
Special event: $2,000–$20,000
Sweepstakes/drawing: $500–$20,000

Website, social media (costs to develop and host the site)
Develop: $800–$1,000
Hosting: $30–$50/month

In addition to the actual cost of the advertisement itself, you may incur production costs. Paid and non-paid media all have production costs associated with them. Usually the production cost is calculated separately from the media cost. Print production can include paid ads and non-paid brochures, media kits, and fliers. Broadcast production includes the cost of producing the television or radio commercial. Broadcast stations in small local markets often will produce television and radio commercials for airing on their station for free, as long as you make a media buy. You just need to send them a script. In larger markets, stations will probably charge for production. Production costs vary widely according to what your concept is and the media you might be using, so it is a good idea to get a quote, or "bid," first so you have a sense of what the cost will be.

Campaign planners draft a media plan or flowchart to lay out when the campaign will begin and end, what types of media will be used, the numbers of people who are estimated to be reached, and the total cost. Flowcharts can be set up on a weekly basis, to cover the 52 weeks in a year or a quarterly basis, to cover the four "quarters" or three-month periods that are commonly used in broadcast media to buy air time and project audience numbers. Table 17-1 is an example of a quarterly flowchart table you can use to begin to develop your campaign plan for media. Follow this up with the detailed campaign media chart and budget (Table 17-2).

Step 11: The Final Project Worksheet: Objectives, Strategies, and Tactics

This final worksheet will represent your best thinking as to the campaign you believe will be most effective to address the problem or opportunity your team has decided to address. Completing each section in the worksheet will provide you with the basis of an effective communications campaign plan (see Figure 17-8).

Step 12: Project Peer Evaluation

One of the ways agency teams stay in touch with their clients is by submitting a conference or "call" report after each meeting and/or when the team gets together to brainstorm and make decisions about the campaign. Use the "Call Report" form to write up your own call reports when working on a campaign (see Figure 17-9).

PROJECT PLAN						
ACTIVITIES	**JAN.–MARCH**	**APRIL–JUNE**	**JULY–SEPT.**	**OCT.–DEC.**	**EST. COST**	**NUMBERS REACHED**

Total Budget:

TABLE 17-1: Example of quarterly flowchart table for a communications campaign.

CAMPAIGN MEDIA CHART AND BUDGET				
Print	Type	# insertions/quantity	Size/placement	Percent of budget
			Total:	
Broadcast	Type	# spots/day	Seconds	Percent of budget
			Total:	
Outdoor billboard	Type	# months	Size/placement	Percent of budget
			Total:	
Direct mail	Type	Quantity	Placement	Percent of budget
			Total:	
Public relations	Type	Production cost	Placement	Percent of budget
			Total:	
Sales promotion	Type	Costs	Placement	Percent of budget
Website		Number of pages	Cost to develop	Percent of budget
Social media elements (Facebook, Twitter, blog, RSS, etc.)		Number of elements/type	Cost to develop	Percent of budget
			Total:	
Total print				
Total broadcast				
Total outdoor				
Total direct marketing				
Total public relations				
Total sales promotion				
Total Web				
Total social media				
Total budget your team is asking for:				

TABLE 17-2: Example of a detailed communications campaign media chart and budget table.

© Cengage Learning 2012

Final Campaign Project Worksheet

- **Team, Client, Need:**

  1. **Team Members:**

  2. **Client:**

  3. **Need:**

  4. **Idea:**

- **Target Audience:**

- **Objectives:**

  1. **Overall Campaign Objectives:**

  2. <u>Marketing</u>: Includes goals for market share, sales, trends, membership increases. Be specific and estimate logical outcomes.

     <u>A.</u>

     <u>B.</u>

  3. <u>Communication</u>: Brand awareness, generate trial, frequency, attitude/image, purchase

     <u>A.</u>

     <u>B.</u>

- **Strategies and Rationale:**

  1. **Overall Campaign Strategies**

     - Creative message:

     - Marketing mix:

  2. **Specific Mass Media Strategies**

     - Media scheduling (type of scheduling, media mix, time of year, length of campaign for both paid and non-paid media):

- **Specific Public Relations Strategies:** Hook or slant to generate newsworthiness, tools to be used (news release, PSA, media kit, special event)

- **Rationale:** Why do you believe it will it work?

- **Tactics:** Be specific and list a tactic for each strategy you are using.

FIGURE 17-8: Final campaign project worksheet.

Call Report

Team: _____

Client: _____

Call report # _____

Date _____

I.   Agency team activities this week:

    a.   Meeting dates:

    b.   Attendees:

II. Topics discussed:

    a.   Agreements:

    b.   Actions taken:

III. Individual activities by team members:

IV. Issues:

V.   Progress status or next steps:

© Cengage Learning 2012

FIGURE 17-9: The Call Report form is used to record team actions and activities, including meetings with the client.

If you are doing a communications campaign for a class, you may want to evaluate yourself and your team members, or your teacher might do so. Here is an example of how a team could be evaluated by its peers:

The _____ team met _____ times to complete the campaign for our client, _____. Below is my evaluation of the contribution of each of the members (include yourself):

1. (name):_____ Grade I would give (A–F): _____

2. (name):_____ Grade I would give (A–F): _____

3. (name):_____ Grade I would give (A–F): _____

4. (name):_____ Grade I would give (A–F): _____

Summarize the mood or morale of the team throughout the campaign process, noting any problems that surfaced and how you dealt with them:

SUMMARY

The chapter provided a comprehensive overview of the communications campaign development process. The chapter introduced integrated marketing communications as a concept, and provided detail review of media objectives, strategies, and tactics as used in a communications campaign plan. The chapter ended with a step-by-step detailed outline of how to develop an integrated communications campaign plan.

CHAPTER EXERCISES

APPLICATIONS

Following are some ways you can apply what you learned in this chapter:

CAMPAIGN MATERIALS

Working alone or in a team, develop the following campaign materials for a client:

- **Marketing mix:** Develop a relevant idea for each element of the marketing mix that could be used for your team's client/campaign. Identify each element of the marketing mix, number it, and describe the idea in a sentence.
- **30-second television or radio script:** Television and radio commercials are first written as scripts, and then produced/filmed and edited for airing. In small local markets, radio and television stations will often produce your "spot" for you if you provide a script. Write a television or radio commercial script (30 or 60 seconds), referring to the instructions in Chapter 5, News Media Writing, and Chapter 10, Video and Audio Production, in this textbook.
- **News release:** Write a one to one-and-a-half page news release for print for your client. A news release is written in the form of a news story. It is usually one or two pages long, and includes contact information so a reporter can contact you if he or she wants to run the story or needs more information. Refer to Chapter 13, Media Relations, for more about writing news releases.
- **Campaign theme and logo:** Develop a campaign theme to be used on subsequent campaign creative materials, as well as a campaign logo. This should incorporate your client's existing logo, if the client has one, and the theme or catchphrase for use on media kits, news releases, and ads. You may use any type of software to develop this, including word processing software.
- **Print ad layout:** Develop a full-page print ad (8.5 by 11 in.) for your client's campaign. Refer to Chapter 6, Document Design. Include the campaign theme and logo. On a separate sheet, include where the ad will be placed (newspaper, magazine).

ANALYZING A CAMPAIGN

Many campaigns, especially those for national consumer brands, strive not only to be persuasive, but also memorable and clever. In these times of media clutter, it is difficult enough for a campaign message to break through; it is even harder to get an audience to remember, and be motivated enough to respond to, a call to action. For this application exercise, reflect on campaigns you have been exposed to recently, using your recollections to answer the following:

- Top three best ads you have seen or heard in the last three months:
- Top three worst ads you have seen or heard in the last three months:
- Most memorable ad, and why:
- Most unusual/shocking/unexpected ad and why:

Pick one ad and describe in detail who you think the target audience is, and what you think the key message was.

CHAPTER QUESTIONS

MATCHING:

Match the component of a business letter below with its definition or characteristics. Not all terms will have a definition associated with them.

_____ **1.** Used to help your team develop a perspective on your client and the issue you are focusing on for your campaign.

_____ **2.** Includes all of the following ways to communicate a campaign message: advertising, public relations, brand marketing, sales promotion, direct marketing, and Internet marketing.

_____ **3.** Provides in-depth background on all facets of your client's business.

_____ **4.** Used in a communications campaign; marketing and communications are examples.

_____ **5.** What the acronym CPM stands for.

a. Marketing mix

b. Situation analysis

c. SWOT analysis

d. Objectives

e. Cost per thousand

MULTIPLE CHOICE

6. _____ involves a variety of tools and techniques to develop incentives to create a perception of greater brand value, while encouraging acceptance or adoption of a product, good, or service right now by a consumer or businessperson.

   a. Sales promotion

   b. Publicity

   c. Advertising

   d. Public relations

7. _____ seeks to generate unpaid "earned" media.

   a. Sales promotion

   b. Publicity

   c. Advertising

   d. Public relations

8. _____ involves integrating a campaign theme or message and delivering it in a format suitable for a specific type of media.

   a. Sales promotion

   b. Publicity

   c. Advertising

   d. Public relations

9. Running your advertising when consumers are most likely to be motivated to pay attention is called _____.

    a. flighting

    b. continuity

    c. consumer aperture

    d. roadblocking

10. Running ads intermittently, with some vehicles running at times different from others, is called _____.

    a. flighting

    b. continuity

    c. consumer aperture

    d. roadblocking

11. _____ are how you will go about achieving your campaign objectives.

    a. Tactics

    b. Goals

    c. Plans

    d. Strategies

FILL IN THE BLANK

12. _____ refers to the size of the audience you are able to reach with a specific media schedule.

13. _____ is the number of times your message is exposed to an audience.

14. _____ is the term used for the process of developing and implementing a set of integrated campaign elements that utilize common and consistent themes across multiple elements.

15. A(n) _____ involves collecting data in the form of surveys or answers to questions from members of your target audience.

REFERENCES AND FURTHER READING

BNET. "Developing a Communications Campaign." Last modified 2007. Accessed November 1, 2010. http://www.bnet.com/article/developing-a-communications-campaign/55157.

Bobbitt, R., and R. Sullivan. *Developing the Public Relations Campaign: A Team-Based Approach.* Boston, MA: Pearson Education, 2005.

"Communicator's Guide for Federal, State Regional and Local Communicators." Accessed November 1, 2010. http://govinfo.library.unt.edu/npr/library/papers/bkgrd/chapter3.html.

O'Guinn, T. C., T. Allen, and R. J. Semenik. *Advertising and Integrated Brand Promotion.* Mason, OH: South-Western College, 2006.

Sayre, S. *Campaign Planner for Integrated Brand Communications.* Mason, OH: Thompson Southwestern, 2005.

The Future

Chapter 18
- Future of Agricultural Communications

18 Future of Agricultural Communications

OBJECTIVES

After completing this chapter, the student will be able to:

- Describe the evolution of agricultural communications and the role of the land-grant university.

- Describe trends in and the future of agricultural media.

- Describe trends in and the future of agricultural communications as a profession.

INTRODUCTION

The everyday routine of agricultural communicators today is very different from what it was even a short few years ago. Today's agricultural communicator may spend his or her day primarily online, updating websites, posting to social networking sites such as Facebook and Twitter, and, of course, handling streams of e-mail from clients, stakeholders, colleagues, media, and the general public (see Figure 18-1).

Corporate video production, which used to require the services of a full production house or studio, can be shot with a personal digital camera in the morning and uploaded to the Web or YouTube before lunchtime. Nearly all images are digital images these days, and even when working with a graphic designer, agricultural communicators have access to desktop layout and image editing software that allows them to manipulate, change the format of, and develop the design for anything that will be printed. It is a far cry from the early days when agricultural publication editors—and they were mostly male—worked on typewriters and shot black-and-white photographs using cameras that contained film, rather than memory cards.

Before the advent of digital media, agricultural communicators were primarily print based, and their focus was on traditional agricultural media outlets such as the **farm press**—publications targeted to famers, their suppliers, and vendors. Today, agricultural communications is making the shift

FIGURE 18-1: Today's communicators employ many types of digital new media when communicating to their audiences.

© Patryk Kosmider/www.Shutterstock.com

from traditional farm press and broadcast to new and converged media, and communications practitioners are looking toward integrating several forms of communications as a way to maintain a consistent look and feel while employing a variety of traditional and non-traditional online media.

The agricultural communications profession was developed primarily to help disseminate information from the agricultural experiment stations at the land-grant institutions of higher education in each state. **Land-grant universities** were established by the **Morrill Land-Grant Act** of 1862 in order to "to teach such branches of learning as are related to agriculture and the mechanic arts, in such manner as the legislatures of the States may respectively prescribe, in order to promote the liberal and practical education of the industrial classes in the several pursuits and professions in life" (Morrill Land-Grant Act, 1862). As a result of the act, each state was given land to develop such an institution. Later extensions of the act added historically black colleges (in 1890) and institutions serving Native Americans (in 1994). These universities were charged with transferring science and technology research, along with educating and informing rural and, later, urban populations (see Figure 18-2).

However, it soon became apparent that editors and journalists were needed to make scientific findings more accessible and easier to disseminate to increasingly diverse audiences. Universities began to offer courses in **agricultural journalism**, which focused on journalistic writing in an agricultural context. The first Bachelor of Science degree in agricultural journalism

FIGURE 18-2: Colleges of agriculture were established in academic institutions in the United States as a result of the Morrill Act, with the goal of extending knowledge in formal and non-formal settings.

was awarded at Iowa State College (now Iowa State University) in 1920. By 1975, agricultural journalism programs were evolving from an exclusive focus on journalism to comprehensive agricultural communications programs that included advertising, public relations, Web technologies, and video production. By 1991, there were more than 30 programs in the United States. Today, the number of undergraduate and graduate programs in agricultural communications is continuing to increase, and many colleges of agriculture find that agricultural communications is an area of enrollment growth.

TRENDS

Traditionally, agricultural media included the "farm press"—print publications such as the *High Plains Journal, Farm Journal, Progressive Farmer,* and *Drovers*—as well as farm radio stations and networks such as Farm TV and RFD-TV. With the advent of the Web and social media, however, agricultural media has expanded to include online media such as websites, social networking sites, and even games like Farmville.com. As an example of convergence between social networking and gaming, Farmville is a farm simulation game that is available as an application on Facebook that, according to Facebook, has more than 44 million active monthly users.

The expansion of **broadband Internet access**, which is defined as high-speed or high-data-rate Internet access via digital subscriber line (DSL) cable modem or wireless transmission, has created many new opportunities for communicators and new information resources and services for users. However, historically, access to the Internet has expanded at a faster rate in urban areas than in rural communities. This has created a "**digital divide**" between urban and rural residents in terms of access to Web-based services. But, even though rural broadband Internet service coverage has always lagged behind that in urban areas, today, many agricultural producers, suppliers, and customers are moving online when they want to communicate, educate, or inform. Print and broadcast farm media are following this trend as well. A monthly magazine like the *Farm Journal* can promote its 24/7 coverage of farm-related news because it has a Web-based "extra" edition that features not only stories from the latest issue, but also additional, continuously updated coverage and features. **Web-based editions** of print and broadcast media can be updated much more

Lisa Lochridge

Director of Public Affairs, Florida Fruit & Vegetable Association

The world of communication has changed immeasurably, it seems like, in the last few years. I'm someone who went to work for a newspaper when they were still using typewriters, so the change that I've seen has been tremendous. In the relatively short time that I've been in public relations, once upon a time, you faxed out a news release and looked for a clip in the newspaper the next day or a story on the news that night. All of that has changed. Those traditional methods still do exist, but there's this whole new world of digital media and the Internet, and that has brought with it things like bloggers. And you have to be monitoring those blogging conversations, for example, and that's tough sometimes. You've got to be out there on Facebook and Twitter. For us it's a matter of figuring out how to use those tools and be effective in using those tools. So it's changed. And who knows what the future holds? What's beyond Twitter? What's beyond YouTube? We have to be able to constantly be learning new things and be adapting to the new ways of communication.

quickly and cost effectively, since there is no need to print more copies or shoot and edit a radio or video news story. In addition, Web and social media sites can link to other types of media.

Because of the trend toward online media, modern agricultural communicators need to be at least somewhat familiar with all of the major media formats. Today, if you work for a farm magazine, you might be expected to know how to shoot and edit video for its Web edition; if you work for a company, you might be managing its brand presence on Twitter and Facebook. Staying on top of all of these forms of communication can be a challenge, especially because the typical agricultural communications practitioner is someone working in a one- or two-person "shop." Many professionals look to their local and national professional associations to provide ongoing professional development training so they can stay abreast of new technologies and learn ways to use both old and new media more effectively. Organizations such as the National Agri-Marketing Association, Cooperative Communicators Association,

American Agricultural Editors Association, North American Agricultural Journalists, Livestock Publications Council, and others offer workshops and training opportunities for members at their annual meetings on a regular basis.

FUTURE

Where are agricultural media and the agricultural communicators who write, edit, and promote in and for agricultural media headed in the future? Will print farm magazines disappear? Will social networking media overtake the Web as the online medium of choice for practitioners and publics? Will agricultural communications jobs become more general in nature or more specialized? Future trends in all media, including agricultural media, indicate that convergence will continue to affect the existing media environment, as media companies merge and entrepreneurs create new applications online.

Some of the new directions for media may involve gaming and virtual reality applications. Gaming

Becky Raulerson

Coordinator, University of Florida/Institute of Food And Agricultural Science's Center for Public Issues Education in Agriculture and Natural Resources

I remember the first time I used the Internet. I even remember the first Web page I designed using HTML codes. That was back when the Internet was just a one-way information source. Even now, with all the engaging and interactive websites available to us, I will occasionally stumble across one of those old HTML-coded pages. I'll stare at it and see the reflection of a 20-year-old student who thought her newly acquired skills were all she would ever need to know about the Internet and its uses. Boy, was she wrong! With today's social networking tools and digital resources, the agriculture and natural resources industries cannot be that 20-year-old student. We must monitor the trends and think ahead to the next step, because there will always be something new. We must be creative in our thinking and find unique ways to use tools when educating the public on issues. Even our research that provides the foundation for education and outreach materials will need to be novel and different. If the agriculture and natural resources industries don't stay ahead of these technologies, others will, and they are likely to be the people providing biased and emotionally charged information to the public.

environments and **virtual reality** can be used to simulate a three-dimensional space where users can interact with each other online. These simulated environments can be used to entertain, communicate, and inform. An example of this is **Second Life**, a virtual world in which individuals, organizations, and companies can buy an "island" in order to have a virtual presence. An example of an innovative educational use of Second Life is a collaboration between Texas A & M University and Texas Tech University to use Second Life to simulate a crisis situation for their agricultural communications students. Students create their own "avatars"—animated representations of themselves—in order to enter the Second Life environment and take on various communications roles as the crisis situation plays out virtually.

As websites become more complex and interactive, users need more bandwidth in order to experience all of these features. It is estimated that rural broadband access, currently estimated at 41% of rural residents versus 57% of residents nationwide, will continue to grow to meet the demands of rural audiences. All of these changes will have a significant effect on the roles and responsibilities of agricultural communicators (Berkes, 2009).

Increasingly, agricultural communicators working in the media or for companies and organizations involved in agriculture will interact with several different forms of media on a regular basis. They will need to wear many "hats," and their training will need to include the ability to write and create communications online, on the Web, and in traditional media at the same time. Further, the agricultural communicator of tomorrow stands to become an even more important member of an organization's executive team. Agricultural communications practitioners will increasingly be seen as **strategic communications** experts, knowledgeable about the form and function of a diverse range of media and communications approaches, and, as such,

FIGURE 18-3: Knowing how to write well will be as important for future agricultural communicators as it was for the earliest agricultural journalists.

will likely be consulted as a routine part of a company's decision-making process. Developing effective strategies for communicating the company's or the organization's message will be more important than ever.

Although **future trends** will bring many changes to the agricultural media landscape, one of the enduring elements of the day-to-day job of an agricultural communicator, however, will be the continued focus on the written word. Media formats may change, and new media emerge, but the basic building block of professional communication remains the same—knowing how to write well will still be the hallmark of an effective agricultural communicator (see Figure 18-3).

SUMMARY

This chapter focused on past, current, and future trends in agricultural communications. The history of the field, starting with the establishment of the land-grant university system and the need for academic programs in agricultural journalism, was outlined, as well as the growth and development of agricultural communications as a specialized field. The chapter concluded with a discussion of future trends, focusing on the role online Web-based media is playing and will continue to play in shaping the functions of communications and agricultural communications practitioners in providing information about agriculture and our natural resources.

CHAPTER EXERCISES

APPLICATIONS

Following are some ways you can apply what you learned in this chapter:

- **In-class discussion**. Form groups of three to four students. Have each group take one of the following questions and prepare a group response as well as two follow-up discussion questions for the rest of the class. When called on, each group should present the question and their response, and lead the class in a discussion.

  Discussion questions:
  What do you think the future holds for agricultural media?
  What changes do you envision?
  What will remain unchanged?
  How will you adapt media use to reach old and new audiences?
  Will agricultural media survive, and if so, what will agricultural media look like in the coming years?

CHAPTER QUESTIONS

MATCHING:

_____ 1. Broadband coverage in the United States being greater for urban residents than for rural residents.

_____ 2. Publications targeted to farmers and their suppliers and vendors.

_____ 3. Established the foundation for land-grant universities in the United States.

_____ 4. The aspect of a weekly or monthly farm magazine that can increase its ability to cover news in a more timely fashion.

_____ 5. These were first needed by land-grant universities to make scientific findings more accessible and easier to disseminate to increasingly diverse audiences.

_____ 6. A 3D virtual world in which individuals, organizations, and companies can buy an "island" in order to have a virtual presence.

a. Farm press

b. Morrill Act

c. Digital divide

d. Second Life

e. Agricultural journalists

f. Web-based edition

MULTIPLE CHOICE:

7. _____ is defined as high-speed or high-data-rate Internet access via digital subscriber line (DSL) cable modem or wireless transmission.

    a. Virtual reality

    b. Broadband

    c. The digital divide

    d. Web-based edition

8. In the future, agricultural communications practitioners will be seen as _____, knowledgeable about the form and function of a diverse range of media and communications approaches.

    a. journalists

    b. technicians

    c. executives

    d. strategic communicators

9. _____ is focused on journalistic writing in an agricultural context.

    a. Strategic communication

    b. Agricultural communication

    c. Agricultural journalism

    d. Technical communication

10. _____ is focused on comprehensive communications programs that include advertising, public relations, Web technologies, and video production.

    a. Strategic communication

    b. Agricultural communication

    c. Agricultural journalism

    d. Technical communication

REFERENCES AND FURTHER READING

Berkes, H. "Stimulus Stirs Debate over Rural Broadband Access, NPR Radio." Last modified 2009. Accessed May 25, 2010. http://www.npr.org/templates/story/story.php?storyId=100739283.

Boone, K., Meisenbach, T., and Tucker, M. *Agricultural Communications Changes and Challenges.* Ames: Iowa State University Press, 2000.

Morrill Land-Grant Act. 1862. U.S. Statutes at Large, 12, 503.

Appendix A:
Associated Press Stylebook Entries to Know

In addition to the following individual entries, beginning agricultural communicators also should know the chapters of *The Associated Press Stylebook* covering punctuation, food, and social media. The following entries are from the 2011 edition of *The Associated Press Stylebook*. Entries may change from year to year, so it is best to purchase updated editions of *The Associated Press Stylebook* each year.

A

Abbreviations and acronyms
Addresses
Adult
Ages
Animals

B

Boy

C

Children
Collective nouns
Company names
Composition titles
Contractions
Courtesy titles

D

Dangling modifiers
Dates
Days of the week
Dimensions
Dollars

E

E. coli
Email
Essential clauses/nonessential clauses
Essential phrases/nonessential phrases
Ethanol

F

Food
Foot-and-mouth disease

G

Girl
Google
Governmental bodies
Grade, grader

H

Highway destinations
Holidays and holy days

I

Initials
Internet
It's/its

J

Jargon

K

L

Legislative titles
-ly

M

Mad cow disease
Millions/billions
Months

N

Names
Nationalities and races
National FFA Organization
No.
Numerals

O

Organizations/institutions
Over

P

Percent
Plants
Plurals
Possessives
President

Q

Quotations in the news

R

Reference works

S

Sentences
Social media

Social networking
Speeds
State names

T

Teen, teenager, teenage
Telephone numbers
Temperatures
Time element
Times
Titles

U

United States

V

Verbs

W

Web
Website

XYZ

Years
ZIP code

Glossary

A

accuracy all text is spelled and punctuated correctly; sentences are grammatically correct; all names are correct.

actuality a term commonly used in radio for the exact words spoken by someone in his or her own voice.

advertising involves integrating a campaign theme or message and delivering it in a format suitable for a specific type of media.

agricultural communications the exchange of information about the agricultural and natural resources industries through effective and efficient media, such as newspapers, magazines, television, radio, and the Web, to reach appropriate audiences.

agricultural journalism focuses on journalistic writing in an agricultural context.

agriculture the process of producing food, feed, and fiber through the raising of plants and animals.

anti-social persuasion warning people about a potential punishment or negative outcome that could result from their action or inaction.

aperture the "iris" of the camera; the opening in the lens through which light passes to the camera sensor. The aperture controls the amount of light that is allowed into the camera.

application letter a special kind of business letter that accompanies a resume for a job; also called a *cover letter*.

argument strategies persuasive strategies designed to create a convincing argument, such as proposition-fact-evidence, common ground, and logical reasoning (induction, deduction, and comparison).

articulation making vocal sounds distinctly; the distinct pronunciation of words.

aspect ratios width-to-height proportions.

Associated Press Style a writing style for news stories developed by the Associated Press, an international organization of professional journalists.

asymmetrical balance occurs when several smaller items on one side of an imaginary line dividing a page either vertically or horizontally are balanced by a large item on the other side, or smaller items are placed further away from the center of the screen than larger items; also called *informal balance*.

attribution telling readers where the information in a news story comes from.

audience the target of the message; may vary dramatically in risk communication situations.

audience analysis describing an audience on the basis of shared characteristics.

B

balance in document design, involves imagining a line dividing a page either vertically or horizontally and then placing visual elements so they are either symmetrically (formally) balanced or asymmetrically (informally) balanced.

benchmarking in special events planning, a type of research that involves looking at what other organizations have done to plan events similar to yours.

blogs (Web logs) shared online journals maintained by the writer and updated regularly, to which readers can also post comments.

brand a product, good, or service that is given an identifiable name and image.

brand loyalty customer preference for a specific brand over other brands.

broadband Internet access high-speed or high-data-rate Internet access via digital subscriber line (DSL) cable modem or wireless transmission.

b-roll any non-narrated video footage shot expressly to "cover" narration or an interview.

business letter a formal document typically sent to people outside of an organization.

C

call to action a specific directive designed to get listeners to act on the advice of the communicator.

campaign development process specific steps and tasks that help you develop a plan that utilizes elements of the marketing mix, a consistent theme, and carefully developed strategies and tactics in order to achieve a communications or marketing objective.

campaign message the theme of the message to be delivered in a media campaign.

caption provides written information underneath a photograph that is necessary for the reader to understand the photograph; also known as a *cutline*.

cascading style sheets (CSS) developed in the 4.0 version of HTML as a way of defining how to display HTML elements in a Web browser.

catchphrase a short, easy-to-understand message; similar to a slogan.

cause-related marketing when one of the partners is a nonprofit or charitable organization; the nonprofit or charity is the "cause" that is being promoted.

channel the means through which a message is sent (e.g., speaking and writing).

chronological resume a resume that is written in reverse chronological order, with headings grouped by what a person has done, such as "education," "employment experience," and "interests/activities."

clarity in visual aid design, the presentation of only one main idea, so that people know—even at a glance—what your poster, display, or computer slide is about.

clichés overused words and phrases, such as "it cost an arm and a leg," "a drop in the bucket," and "on the cutting edge."

CMYK colors the inks used for printed documents (C = cyan, M = magenta, Y = yellow, and K = black).

coding sheet provides explicit definitions, examples, and coding rules for each category in a content analysis study.

cognitive dissonance the basic incompatibility of holding two or more beliefs simultaneously (dissonance).

color temperature different color qualities of light; light sources contain different amounts of red, green, and blue light.

commodities products such as corn, wheat, cotton, or citrus that are not sold on the basis of a brand or product name.

common ground the act of taking polarizing viewpoints and showing where they agree.

communication points the most important issues or points you hope to address and get across to the reporter during the interview.

communications the process of transferring information from a sender to a receiver with the use of a medium—such as newspapers, your voice, radio, television, the Web, or other media—in which the communicated information is understood by both the sender and receiver.

communications campaign a strategic, structured plan consisting of a mix of media and message strategies and tactics with a consistent, unified theme.

comparison the form of logical reasoning in which writers and speakers choose between or among best alternatives based on a set of standards or criteria.

composition organizing the subject—the person or object of the photograph—through the viewfinder of the camera.

content analysis using this tool, the researcher examines the content itself. This can take the form of analyzing messages or words in newspaper articles, violent images in pictures or videos, key phrases in advertisements that persuade people to buy items, or inflammatory messages in political speeches. Content analysis can be conducted with any type of recorded communication.

contingency amount added to an event budget in case costs are more than anticipated.

continuity when each shot in a video logically flows from the one before it.

contrast the dominant focus or element on a page in document design.

converged media the trend that technological developments in communications media will eventually lead to the point where all existing media become one medium.

co-sponsorship in special events, arrangement in which more than one organization shares sponsorship of the event.

cost per thousand (CPM) a measure that looks at the total cost divided by the total estimated audience for using a particular media vehicle; a way of making all media equivalent in some way.

crisis communication communications that take place when things go wrong.

crisis communication plan advance preparation by an organization to be ready to respond when a crisis occurs.

cross promotion when an event is co-sponsored by more than one organization, or an organization and another type of sponsor, such as a media outlet.

cut a direct transition from one shot to the next in a video composition.

D

decoding interpretation by the receiver of the message sent by the sender.

deduction when the writer or speaker moves from the general to the particular; starts with a general principle, then applies the principle to a fact, and finally draws a conclusion concerning the fact.

demographics characteristics about audience members that are hard to change, such as gender, age, income, education level, and place of residence; used in audience analysis.

depth of field the portion of the scene in focus in the camera; can be long or short.

digital divide exists between urban and rural residents in terms of access to Web-based services.

digital immigrants those not born into the digital world but who have *migrated* into digital technologies.

digital natives people for whom digital technologies already existed before they were born.

direct mail when companies use the postal system to deliver a direct offer.

direct marketing focuses directly on the end users; includes use of direct offers to consumers, often through either a form of advertising media, such as a television infomercial, or through postal mail.

dissolve a gradual change from one shot to the next in a video composition.

document design the process of choosing how to present all of the basic document elements so your document s message is clear and effective.

dyadic communication involves only two people; also called *one-on-one communication*.

E

earned (non-paid) media free to the organization; includes news releases, media kits, and public service announcements; used in a public relations campaign.

editorializing allowing the writer's opinions, prejudices, and biases to enter a story; should be avoided in news writing.

emphasis key component of visual aid design; focuses the audience's attention to a specific place on a poster, exhibit, or slide.

encoding the use of language, symbols, and metaphors by the sender to put a message in a specific form.

evaluation the process of determining how effective a special event was.

event budget all of the income and expenses of the event; based on income and expenses you reasonably believe you can expect with the resources available.

event managers the people responsible for organizing and conducting a special event.

evidence any proof you have that helps you argue your main points.

exclusive sponsor in special events, the one who pays extra to have an exclusive role, that is, to be the only sponsor of the event.

extemporaneous speech requires considerable preparation before the speech is given, but the speaker waits until the actual presentation to select the exact wording of the speech.

F

facts statements of what is known to be true in a given situation.

fade a special form of the dissolve; any shot that dissolves to black in a video composition.

false argument a form of persuasion based on illogical or poor reasoning.

farm press publications targeted to famers, their suppliers, and vendors.

fear appeals raise fear and anxiety in the audience of a persuasive communication.

feature story more relaxed in style than a traditional news story; set apart from a news story because of the greater amount of detail and description it contains.

feedback the receiver's response to the sender's message.

five P's of marketing product, promotion, price, place (location), and public relations.

five Ws and H the key components of any news stories: *who, what, when, where, why,* and *how.*

five-part argument the division of a persuasive essay or speech into five parts: introduction, background, lines of argument, refuting objections, and conclusions.

focus groups allow small groups of people to interact and trigger responses from each other; essentially a group interview, comprised of between six to eight persons who are asked questions by an objective moderator.

frame of reference shared understanding in a communication situation.

framing (1) the way the various elements within the video screen are arranged; (2) how a message conveyed in the media provides contextual cues that can indicate to receivers how to think about what is being communicated.

frequency the number of times your targeted audience members are exposed to your message.

FTP (file transfer protocol) function of an HTML editor that allows you to publish or upload a site to a Web server.

full block letter has all of the components of the letter (heading, date line, inside address, salutation, paragraphs, and the signature block) flush left, meaning that the components are all the way to the left side of the page. The paragraphs are either flush left or full justified, and the paragraphs are not indented.

functional resume classifies the experiences that demonstrate your skills and capabilities into categories, such as "professional," "technical," "communication," "leadership," "management," and "sales"; usually finishes with a reverse chronological listing of your job experiences.

future trends events that will bring many changes to the agricultural media landscape.

G

gatekeeper in the news media, determines what you read, view, and hear about the world around you.

H

head room the space between the person's head and the top edge of the screen in a video composition.

hexadecimal color number a six-number system that converts the colors your computer monitor displays to mathematical equivalents of other colors.

HTML (hypertext markup language) the programming language or code used to create pages on the Web.

hyperlink (link) a file name or address to another Web page.

hypothesis the assumption or induction a scientist makes.

I

impromptu speech unpracticed, spontaneous, or improvised speech.

induction a method of logical reasoning that is associated with the sciences; involves the writer or speaker moving from particular facts to general conclusions.

information can be neutral or can be biased, but it is generally intended to show evidence, facts, and details about something.

informative speech the purpose of the speech is to define, explain, describe, or demonstrate.

in-kind sponsors in a special event, sponsors that do not financially support an event, but donate products or services.

integrated marketing campaign one in which your chief goal is to communicate a message, using appropriate tools packaged for specifically targeted audiences and a strategically framed message.

integrated marketing communications (IMC) the term used for the process of developing and implementing a set of integrated campaign elements that utilize common and consistent themes across multiple elements.

Internet the global computer information network that also includes e-mail, listservs, blogs, social media applications such as Facebook and Twitter, and search engines such as Google.

interpersonal communication the process of sending and receiving information between two or more people; usually divided into dyadic communication, small-group communication, public speaking, and mass communication.

inverted pyramid structure the most commonly used structure for news writing; presents the most important information in a news story first, followed in descending order by less-important information.

IP (Internet protocol) address a series of numbers that connects a computer to the Internet.

ISO stands for International Standards Organization; scale for measuring light sensitivity.

J

jargon technical language; generally avoided in news writing so that the message is easily understood.

JPG file format developed by the Joint Photographic Experts Group; all Web browsers and e-mail programs can display a JPG image. JPGs are smaller in file size than other formats. If you record your images as JPGs, you should try to save the image at the highest resolution possible to minimize compression.

jump cut occurs when a video shot shows the same prominent person or object in different angles or different locations in back-to-back shots. This makes the two shots appear to "jump," due to the way the shots are framed in relation to each other.

L

land-grant universities established by the Morrill Land-Grant Act of 1862 in order to "to teach such branches of learning as are related to agriculture and the mechanic arts, in such manner as the legislatures of the States may, respectively, prescribe, in order to promote the liberal and practical education of the industrial classes in the several pursuits and professions in life" (Morrill Land-Grant Act, 1862).

lead paragraph the first paragraph in the news story; grabs the reader's attention and answers the most important of the five Ws and H.

lead room the space in front of a person in a video composition.

lines of argument the claims, reasons, and supporting evidence that help make your points in the body of a speech.

logical reasoning a way of thinking; includes induction, deduction, and comparison.

logistics in special events planning, working out everything involved in selecting and developing the right type of event to achieve your objectives.

M

margin of error an estimate of how much the results of a given sample might differ from the results of the "true" population.

marketing mix includes the use of advertising, public relations, brand marketing, sales promotion, direct marketing, and Internet marketing to communicate a campaign message.

mass communication involves communicating to an audience through a media channel that reaches large numbers of people at the same time; the audience is not present with the speaker or is so large that there is no interaction between the speaker and the audience.

media clutter the situation that exists when there are so many media channels and messages out there that it becomes difficult for your message to break through.

media kit package of news releases, fact sheets, photos, brochures, and other items in an attractive, logo-identified package for use by news editors and journalists.

media promotion in a media sales promotion, when an organization partners with a media outlet to jointly promote something. The media outlet uses its channel to convey information to its audience, while the organization includes the media outlet as a sponsor on its materials.

media relations a strategy of working with the news media in order to get out information about an organization's events and activities in local news outlets, such as newspapers, television and radio newscasts, and magazines.

media sponsor when the sponsor of a special event is a media outlet, such as a local newspaper or radio station.

media vehicle media outlets or channels (e.g., Web, television, and print); the term is frequently used in the development of a communications campaign.

medium in risk communication, the manner in which the message is conveyed; depends on the specifics of the situation—it could be a brochure, a billboard, or a television commercial, for example.

megapixel one million pixels.

memorandum a written reminder of something important that has occurred or will occur.

message (1) the "what" and "how" of communication; (2) in risk communication, the overall information an organization wants its audience to comprehend, even if the audience forgets the details. The message should inform and persuade; it should help audiences understand the message and get them to take certain action.

message sidedness the degree to which communications messages present only one side, both sides, or both sides plus an evaluation of the arguments or claims.

message ordering the placement of elements of the persuasive argument.

message strategy a strategically framed message based on the audience of the message and the desired outcome of the message.

moderator objective leader of a focus group.

modified block letter similar to the full block letter, in that the sender's address, salutation, and paragraphs are flush left, with the paragraphs not indented; however, the date line, return address, and signature block are moved to the right.

Morrill Land-Grant Act act of 1862 giving each state land to develop agricultural higher-education institutions. Later extensions of the act added historically black colleges (in 1890) and institutions serving Native Americans (in 1994). These universities were charged with transferring science and technology research, along with educating and informing rural, and later urban, populations.

N

natural sound the audio that is naturally in the environment where the video is being shot; also called *nat sound*.

navigation how users get around the pages that make up the website to find information.

netiquette the set of common rules that govern how to send and receive e-mail properly; essentially, "Internet etiquette."

new media a term often used to describe Web-based communications technologies based on the concept of file sharing, including such technologies as social networking sites, wikis, podcasts, video podcasts (vodcasts), and even cell phone videos.

news media radio and television stations, newspapers, magazines, and Internet news outlets from which people receive information.

news release information about an organization's activities, interesting news, or important events; sent to news media to announce something that has news value; provides reporters with the basics they need to develop a news story; also called a *press release*.

news writing similar to, yet slightly different than, "traditional" writing; it is a more concise form of communication; also called *journalistic writing*.

newsworthiness news value; what stories the media will cover in their newscasts and newspapers.

noise anything that interferes with the receiver's reception of the sender's message, reducing its clarity.

non-random sample technique—such as street-corner interviews and telephone call-ins—that does not allow you to infer results back to the population because the sample was not generated randomly.

O

objective risk the risk that is calculated by scientists based on research.

optimizing resampling or re-creating an image to use fewer colors and pixels in order to compress the file size.

order how you show sequence and importance in document design.

organization in visual aid design, use of a logical visual pattern that is easy to comprehend.

outcue the last thing a reporter says; usually, the outcue gives the reporter's name and television station.

P

pacing the speed or rhythm of a video program, as perceived by the audience.

pan horizontal movement of the video camera; moves the camera left or right.

parallelism items in a group match one another; for example, if your first bulleted item begins with a verb, so must all the others.

permits necessary to conduct activities at special events that are regulated and for which you need permission, such as having a fireworks display or serving food to the public.

personal statement a statement about how your personal, familial, academic, and professional experiences and background qualify you for a particular job, scholarship, or college program; also called *application essays* or *statements of purpose.*

persuasion attempting to influence or convince others to take a specific action or to reach a certain conclusion about an issue.

persuasive communication includes the use of strategies that are designed to create a convincing argument.

photography literally means "drawing or writing with light"; a process for creating images.

pitch the high and low sounds of your voice.

pixels picture elements.

pixels per inch (ppi) measure of an image's resolution.

podcasts series of digital media files (audio or video) that are available for download by users.

population the overall group you want information about in a research study.

positioning describing your subject in a way that influences how others "see" or perceive it.

positive space areas occupied by text or images in document design.

presentation software tools for developing slide presentations, such as Microsoft's PowerPoint or Apple's Keynote.

presentation template an already-thought-out color scheme and formatting pattern; often available as part of a presentation software package, or they can be downloaded from Web sites.

primary research in special events, involves activities such as talking to people who might come to a meeting to find out the best day and time for them or sending out a survey in the case of a large event, takes place before planning begins.

proportion the spatial relationship between each design element in a document; the eye visually compares the relationship of each element's area, size, weight, and location to all of the others on the page.

proposition a claim indicating your stance or position on an issue, or a proposed solution to a problem.

pro-social persuasion when a person stands to gain something positive or be rewarded by agreeing with the persuasive argument.

psychographics combine the attitudes and values people have with their lifestyle choices; used in audience analysis.

public communication takes place when someone communicates directly to a large audience, such as a motivational speaker at a conference, a pastor during a church service, or an announcer at a football game.

public relations the attempt to promote goodwill and garner publicity on behalf of a company, organization, or individual through earned (non-paid) media; one of the most effective ways to promote a special event; includes what you say about an event as well as what others say about it.

public service announcement (PSA) a free advertisement that radio and television stations air or newspapers and magazines print to highlight information about nonprofit organizations' programs, activities, or events.

public speaking one of the many ways you can communicate a message to an audience; valuable because people still like a "human touch" provided through interpersonal communication.

Q

questionnaire a series of questions that includes close-ended questions, open-ended questions, or a mixture of both, used to collect data.

R

random sample allows the researcher to infer results back to the larger population; the sample is generated using random methods, such as selecting every 10th person from a list.

raster graphics images made up of pixels, such images as photographs; they are limited in their size and shape and are linked to the image's resolution.

rate the speed at which someone talks.

RAW file format specified by each camera manufacturer; records data straight from the camera's sensor, just the way it looks on the sensor, onto the camera's data card. The files are uncompressed, meaning they are larger than JPG files.

reach the number of members of your target audience who are reached one time by the media vehicles you have chosen.

readability the ease or difficulty with which the reader can understand your writing.

receiver the person receiving the message, which must be decoded.

refuting objections disproving, ruling out, and countering any potential objections before readers and listeners can think of reasons not to be persuaded.

resampling changing a picture's pixel dimensions by adding or subtracting pixels from an original image.

resolution a measurement of how closely packed together the pixels are in an image; usually measured as *pixels per inch (ppi).*

resume a summary of your education, job experience, and job-related skills that you send to potential employers.

return on investment expenses versus income in a special event.

RGB stands for "red, green, blue"; the color format used by televisions and computer monitors. If a photograph s final destination is the Web or a television monitor, the final color format needs to be RGB.

risk communication the knowledge and skills used by professional communicators to inform people about hazards to their environment or their health, to manage issues, to disseminate information, and to communicate effectively about potential crisis and emergency situations.

rule of thirds basic principle of video and photographic composition in which you imagine dividing an image into thirds horizontally and vertically so that you have nine parts; you position the main subject elements where the dividing lines intersect.

s

sales promotion involves a variety of tools and techniques to develop incentives to create a perception of greater brand value, while encouraging immediate acceptance or adoption of a product, good, or service by a consumer or businessperson.

sampling used to represent an entire population in a way that allows you to collect data from a smaller subset that shares the same characteristics as the larger population.

scaling resizing an image.

search engines Internet tools that use powerful programming languages to develop applications that search through all the pages on the Internet and automatically compile a list of site pages that contain your keyword(s).

Second Life a virtual world in which individuals, organizations, and companies can buy an "island" in order to have a virtual presence.

secondary research in special events, involves reviewing published sources of information when planning for the event.

semi-block letter the return address and signature block are on the right side of the page, and the paragraphs are indented.

sender the person sending the message in communication.

Shannon–Weaver model model of the communication process in which a source of information (such as a speaker) transmits a message through a channel to a receiver who decodes the message.

shared mail when information materials from several different organizations are included in one mailing.

shot outline a detailed written description of the video you plan to shoot.

similarity shows that design elements are alike in a document design; when people see things that look similar, they assume they have similar functions.

simplicity key component of visual aid design; the fewer the elements (text, images, and illustrations) on the page or screen, the easier it is for your audience to understand and recall your message.

site diagrams in special events planning, sketches of the facility, usually with dimensions included.

situation analysis identifies the problem and provides a background and rationale for the suggested media campaign elements.

slug provides the title of the script, the running time (how long the news story is, measured in minutes and seconds), and the date that the story is to be aired or when it was written.

small-group communication takes place among members of a group or team, such as the members of a student organization, an athletic team, a church youth group, or a work team.

social media a form of new media that includes primarily Internet- and mobile-based tools for sharing and discussing information using two-way communication; most often refers to Web-based activities that integrate technology, telecommunications, and social interaction.

social networking sites Internet sites that let users interact with each other, upload files, and post messages, photos, and videos that visitors can see.

sound on tape (SOT) any audio accompanying a video; most frequently describes when any person talking is shown speaking.

soundbite the exact words spoken by someone in his or her own recorded voice.

source a source of information in the communication process, such as a speaker.

special event a planned gathering of people who get together to engage in some shared activity.

splash page an opening, attention-getting page with graphics that either automatically refreshes to your home page or is clickable.

stand-up when a reporter narrates a portion of a story on camera.

storyboard a series of drawings with captions that describe video shots and their accompanying audio or narration.

strategic communications involves being knowledgeable about the form and function of a diverse range of media and communications approaches.

subjective risk the risk that the public perceives to be hazardous; affected by issues of familiarity, dread, and personal control.

superimposition on-screen text that appears below a person's face; also called a *super*.

survey the act of conducting a study to collect data using a questionnaire.

syllogism a logical argument with three propositions or claims: a major premise, a minor premise, and a conclusion.

symmetrical balance mirror-image balance; occurs when all the visual elements on one side of the page are mirrored on the other side; also called *formal balance*.

T

tagline a short, easy-to-understand message, similar to a slogan.

target audience the intended recipient of a specific message.

text any size, shape, and placement of the printed word in your document.

thesis statement (1) states your main point and purpose for writing; a way of getting your key message across to the reader; (2) the main idea of the speech; should be composed as a single, brief sentence that illustrates what you will attempt to demonstrate in your speech.

thumbnail a hand-drawn sketch of what the finished document will look like, including all text, images, headlines, and borders.

TIFF stands for *tagged image file format*. TIFF files are much larger than JPGs because TIFFs do not compress files very much. TIFFs cannot be displayed on some Web browsers. A TIFF file must usually be opened in a photo editing program and converted to a JPG before the image can be shared on the Web. TIFF is the preferred format for image files for print publications.

tilt vertical movement of the video camera.

title sponsor in special events, pays a premium fee to have its name become a part of the event itself.

Twitter a micro-blog that lets its users send and view other users 140-character updates, called "tweets."

typeface the actual look of the letters in document design; also called *type* or *font*. Type usually is classified into two categories: serif type (also called serif font) and sans serif type (or sans serif font).

U

unity how all of a message's elements tie together visually; a message with good unity is one where all of its visual elements complement each other.

URL (uniform resource locator) the formal name for a Web address.

V

vector graphics images composed of mathematically defined shapes created by illustration programs, such as CorelDraw and Adobe Illustrator; usually lines, curves, and geometric shapes.

virtual reality simulates a three-dimensional space where users can interact with each other online.

visual aid anything used to communicate visually.

visual communication any attempt to reach us through our eyes, whether it is a roadside billboard or a flier posted for a meeting, or gestures and body language.

visual weight the amount of "weight" or prominence an image has in a given document.

VO (voice over) video that will be shown, with a newscaster narrating the script; the newscaster is providing voice *over* the video.

VO/SOT (voice over/sound on tape) when a newscaster narrates a script on air while video runs over the narrator's voice (VO), followed immediately by someone talking (a soundbite) about a topic related to the voice over.

vocal delivery how you use your voice to communicate your message in a public speech.

volume the loudness or softness of your voice.

W

Web-based editions Internet versions of print and broadcast media; can be updated much more quickly and cost effectively; Web and social media sites can link to other types of media.

Web browser a graphical interface on your computer that lets you type in a URL.

Web marketing the marketing of products, goods, and services online; advertisement on Web pages.

white balancing tells the camera what combination of red, green, and blue light it should perceive as white, given a particular lighting condition.

white space the area not taken up with text or images in document design; used to create a sense of openness; also called *blank space* or *negative space*.

wikis pages or collections of Web pages designed to enable anyone who accesses them to contribute to or modify the content.

wipe when one video picture "wipes" off and another appears in a video composition.

World Wide Web or **Web** the system of specially formatted interlinked documents accessed through the Internet.

Y

YouTube Internet site that allows users to upload and download media files and share these with others.

Z

zoom a change in the focal length of the video camera lens. When you "zoom in," the shot gets tighter on the subject. When you "zoom out," the shot gets wider.

Index